Secure Communication Protocols

Contents

Chapter 1

Secure communication

Secure communication is when two entities are communicating and do not want a third party to listen in. For that they need to communicate in a way not susceptible to eavesdropping or interception.[1][2] Secure communication includes means by which people can share information with varying degrees of certainty that third parties cannot intercept what was said. Other than spoken face-to-face communication with no possible eavesdropper, it is probably safe to say that no communication is guaranteed secure in this sense, although practical obstacles such as legislation, resources, technical issues (interception and encryption), and the sheer volume of communication serve to limit surveillance.

With many communications taking place over long distance and mediated by technology, and increasing awareness of the importance of interception issues, technology and its compromise are at the heart of this debate. For this reason, this article focusses on communications mediated or intercepted by technology.

Also see *Trusted Computing*, an approach under present development that achieves security in general at the potential cost of compelling obligatory trust in corporate and government bodies.

1.1   History

In 1898, Nikola Tesla demonstrated a radio controlled boat in Madison Square Garden that allowed secure communication between transmitter and receiver.[3]

One of the most famous systems of secure communication was the Green Hornet. During WWII, Winston Churchill had to discuss vital matters with Franklin D. Roosevelt. At first, the calls were made using a voice scrambler as this was thought to be secure. When this was found to be untrue the engineers started work on a whole new system, the Green Hornet or SIGSALY. Anyone listening in would just hear white noise but the conversation was clear to the parties. As secrecy was paramount, the location of the Green Hornet was only known by the people who built it and Winston

Churchill, and if anyone did see him entering the room it was kept in, all they would see was the Prime Minister entering a closet labeled 'Broom Cupboard.' It is said that because the Green Hornet works by a one-time pad it cannot be beaten.

1.2   Nature and limits of security

1.2.1   Types of security

Security can be broadly categorised under the following headings, with examples:

- Hiding the content or nature of a communication

  - Code – a rule to convert a piece of information (for example, a letter, word, phrase, or gesture) into another form or representation (one sign into another sign), not necessarily of the same type. In communications and information processing, encoding is the process by which information from a source is converted into symbols to be communicated. Decoding is the reverse process, converting these code symbols back into information understandable by a receiver. One reason for coding is to enable communication in places where ordinary spoken or written language is difficult or impossible. For example, semaphore, where the configuration of flags held by a signaler or the arms of a semaphore tower encodes parts of the message, typically individual letters and numbers. Another person standing a great distance away can interpret the flags and reproduce the words sent.

  - Encryption

  - Steganography

  - Identity Based

- Hiding the parties to a communication – preventing identification, promoting anonymity

- "Crowds" and similar anonymous group structures – it is difficult to identify who said what when it comes from a "crowd"
- Anonymous communication devices – unregistered cellphones, Internet cafes
- Anonymous proxies
- Hard to trace routing methods – through unauthorized third-party systems, or relays

- Hiding the fact that a communication takes place

  - "Security by obscurity" – similar to needle in a haystack
  - Random traffic – creating random data flow to make the presence of genuine communication harder to detect and traffic analysis less reliable

Each of the three is important, and depending on the circumstances any of these may be critical. For example, if a communication is not readily identifiable, then it is unlikely to attract attention for identification of parties, and the mere fact a communication has taken place (regardless of content) is often enough by itself to establish an evidential link in legal prosecutions. It is also important with computers, to be sure where the security is applied, and what is covered.

1.3   Borderline cases

A further category, which touches upon secure communication, is software intended to take advantage of security openings at the end-points. This software category includes trojan horses, keyloggers and other spyware.

These types of activity are usually addressed with everyday mainstream security methods, such as antivirus software, firewalls, programs that identify or neutralize adware and spyware, and web filtering programs such as Proxomitron and Privoxy which check all web pages being read and identify and remove common nuisances contained. As a rule they fall under computer security rather than secure communications.

1.4   Tools used to obtain security

1.4.1   Encryption

Main article: Encryption

Encryption is where data is rendered hard to read by an unauthorized party. Since encryption can be made extremely hard to break, many communication methods either use deliberately weaker encryption than possible, or have backdoors inserted to permit rapid decryption. In some cases government authorities have required backdoors be installed in secret. Many methods of encryption are also subject to "man in the middle" attack whereby a third party who can 'see' the establishment of the secure communication is made privy to the encryption method, this would apply for example to interception of computer use at an ISP. Provided it is correctly programmed, sufficiently powerful, and the keys not intercepted, encryption would usually be considered secure. The article on key size examines the key requirements for certain degrees of encryption security.

The encryption can be implemented in a way to require the use of encryption, i.e. if encrypted communication is impossible then no traffic is sent, or opportunistically. Opportunistic encryption is a lower security method to generally increase the percentage of generic traffic which is encrypted. This is analogous to beginning every conversation with "Do you speak Navajo?" If the response is affirmative, then the conversation proceeds in Navajo, otherwise it uses the common language of the two speakers. This method does not generally provide authentication or anonymity but it does protect the content of the conversation from eavesdropping.

An Information-theoretic security technique known as physical layer encryption ensures that a wireless communication link is provably secure with communications and coding techniques.

1.4.2   Steganography

Steganography ("hidden writing") is the means by which data can be hidden within other more innocuous data. Thus a watermark proving ownership embedded in the data of a picture, in such a way it is hard to find or remove unless you know how to find it. Or, for communication, the hiding of important data (such as a telephone number) in apparently innocuous data (an MP3 music file). An advantage of steganography is plausible deniability, that is, unless one can prove the data is there (which is usually not easy), it is deniable that the file contains any. *(Main article: Steganography)*

1.4.3   Identity based networks

Unwanted or malicious behavior is possible on the web since it is inherently anonymous. True identity based networks replace the ability to remain anonymous and are inherently more trustworthy since the identity of the sender and recipient are known. (The telephone system is an example of an identity based network.)

1.4.4 Anonymized networks

Recently, anonymous networking has been used to secure communications. In principle, a large number of users running the same system, can have communications routed between them in such a way that it is very hard to detect what any complete message is, which user sent it, and where it is ultimately going from or to. Examples are Crowds, Tor, I2P, Mixminion, various anonymous P2P networks, and others.

1.4.5 Anonymous communication devices

In theory, an unknown device would not be noticed, since so many other devices are in use. This is not altogether the case in reality, due to the presence of systems such as Carnivore and Echelon which can monitor communications over entire networks, and the fact that the far end may be monitored as before. Examples include payphones, Internet cafe, etc.

1.5 Methods used to "break" security

1.5.1 Bugging

Main article: Covert listening device

The placing covertly of monitoring and/or transmission devices either within the communication device, or in the premises concerned.

1.5.2 Computers (general)

Main article: Computer security

Any security obtained from a computer is limited by the many ways it can be compromised – by hacking, keystroke logging, backdoors, or even in extreme cases by monitoring the tiny electrical signals given off by keyboard or monitors to reconstruct what is typed or seen (TEMPEST, which is quite complex).

1.5.3 Laser audio surveillance

Main article: Laser microphone

Sounds, including speech, inside rooms can be sensed by bouncing a laser beam off a window of the room where a conversation is held, and detecting and decoding the vibrations in the glass caused by the sound waves.[4]

1.6 Systems offering partial security

1.6.1 Anonymous cellphones

Cellphones can easily be obtained, but are also easily traced and "tapped". There is no (or only limited) encryption, the phones are traceable – often even when switched off – since the phone and SIM card broadcast their International Mobile Subscriber Identity (IMSI). It is possible for a cellphone company to turn on some cellphones when the user is unaware and use the microphone to listen in on you, and according to James Atkinson, a counter-surveillance specialist cited in the same source, "Security-conscious corporate executives routinely remove the batteries from their cell phones" since many phones' software can be used "as-is", or modified, to enable transmission without user awareness and the user can be located within a small distance using signal triangulation and now using built in GPS features for newer models. Transceivers may also be defeated by jamming or Faraday cage.

Some cellphones (Apple's iPhone, Google's Android) track and store users' position information, so that movements for months or years can be determined by examining the phone.[5]

1.6.2 Landlines

Analogue landlines are not encrypted, and it is very easy to tap them. Such tapping requires physical access to the line, easily obtained from a number of places, e.g. the phone location, distribution points, cabinets and the exchange itself. Tapping a landline in this way can enable an attacker to make calls which appear to originate from the tapped line.

1.6.3 Anonymous Internet

Main article: Anonymity

Using a third party system of any kind (payphone, Internet cafe) is often quite secure, however if that system is used to access known locations (a known email account or 3rd party) then it may be tapped at the far end, or noted, and this will remove any security benefit obtained. Some countries also impose mandatory registration of Internet cafe users.

Anonymous proxies are another common type of protection, which allow one to access the net via a third party (often in a different country) and make tracing difficult. Note that there is seldom any guarantee that the plaintext is not tappable, nor that the proxy does not keep its own records of users or entire dialogs. As a result, anonymous proxies are a generally useful tool but may not be as secure as other systems whose security can be better assured. Their most common use is to prevent a record of the originating IP, or address, being left on the target site's own records. Typical anonymous proxies are found at both regular websites such as Anonymizer.com and spynot.com, and on proxy sites which maintain up to date lists of large numbers of temporary proxies in operation.

A recent development on this theme arises when wireless Internet connections ("Wi-Fi") are left in their unsecured state. The effect of this is that any person in range of the base unit can piggyback the connection – that is, use it without the owner being aware. Since many connections are left open in this manner, situations where piggybacking might arise (willful or unaware) have successfully led to a defense in some cases, since it makes it difficult to prove the owner of the connection was the downloader, or had knowledge of the use to which unknown others might be putting their connection. An example of this was the Tammie Marson case, where neighbours and anyone else might have been the culprit in the sharing of copyright files.[6] Conversely, in other cases, people deliberately seek out businesses and households with unsecured connections, for illicit and anonymous Internet usage, or simply to obtain free bandwidth.[7]

1.6.4   Programs offering more security

- Cryptocat is a simple, open source, online encrypted chat service and software project supporting encrypted file transfers and up to 10 users in a room. The service can be used any modern browser, including those on mobile devices. A local app for Chrome exists, with more platforms in development.

- I2P-Messenger is a simple secure (end-to-end encrypted), anonymous, and serverless instant messenger with file transfer support.

- Jitsi is another open source voice and chat client implementing ZRTP and SRTP encryption for calls, as well as OTR security for chat.

- Off-the-Record Messaging (OTR) is a plugin which adds end-to-end encryption, authentication and perfect forward secrecy (PFS) to instant messaging. It is not a separate protocol but runs under most every instant messaging (IM) protocol.

- pbxnsip is a SIP-based PBX that uses TLS and SRTP to encrypt the voice traffic. In contrast to other proprietary protocols, the protocol is open so that devices from independent vendors can be used. The encryption includes the relay of instant messaging, presence information, and the management interface.

- Secure email – some email networks such as "hushmail" or Opolis Secure Mail, are designed to provide encrypted and/or anonymous communication. They authenticate and encrypt on the users own computer, to prevent transmission of plain text, and mask the sender and recipient. Mixminion and I2P-Bote provide a higher level of anonymity by using a network of anonymizing intermediaries, (similar to how Tor and crowds work above, but at a higher latency).

- Secure IRC and web chat – Some IRC clients and systems use security such as SSL. This is not standardised. Likewise some web chat clients such as Yahoo Messenger use secure communications on their web based program. Again the security of these is unverified, and it is likely the communication is not secured other than to and from the client.

- Signal is a free and open-source encrypted voice calling and instant messaging application developed by Open Whisper Systems for iOS. Signal communications are compatible with RedPhone and TextSecure on Android. It uses end-to-end encryption with forward secrecy and deniable authentication to secure all communications to RedPhone, TextSecure, and other Signal users.

- Skype – secure voice over Internet, secure chat. Uses 128-bit AES (256-bit is the standard) and 1024-bit asymmetrical protocols to exchange initial keys (which is considered relatively weak by NIST[8]). Proprietary. An article in 2004 suggested that Skype has relatively weak encryption. Criticism focuses upon its proprietary "black box" design, its relatively short (1536 bit) keys, excessive bandwidth use of user supernodes, and excessive trust of other computers able to "speak Skype". The 2013 mass surveillance disclosures revealed that agencies such as the NSA and the FBI have the ability to eavesdrop on Skype, including the monitoring and storage of text and video calls and file transfers. *(See Skype#Security and privacy)*

- Syme is a secure, encrypted social network. It uses end-to-end encryption and provides confidentiality, integrity and authentication security features.[9] AES-256 is used for symmetric encryption and ECC-384 is used for asymmetric encryption.[10]

- Tox – A New Kind of Instant Messaging. With the rise of government monitoring programs, Tox provides an easy to use application that allows you to connect with friends and family without anyone else listening in. While other big-name services require you to pay for features, Tox is totally free, and comes without advertising.

- Trillian – offers secure IM facility, however appears to have weaknesses in key exchange which would enable a "man in the middle" attack with ease. Proprietary, no information on backdoors.

- TrueConf Server is an on-premises unified communications software server for group video conferencing. It is originally designed to work in private corporate networks without Internet connection. In TrueConf Server connections are secured with TLS/SSL protocol, data is encrypted using AES-256, proprietary transport layer is used. The solution can be deployed within LAN/VPN/DMZ networks and is fully compatible with 3rd party data encryption systems.

- WASTE – open source secure IM, high strength "end to end" encryption, within an anonymised network.

- Zfone is an open source secure voice over Internet program implementing the ZRTP encryption protocol, by Phil Zimmermann, the creator of PGP.

1.7   See also

1.7.1   General background

- Computer security

- Opportunistic encryption

- Communications security

- Secure messaging

1.7.2   Software selections and comparisons

- Comparison of VoIP software

- Comparison of instant messaging clients

- Anonymous P2P

1.7.3   Other

- I2P

- Freenet

- Hepting vs. AT&T, a 2006 lawsuit in which the Electronic Frontier Foundation alleges AT&T Inc. allowed the NSA to tap all of its clients' Internet and Voice over IP communications

- NSA warrantless surveillance controversy

- Secret cell phone

1.8   References

[1] D. P. Agrawal and Q-A. Zeng, *Introduction to Wireless and Mobile Systems* (2nd Edition, published by Thomson, April 2005) ISBN 978-0-534-49303-5

[2] J.K. and K. Ross, *Computer Networking* (2nd Ed, Addison Wesley, 2003) ISBN 978-0-321-17644-8

[3] The schematics are illustrated in U.S. Patent 613,809 and describes "rotating coherers".

[4] "High-tech bugging techniques, and a costly fix". *Popular Science*. Google. August 1987.

[5] Wall Street Journal: How concerned are you that the iPhone tracks and stores your location?

[6] Open Wi-Fi proves no defence in child porn case, The Register

[7] 'Extortionist' turns Wi-Fi thief to cover tracks, The Register

[8] NIST recommendations from 2005 state that 1024-bit asymmetric ciphers are the lowest standard it considers when evaluating ciphers, and that this is expected to be categorized as technically weak or breakable from around 2006–2010. (click on NIST and scroll to the last table, or see article: Key size).

[9] https://getsyme.com/faq

[10] https://github.com/symeapp/syme

1.9   External links

- X. Y. Wang, S. Chen, S. Jajodia. "Tracking Anonymous Peer-to-Peer VoIP Calls on the Internet". In Proceedings of the 12th ACM Conference on Computer Communications Security (CCS 2005), November 2005.

Chapter 2

Automated Certificate Management Environment

The **Automated Certificate Management Environment** (**ACME**) protocol is a communications protocol for automating interactions between certificate authorities and their users' web servers, allowing the automated deployment of public key infrastructure at very low cost.[1][2] It was designed by the Internet Security Research Group for their Let's Encrypt service.[1]

The protocol, based on passing JSON-formatted messages over HTTPS,[3][2] has been published as an Internet-Draft.[4]

The Internet Security Research Group provides reference open source software implementations for ACME: letsencrypt-preview is a Python-based test implementation of server certificate management software using the ACME protocol,[5] and boulder is a CA implementation, written in the Go programming language.[6]

2.1   References

[1] Steven J. Vaughan-Nichols (9 April 2015). "Securing the web once and for all: The Let's Encrypt Project". ZDNet.

[2] "letsencrypt/acme-spec". github.com. Retrieved 2014-11-20.

[3] Chris Brook (18 November 2014). "EFF, Others Plan to Make Encrypting the Web Easier in 2015". ThreatPost.

[4] R. Barnes, P. Eckersley, S. Schoen, A. Halderman, J. Kasten (January 28, 2015). "Automatic Certificate Management Environment (ACME) draft-barnes-acme-01".

[5] "letsencrypt/lets-encrypt-preview". github.com. Retrieved 2014-11-20.

[6] "letsencrypt/boulder". github.com. Retrieved 2015-06-22.

Chapter 3

Automatic Secure Voice Communications Network

The **Automatic Secure Voice Communications Network** (**AUTOSEVOCOM**) was a worldwide, switched, secure voice network for the United States Armed Forces, which was operational from the late 1960s to the end of the 1980s. It was closely related to the Automatic Voice Network or AUTOVON, which was the main non-secure switched telephone network for the military.

3.1 Phase I

During the mid 1960s, the United States Government decided to implement a worldwide secure voice network. This was named Automatic Secure Voice Communications Network, or by its acronym AUTOSEVOCOM, and was the National Security Agency's first program for DoD telephone protection. It was·a cumbersome and expensive system, that was available only for high-level users. Because of its inadequacies, the Defense Department capped it at 1850 terminals, and in the late 1960s, hoping for something better, decided not to continue with the expansion of AUTOSEVOCOM.[1]

Phase I of the network was approved by the Deputy Secretary of Defense in July 1967 and after that it took several years to implement AUTOSEVOCOM within the continental United States. AUTOSEVOCOM-I was a non-tactical network, that enabled users to discuss classified or sensitive information over the telephone. The network consisted of switching centres, transmission facilities and subscriber terminals. Subscribers were homed either on an AUTOSEVOCOM switch, on an Automatic Voice Network (AUTOVON) switch, or a Joint Overseas Switchboard (JOSS), which were for example operated and maintained by numerous Signal Battalions in Vietnam.

The AUTOSEVOCOM switches provided for wideband secure voice communications between local subscribers and enabled them to establish long distance secure voice calls.

The majority of the long distance calls were routed by the AUTOVON.

3.2 Phase II

Difficulties with speech intelligibility, requirements for voice recognition,[2] the holding of telephone conferences, speedier service, and simpler calling procedures led Defense officials to approve the development of an improved system, called AUTOSEVOCOM II. The Army was designated as the agency with the primary responsibility of developing the system. In May 1976, the Deputy Secretary of Defense approved the full-scale development of the AUTOSEVOCOM II programme.

AUTOSEVOCOM II incorporated technological advances and furnished higher quality communications for the several thousand subscribers who were expected to use it when put into operation during the years 1980 to 1985. The U.S. Army Communications Command acted as program manager for AUTOSEVOCOM II.

The Automatic Secure Voice Communications Network was succeeded by the Defense Red Switch Network (DRSN) and the STU-III secure phones. The last AUTOSEVOCOM secure voice switch in the world was deactivated at the Pentagon in 1994.[3]

3.3 External links

- Crypto Machines: AUTOSEVOCOM I and II

- Janes Defense: Automatic Secure Voice Communications (AUTOSEVOCOM)

3.4   References

[1] Thomas R. Johnson, American Cryptology during the Cold War, 1945-1989, 1998, Chapter 17, p. 142.

[2] How to Wreck a Nice Beach: The Vocoder from World War II to Hip-Hop, By Dave Tompkins

[3] Christopher H. Sterling, Military Communications: From Ancient Times to the 21st Century, 2008, p. 48-49.

Chapter 4

BID 150

A British voice encryption device used with (for example) Larkspur radio system sets.

BID means 'British Inter Departmental'. These systems or equipment types were generally used by more than one single governmental agency or department. The authority for BID's are the Communications-Electronics Security Group (CESG) who are part of Government Communications Headquarters (GCHQ).

The BID/150 speech encryption key generator is a single channel device for use with the British Army C42 and C45 Larkspur radio system. This was the first Combat Net secure speech system whose key was set through the use of punch cards within the device. Examples of the BID/150 are on display at the Royal Signals Museum, Blandford Forum.

It was first used operationally by the 15th Signal Regiment during the Aden crisis in the late 1960s and was widely used from Battalion level up to Corps Headquarters. It remained in use until the early 1980s.

In secure mode the analogue speech signal is digitized, then encrypted by combination with the digital key generated by the BID/150; the digital stream is then fed to the transmitter, to pulse modulate the carrier. On receive, the sequence is reversed.

Good radio performance was needed for reliable secure working. To this end, vehicles carrying the C42-DM-BID/150 system were issued with a 27-foot mast and an elevated antenna. The set carried a 'Goodman Box', to check signal strengths and antenna performance. The operator had searching set-up and performance drills to be strictly followed.

4.1 External links

- Crypto Machine Website
- Royal Signals Museum, Blandford Forum

Chapter 5

Blacker (security)

Blacker (styled **BLACKER**) is a U.S. Department of Defense computer network security project designed to achieve A1 class ratings of the Trusted Computer System Evaluation Criteria (TCSEC).[1][2] The project was implemented by SDC and Burroughs. It was the first secure system with trusted End-to-end encryption on the United States' Defense Data Network.[3]

5.1 See also

- RED/BLACK concept for segregation of sensitive *plaintext* information (RED signals) from *encrypted* ciphertext (BLACK signals)

5.2 References

[1] Weissman, Clark. "BLACKER: security for the DDN examples of A1 security engineering trades". Retrieved 2007-12-02.

[2] Weissman, Clark (1995-01-24). "Handbook for the Computer Security Certification of Trusted Systems". Retrieved 2007-12-02.

[3] Pike, John (2000-02-11). "BLACKER, an article at the Intelligence Resource Program". Retrieved 2007-12-02.

Chapter 6

CA/Browser Forum

The Certification Authority Browser Forum, also known as **CA/Browser Forum**, is a voluntary consortium of certification authorities, vendors of Internet browser software, operating systems, and other PKI-enabled applications that promulgates industry guidelines governing the issuance and management of X.509 v.3 digital certificates that chain to a trust anchor embedded in such applications. Its guidelines cover certificates used for the SSL/TLS protocol and code signing, as well as system and network security of certificate authorities.

As of October 2014, the CA/Browser Forum includes over 40 Certificate Authority members and the following six Internet Browser Software Vendors: Microsoft (Internet Explorer), Apple (Safari), Mozilla (Firefox), Google (Chrome), Opera, and Qihoo 360 (360 Secure Browser).[1]

The CA/Browser Forum maintains "Guidelines For The Issuance And Management Of Extended Validation (EV) Certificates". The EV SSL standard improves security for Internet transactions and creates a more intuitive method of displaying secure sites to Internet users. In order for Certification Authorities to issue EV SSL Certificates, they must be audited for compliance with the Forum's EV Guidelines[2] in accordance with either WebTrust or ETSI audit criteria.

The CA/Browser Forum adopted the "Baseline Requirements for the Issuance and Management of Publicly-Trusted Certificates" in 2011. These Guidelines, which are binding on members of the CA/Browser Forum, took effect July 1, 2012. These guidelines cover all CA-issued certificates. Certificates are now classified as "DV" (Domain Validated), "OV" (Organization Validated), "IV" (Individual Validated), and "EV" (Extended Validation), and a method is defined within the specification to distinguish the types of certificates.[3]

6.1 History

In 2005, Melih Abdulhayoglu of the Comodo Group organized[4] and arranged the first meeting of CA/Browser Forum. The first meeting was held in New York City. This was followed by a meeting in November 2005 in Kanata, Ontario, and a meeting in December, 2005, in Scottsdale, Arizona with the main objective to enable secure connections between users and websites.

In addition to CA/Browser Forum members, representatives of the Information Security Committee of the American Bar Association Section of Science & Technology, Law and the Canadian Institute of Chartered Accountants participated in developing the standards for issuing and managing Extended Validation SSL certificates.

Version 1.0 of the EV Guidelines was adopted on 7 June 2007.[5]

Version 1.1 was adopted by the CA/Browser Forum on 10 April 2008.[6]

Version 1.2 was adopted by the CA/Browser Forum on 1 Oct 2009.[7]

It is a great step forward in establishing verified identity for websites considers MSDN in its blog post.[8] Also, Microsoft's vision is that the backbone of an Internet identity system is composed of Extended Validation SSL Certificates intimately integrated with the users' browsing experience.[9]

The tougher certificates, coupled with browser developments,[10] could help fight phishing, which threatens the multibillion-dollar online retail market.

In November 2011, the CA/Browser Forum adopted version 1.0 of the "Baseline Requirements for the Issuance and Management of Publicly-Trusted Certificates."[3]

In February 2013 a new industry group, the Certificate Authority Security Council (CASC), was formed with a mission that includes promoting CA/Browser Forum standards. Membership requires adherence to

CA/Browser Forum standards.[11] The CASC's founding members consist of the 7 largest Certificate Authorities: Comodo,[12][13] Symantec,[14] Trend Micro, DigiCert, Entrust,[15] GlobalSign [16] and GoDaddy.[17][18][18] [19] [20][21]

6.2   See also

- Certification Authority
- Transport Layer Security (TLS)
- Comparison of SSL certificates for web servers
- Extended Validation Certificate

6.3   References

[1] "Members of the CA - Browser Forum - Over 30 CAs and All Major Browsers". CA/Browser Forum. Archived from the original on 2015-01-24. Retrieved 23 January 2015.

[2] CA/Browser Extended Validation Guidelines

[3] "Baseline Requirements Documents". CA/Browser Forum. Retrieved 2014-10-26.

[4] eWeek Article about Origins of CA/Browser Forum and EV SSL

[5] "GUIDELINES FOR THE ISSUANCE AND MANAGEMENT OF EXTENDED VALIDATION CERTIFICATE" (PDF). The CA/Browser Forum.

[6] "GUIDELINES FOR THE ISSUANCE AND MANAGEMENT OF EXTENDED VALIDATION CERTIFICATES" (PDF). The CA/Browser Forum.

[7] "Guidelines For The Issuance And Management Of Extended Validation Certificates". The CA/Browser Forum.

[8] Extended Validation Guidelines v1 Released

[9] Microsoft information on EV in IE7

[10] CNet News - Browsers to get sturdier padlocks

[11] https://casecurity.org/casc/

[12] *SSL Certificate Types*, retrieved 2015-07-02

[13] *SSL Certificate*, retrieved 2015-07-02

[14] http://www.symantec.com/connect/blogs/let-s-build-more-secure-future

[15] http://www.entrust.com/news/2013-02-14-Entrust-Joins-Worlds-Leading-CAs-to-Form-Certificate-Authority-Security-Council-Advance-Internet-Security-and-Trusted-SSL-Ecosystem

[16] http://www.thepaypers.com/news/e-identity-security-online-fraud/globalsign-joins-the-certificate-authority-security-council-to-upgrade-internet-security/750211-26

[17] http://inside.godaddy.com/announcing-certificate-authority-security-council/

[18] http://www.darkreading.com/authentication/167901072/security/news/240148546/major-certificate-authorities-unite-in-the-name-of-ssl-security.html

[19] http://www.networkworld.com/news/2013/021413-council-digital-certificate-266728.html

[20] http://www.cmswire.com/cms/customer-experience/website-certificate-authorities-set-up-security-council-for-advocacy-research-019619.php

[21] http://electronicstaff.com/2013/ssl-certificate-authority-security-council-takes-root

6.4   External links

- CA/Browser Forum Website
- CAs approved for EV in Microsoft IE7
- Microsoft's Internet Identity Technology Gets Certified

Chapter 7

CCMP

This article is about a data encryption protocol. For the private equity investment firm, see CCMP Capital.

Counter Mode Cipher Block Chaining Message Authentication Code Protocol, **Counter Mode CBC-MAC Protocol** or simply **CCMP** (**CCM mode Protocol**) is an encryption protocol designed for Wireless LAN products that implement the standards of the IEEE 802.11i amendment to the original IEEE 802.11 standard. CCMP is an enhanced data cryptographic encapsulation mechanism designed for data confidentiality and based upon the Counter Mode with CBC-MAC (CCM) of the AES standard.[1] It was created to address the vulnerabilities presented by WEP, a dated, insecure protocol.[1]

7.1 Technical details

CCMP uses CCM that combines CTR for data confidentiality and CBC-MAC for authentication and integrity. CCM protects the integrity of both the MPDU data field and selected portions of the IEEE 802.11 MPDU header. CCMP is based on AES processing and uses a 128-bit key and a 128-bit block size. CCMP uses CCM with the following two parameters:

- M = 8; indicating that the MIC is 8 octets (eight bytes).

- L = 2; indicating that the Length field is 2 octets.

A CCMP Medium Access Control Protocol Data Unit (MPDU) comprises five sections. The first is the MAC header which contains the destination and source address of the data packet. The second is the CCMP header which is composed of 8 octets and consists of the packet number (PN), the Ext IV, and the key ID. The packet number is a 48-bit number stored across 6 octets. The PN codes are the first two and last four octets of the CCMP header and are incremented for each subsequent packet. Between the PN codes are a reserved octet and a Key ID octet. The Key ID octet contains the Ext IV (bit 5), Key ID (bits 6-7), and a reserved subfields (bits 0-4). CCMP uses these values to encrypt the data unit and the MIC. The third section is the data unit which is the data being sent in the packet. Lastly are the Message Integrity Code (MIC) which protects the integrity and authenticity of the packet and the frame check sequence (FCS) which is used for error detection and correction. Of these sections only the data unit and MIC are encrypted.[1]

7.2 Security

CCMP is the standard encryption protocol for use with the WPA2 standard and is much more secure than the WEP protocol and TKIP protocol of WPA. CCMP provides the following security services:[2]

- Data confidentiality; ensures only authorized parties can access the information

- Authentication; provides proof of genuineness of the user

- Access control in conjunction with layer management

Because CCMP is a block cipher mode using a 128-bit key, it is secure against attacks to the 2^{64} steps of operation. Generic meet-in-the-middle attacks do exist and can be used to limit the theoretical strength of the key to $2^{n/2}$ (where n is the number of bits in the key) operations needed.[3]

7.2.1 Known attacks

Main article: Advanced Encryption Standard § Known attacks

7.3 References

[1] Cole, Terry (12 June 2007). "IEEE Std 802.11-2007" (PDF). New York, New York: The Institute of Electrical and Electronics Engineers, Inc. Retrieved 11 April 2011.

[2] Ciampa, Mark (2009). *Security Guide To Network Security Fundamentals* (3 ed.). Boston, MA: Course Technology. pp. 205, 380, 381. ISBN 1-4283-4066-1.

[3] Whiting, Doug; R. Housley; N. Ferguson (September 2003). "Counter with CBC-MAC (CCM)". The Internet Society. Retrieved 11 April 2011.

Chapter 8

Cipher suite

A **cipher suite** is a named combination of authentication, encryption, message authentication code (MAC) and key exchange algorithms used to negotiate the security settings for a network connection using the Transport Layer Security (TLS) / Secure Sockets Layer (SSL) network protocol.

The structure and use of the cipher suite concept is defined in the documents that define the protocol.[1] A reference for named cipher suites is provided in the TLS Cipher Suite Registry.[2]

8.1 Use

Main article: Transport Layer Security

When a TLS connection is established, a handshaking, known as the TLS Handshake Protocol, occurs. Within this handshake, a *client hello* (ClientHello) and a *server hello* (ServerHello) message are passed.[3] First, the client sends a cipher suite list, a list of the cipher suites that it supports, in order of preference. Then the server replies with the cipher suite that it has selected from the client cipher suite list.[4] In order to test which TLS ciphers that a server supports an SSL/TLS Scanner may be used.

8.2 Detailed description

Each named cipher suite defines a key exchange algorithm, a bulk encryption algorithm, a message authentication code (MAC) algorithm, and a pseudorandom function (PRF).[5][6][7]

8.3 References

[1] RFC 5246

[2] TLS Cipher Suite Registry

[3] RFC 5246, p. 37

[4] RFC 5246, p. 40

[5] "CipherSuites and CipherSpecs". IBM. Retrieved 20 November 2009.

[6] "Cipher Suites in Schannel". Microsoft MSDN. Retrieved 20 November 2009.

[7]
- The **key exchange algorithm** is used to determine if and how the client and server will authenticate during the handshake.<ref>RFC 5246, p. 47</ref>
- The **bulk encryption algorithm** is used to encrypt the message stream. It also includes the key size and the lengths of explicit and implicit initialization vectors (cryptographic nonces).<ref name="RFC5246p17">RFC 5246, p. 17</ref>
- The **message authentication code** (MAC) algorithm is used to create the message digest, a cryptographic hash of each block of the message stream.<ref name="RFC5246p17"/>
- The **pseudorandom function** (PRF) is used to create the **master secret**, a 48-byte secret shared between the two peers in the connection. The master secret is used as a source of entropy when creating session keys, such as the one used to create the MAC.<ref>RFC 5246, p. 16-17, 26</ref>

8.3.1 Examples of algorithms used

See also: Transport Layer Security § Applications and adoption

key exchange/agreement RSA, Diffie-Hellman, ECDH, SRP, PSK

authentication RSA, DSA, ECDSA

bulk ciphers RC4, Triple DES, AES, IDEA, DES, or Camellia. In older versions of SSL, RC2 was also used.

message authentication for TLS, a Hash-based Message Authentication Code using MD5 or one of the SHA hash functions is used. For SSL, SHA, MD5, MD4, and MD2 are used.

8.4   Programming references

Programatically, a cipher suite is referred to as:

CipherSuite cipher_suites a list of the cryptographic options supported by the client<ref>RFC 5246, p. 41</ref>

CipherSuite cipher_suite the cipher suite selected by the server and revealed in the ServerHello message<ref>RFC 5246, p. 42-43, 64

Chapter 9

Common Data Link

Common Data Link (CDL) is a secure U.S. military communications protocol. It was established by the U.S. Department of Defense in 1991 as the military's primary protocol for imagery and signals intelligence.[1][2] CDL operates within the K_u band at data rates up to 274 Mbit/s. CDL allows for full duplex data exchange. CDL signals are transmitted, received, synchronized, routed, and simulated by **Common data link (CDL) Interface Boxes (CIBs)**.

The FY06 Authorization Act (Public Law 109-163) requires use of CDL for all imagery, unless waiver is granted. The primary reason waivers are granted is from the inability to carry the 300 pound radios on a small (30 pound) aircraft. Emerging technology expects to field a 2-pound version by the end of the decade (2010).

The **Tactical Common Data Link (TCDL)** is a secure data link being developed by the U.S. military to send secure data and streaming video links from airborne platforms to ground stations. The TCDL can accept data from many different sources, then encrypt, multiplex, encode, transmit, demultiplex, and route this data at high speeds. It uses a K_u narrowband uplink that is used for both payload and vehicle control, and a wideband downlink for data transfer.

The TCDL uses both directional and omnidirectional antennas to transmit and receive the K_u band signal. The TCDL was designed for UAVs, specifically the MQ-8B Fire Scout, as well as manned non-fighter environments. The TCDL transmits radar, imagery, video, and other sensor information at rates from 1.544 Mbit/s to 10.7 Mbit/s over ranges of 200 km. It has a bit error rate of 10e-6 with COMSEC and 10e-8 without COMSEC. It is also intended that the TCDL will in time support the required higher CDL rates of 45, 137, and 274 Mbit/s.

- L-3 business segments
- Avionics Systems Standardisation Committee

9.1 References

[1] http://directory.eoportal.org/auth/8127/pres_
    TacSat2Roadrunner.html Accessed 28 June 2007.

[2] L-3 Communication Systems - West Homepage

Chapter 10

Computer security

Computer security, also known as **cybersecurity** or **IT security**, is the protection of information systems from theft or damage to the hardware, the software, and to the information on them, as well as from disruption or misdirection of the services they provide.[1] It includes controlling physical access to the hardware, as well as protecting against harm that may come via network access, data and code injection,[2] and due to malpractice by operators, whether intentional, accidental, or due to them being tricked into deviating from secure procedures.[3]

The field is of growing importance due to the increasing reliance of computer systems in most societies.[4] Computer systems now include a very wide variety of "smart" devices, including smartphones, televisions and tiny devices as part of the Internet of Things - and networks include not only the Internet and private data networks, but also Bluetooth, Wi-Fi and other wireless networks.

Computer security covers all the processes and mechanisms by which digital equipment, information and services are protected from unintended or unauthorized access, change or destruction and the process of applying security measures to ensure confidentiality, integrity, and availability of data both in transit and at rest.[5]

10.1 Vulnerabilities and attacks

Main article: Vulnerability (computing)

A vulnerability is a system susceptibility or flaw, and many vulnerabilities are documented in the Common Vulnerabilities and Exposures (CVE) database and vulnerability management is the cyclical practice of identifying, classifying, remediating, and mitigating vulnerabilities as they are discovered. An *exploitable* vulnerability is one for which at least one working attack or "exploit" exists.

To secure a computer system, it is important to understand the attacks that can be made against it, and these threats can typically be classified into one of the categories below:

10.1.1 Backdoors

A backdoor in a computer system, a cryptosystem or an algorithm, is any secret method of bypassing normal authentication or security controls. They may exist for a number of reasons, including by original design or from poor configuration. They may also have been added later by an authorized party to allow some legitimate access, or by an attacker for malicious reasons; but regardless of the motives for their existence, they create a vulnerability.

10.1.2 Denial-of-service attack

Main article: Denial-of-service attack

Denial of service attacks are designed to make a machine or network resource unavailable to its intended users. Attackers can deny service to individual victims, such as by deliberately entering a wrong password enough consecutive times to cause the victim account to be locked, or they may overload the capabilities of a machine or network and block all users at once. While a network attack from a single IP address can be blocked by adding a new firewall rule, many forms of Distributed denial of service (DDoS) attacks are possible, where the attack comes from a large number of points - and defending is much more difficult. Such attacks can originate from the zombie computers of a botnet, but a range of other techniques are possible including reflection and amplification attacks, where innocent systems are fooled into sending traffic to the victim.

10.1.3 Direct-access attacks

An unauthorized user gaining physical access to a computer is often able to directly download data from it. They

Common consumer devices that can be used to transfer data surreptitiously.

may also compromise security by making operating system modifications, installing software worms, keyloggers, or covert listening devices. Even when the system is protected by standard security measures, these may be able to be by passed by booting another operating system or tool from a CD-ROM or other bootable media. Disk encryption and Trusted Platform Module are designed to prevent these attacks.

10.1.4 Eavesdropping

Eavesdropping is the act of surreptitiously listening to a private conversation, typically between hosts on a network. For instance, programs such as Carnivore and NarusInsight have been used by the FBI and NSA to eavesdrop on the systems of internet service providers. Even machines that operate as a closed system (i.e., with no contact to the outside world) can be eavesdropped upon via monitoring the faint electro-magnetic transmissions generated by the hardware; TEMPEST is a specification by the NSA referring to these attacks.

10.1.5 Spoofing

Spoofing of user identity describes a situation in which one person or program successfully masquerades as another by falsifying data.

10.1.6 Tampering

Tampering describes a malicious modification of products. So-called "Evil Maid" attacks and security services planting of surveillance capability into routers[6] are examples.

10.1.7 Privilege escalation

Privilege escalation describes a situation where an attacker with some level of restricted access is able to, without authorization, elevate their privileges or access level. So for example a standard computer user may be able to fool the system into giving them access to restricted data; or even to "become root" and have full unrestricted access to a system.

10.1.8 Phishing

Phishing is the attempt to acquire sensitive information such as usernames, passwords, and credit card details. Phishing is typically carried out by email spoofing or instant messaging, and it often directs users to enter details at a fake website whose look and feel are almost identical to the legitimate one.

10.1.9 Clickjacking

Clickjacking, also known as "UI redress attack or User Interface redress attack", is a malicious technique in which an attacker tricks a user into clicking on a button or link on another webpage while the user intended to click on the top level page. This is done using multiple transparent or opaque layers. The attacker is basically "hijacking" the clicks meant for the top level page and routing them to some other irrelevant page, most likely owned by someone else. A similar technique can be used to hijack keystrokes. Carefully drafting a combination of stylesheets, iframes, buttons and text boxes, a user can be led into believing that they are typing the password or other information on some authentic webpage while it is being channeled into an invisible frame controlled by the attacker.

10.1.10 Social engineering and trojans

Main article: Social engineering (security)
See also: Category:Cryptographic attacks

Social engineering aims to convince a user to disclose secrets such as passwords, card numbers, etc. by, for example, impersonating a bank, a contractor, or a customer.[7]

10.2 Systems at risk

Computer security is critical in almost any industry which uses computers.[8]

10.2.1   Financial systems

Web sites that accept or store credit card numbers and bank account information are prominent hacking targets, because of the potential for immediate financial gain from transferring money, making purchases, or selling the information on the black market. In-store payment systems and ATMs have also been tampered with in order to gather customer account data and PINs.

10.2.2   Utilities and industrial equipment

Computers control functions at many utilities, including co-ordination of telecommunications, the power grid, nuclear power plants, and valve opening and closing in water and gas networks. The Internet is a potential attack vector for such machines if connected, but the Stuxnet worm demonstrated that even equipment controlled by computers not connected to the Internet can be vulnerable to physical damage caused by malicious commands sent to industrial equipment (in that case uranium enrichment centrifuges) which are infected via removable media. In 2014, the Computer Emergency Readiness Team, a division of the Department of Homeland Security, investigated 79 hacking incidents at energy companies.[9]

10.2.3   Aviation

The aviation industry is very reliant on a series of complex system which could be attacked.[10] A simple power outage at one airport can cause repercussions worldwide,[11] much of the system relies on radio transmissions which could be disrupted,[12] and controlling aircraft over oceans is especially dangerous because radar surveillance only extends 175 to 225 miles offshore.[13] There is also potential for attack from within an aircraft.[14]

The consequences of a successful attack range from loss of confidentiality to loss of system integrity, which may lead to more serious concerns such as exfiltration of data, network and air traffic control outages, which in turn can lead to airport closures, loss of aircraft, loss of passenger life, damages on the ground and to transportation infrastructure. A successful attack on a military aviation system that controls munitions could have even more serious consequences.

10.2.4   Consumer devices

Desktop computers and laptops are commonly infected with malware either to gather passwords or financial account information, or to construct a botnet to attack another target. Smart phones, tablet computers, smart watches, and other mobile devices such as Quantified Self devices like activity trackers have also become targets and many of these have sensors such as cameras, microphones, GPS receivers, compasses, and accelerometers which could be exploited, and may collect personal information, including sensitive health information. Wifi, Bluetooth, and cell phone network on any of these devices could be used as attack vectors, and sensors might be remotely activated after a successful breach.[15]

Home automation devices such as the Nest thermostat are also potential targets.[15]

10.2.5   Large corporations

Large corporations are common targets. In many cases this is aimed at financial gain through identity theft and involves data breaches such as the loss of millions of clients' credit card details by Home Depot,[16] Staples,[17] and Target Corporation.[18]

Large corporations are common targets. In many cases this is aimed at financial gain through identity theft and involves data breaches such as the loss of millions of clients' credit card details by Home Depot,[19] Staples,[20] and Target Corporation.[18]

Not all attacks are financially motivated however; for example security firm HBGary Federal suffered a serious series of attacks in 2011 from hacktivist goup Anonymous in retaliation for the firm's CEO claiming to have infiltrated their group, [21][22] and Sony Pictures was attacked in 2014 where the motive appears to have been to embarrass with data leaks, and cripple the company by wiping workstations and servers.[23][24]

10.2.6   Automobiles

If access is gained to a car's internal controller area network, it is possible to disable the brakes and turn the steering wheel.[25] Computerized engine timing, cruise control, anti-lock brakes, seat belt tensioners, door locks, airbags and advanced driver assistance systems make these disruptions possible, and self-driving cars go even further. Connected cars may use wifi and bluetooth to communicate with onboard consumer devices, and the cell phone network to contact concierge and emergency assistance services or get navigational or entertainment information; each of these networks is a potential entry point for malware or an attacker.[25] Researchers in 2011 were even able to use a malicious compact disc in a car's stereo system as a successful attack vector,[26] and cars with built-in voice recognition or remote assistance features have onboard microphones which could be used for eavesdropping.

A 2015 report by U.S. Senator Edward Markey criticized manufacturers' security measures as inadequate, and also highlighted privacy concerns about driving, location, and diagnostic data being collected, which is vulnerable to abuse by both manufacturers and hackers.[27]

10.2.7 Government

Government and military computer systems are commonly attacked by activists[28][29][30][31] and foreign powers.[32][33][34][35] Local and regional government infrastructure such as traffic light controls, police and intelligence agency communications, personnel records and financial systems are also potential targets as they are now all largely computerized.

10.3 Impact of security breaches

Serious financial damage has been caused by security breaches, but because there is no standard model for estimating the cost of an incident, the only data available is that which is made public by the organizations involved. "Several computer security consulting firms produce estimates of total worldwide losses attributable to virus and worm attacks and to hostile digital acts in general. The 2003 loss estimates by these firms range from $13 billion (worms and viruses only) to $226 billion (for all forms of covert attacks). The reliability of these estimates is often challenged; the underlying methodology is basically anecdotal."[36]

However, reasonable estimates of the financial cost of security breaches can actually help organizations make rational investment decisions. According to the classic Gordon-Loeb Model analyzing the optimal investment level in information security, one can conclude that the amount a firm spends to protect information should generally be only a small fraction of the expected loss (i.e., the expected value of the loss resulting from a cyber/information security breach).[37]

10.4 Attacker motivation

As with physical security, the motivations for breaches of computer security vary between attackers. Some are thrill-seekers or vandals, others are activists; or criminals looking for financial gain. State-sponsored attackers are now common and well resourced, but started with amateurs such as Markus Hess who hacked for the KGB, as recounted by Clifford Stoll, in *The Cuckoo's Egg*.

A standard part of threat modelling for any particular system is to identify what might motivate an attack on that system, and who might be motivated to breach it. The level and detail of precautions will vary depending on the system to be secured. A home personal computer, bank and classified military network all face very different threats, even when the underlying technologies in use are similar.

10.5 Computer protection (countermeasures)

In computer security a countermeasure is an action, device, procedure, or technique that reduces a threat, a vulnerability, or an attack by eliminating or preventing it, by minimizing the harm it can cause, or by discovering and reporting it so that corrective action can be taken.[38][39][40]

Some common countermeasures are listed in the following sections:

10.5.1 Security measures

A state of computer "security" is the conceptual ideal, attained by the use of the three processes: threat prevention, detection, and response. These processes are based on various policies and system components, which include the following:

- User account access controls and cryptography can protect systems files and data, respectively.

- Firewalls are by far the most common prevention systems from a network security perspective as they can (if properly configured) shield access to internal network services, and block certain kinds of attacks through packet filtering. Firewalls can be both hardware- or software-based.

- Intrusion Detection System (IDS) products are designed to detect network attacks in-progress and assist in post-attack forensics, while audit trails and logs serve a similar function for individual systems.

- "Response" is necessarily defined by the assessed security requirements of an individual system and may cover the range from simple upgrade of protections to notification of legal authorities, counter-attacks, and the like. In some special cases, a complete destruction of the compromised system is favored, as it may happen that not all the compromised resources are detected.

Today, computer security comprises mainly "preventive" measures, like firewalls or an exit procedure. A firewall can be defined as a way of filtering network data between a host or a network and another network, such as the Internet, and can be implemented as software running on the machine, hooking into the network stack (or, in the case of most UNIX-based operating systems such as Linux, built into the operating system kernel) to provide real time filtering and blocking. Another implementation is a so-called physical firewall which consists of a separate machine filtering network traffic. Firewalls are common amongst machines that are permanently connected to the Internet.

However, relatively few organisations maintain computer systems with effective detection systems, and fewer still have organised response mechanisms in place. As result, as Reuters points out: "Companies for the first time report they are losing more through electronic theft of data than physical stealing of assets".[41] The primary obstacle to effective eradication of cyber crime could be traced to excessive reliance on firewalls and other automated "detection" systems. Yet it is basic evidence gathering by using packet capture appliances that puts criminals behind bars.

10.5.2   Reducing vulnerabilities

While formal verification of the correctness of computer systems is possible,[42][43] it is not yet common. Operating systems formally verified include seL4,[44] and SYSGO's PikeOS[45][46] - but these make up a very small percentage of the market.

Cryptography properly implemented is now virtually impossible to directly break. Breaking them requires some non-cryptographic input, such as a stolen key, stolen plaintext (at either end of the transmission), or some other extra cryptanalytic information.

Two factor authentication is a method for mitigating unauthorized access to a system or sensitive information. It requires "something you know"; a password or PIN, and "something you have"; a card, dongle, cellphone, or other piece of hardware. This increases security as an unauthorized person needs both of these to gain access.

Social engineering and direct computer access (physical) attacks can only be prevented by non-computer means, which can be difficult to enforce, relative to the sensitivity of the information. Even in a highly disciplined environment, such as in military organizations, social engineering attacks can still be difficult to foresee and prevent.

It is possible to reduce an attacker's chances by keeping systems up to date with security patches and updates, using a security scanner or/and hiring competent people responsible for security. The effects of data loss/damage can be reduced by careful backing up and insurance.

10.5.3   Security by design

Main article: Secure by design

Security by design, or alternately secure by design, means that the software has been designed from the ground up to be secure. In this case, security is considered as a main feature.

Some of the techniques in this approach include:

- The principle of least privilege, where each part of the system has only the privileges that are needed for its function. That way even if an attacker gains access to that part, they have only limited access to the whole system.

- Automated theorem proving to prove the correctness of crucial software subsystems.

- Code reviews and unit testing, approaches to make modules more secure where formal correctness proofs are not possible.

- Defense in depth, where the design is such that more than one subsystem needs to be violated to compromise the integrity of the system and the information it holds.

- Default secure settings, and design to "fail secure" rather than "fail insecure" (see fail-safe for the equivalent in safety engineering). Ideally, a secure system should require a deliberate, conscious, knowledgeable and free decision on the part of legitimate authorities in order to make it insecure.

- Audit trails tracking system activity, so that when a security breach occurs, the mechanism and extent of the breach can be determined. Storing audit trails remotely, where they can only be appended to, can keep intruders from covering their tracks.

- Full disclosure of all vulnerabilities, to ensure that the "window of vulnerability" is kept as short as possible when bugs are discovered.

10.5.4   Security architecture

The Open Security Architecture organization defines IT security architecture as "the design artifacts that describe how the security controls (security countermeasures) are positioned, and how they relate to the overall information technology architecture. These controls serve the purpose to maintain the system's quality attributes: confidentiality, integrity, availability, accountability and assurance services".[47]

Techopedia defines security architecture as "a unified security design that addresses the necessities and potential risks involved in a certain scenario or environment. It also specifies when and where to apply security controls. The design process is generally reproducible." The key attributes of security architecture are:[48]

- the relationship of different components and how they depend on each other.

- the determination of controls based on risk assessment, good practice, finances, and legal matters.

- the standardization of controls.

10.5.5 Hardware protection mechanisms

See also: Computer security compromised by hardware failure

While hardware may be a source of insecurity, such as with microchip vulnerabilities maliciously introduced during the manufacturing process,[49][50] hardware-based or assisted computer security also offers an alternative to software-only computer security. Using devices and methods such as dongles, trusted platform modules, intrusion-aware cases, drive locks, disabling USB ports, and mobile-enabled access may be considered more secure due to the physical access (or sophisticated backdoor access) required in order to be compromised. Each of these is covered in more detail below.

- USB dongles are typically used in software licensing schemes to unlock software capabilities,[51] but they can also be seen as a way to prevent unauthorized access to a computer or other device's software. The dongle, or key, essentially creates a secure encrypted tunnel between the software application and the key. The principle is that an encryption scheme on the dongle, such as Advanced Encryption Standard (AES) provides a stronger measure of security, since it is harder to hack and replicate the dongle than to simply copy the native software to another machine and use it. Another security application for dongles is to use them for accessing web-based content such as cloud software or Virtual Private Networks (VPNs).[52] In addition, a USB dongle can be configured to lock or unlock a computer.[53]

- Trusted platform modules (TPMs) secure devices by integrating cryptographic capabilities onto access devices, through the use of microprocessors, or so-called computers-on-a-chip. TPMs used in conjunction with server-side software offer a way to detect and authenticate hardware devices, preventing unauthorized network and data access.[54]

- Computer case intrusion detection refers to a push-button switch which is triggered when a computer case is opened. The firmware or BIOS is programmed to show an alert to the operator when the computer is booted up the next time.

- Drive locks are essentially software tools to encrypt hard drives, making them inaccessible to thieves.[55] Tools exist specifically for encrypting external drives as well.[56]

- Disabling USB ports is a security option for preventing unauthorized and malicious access to an otherwise secure computer. Infected USB dongles connected to a network from a computer inside the firewall are considered by Network World as the most common hardware threat facing computer networks.[57]

- Mobile-enabled access devices are growing in popularity due to the ubiquitous nature of cell phones. Built-in capabilities such as Bluetooth, the newer Bluetooth low energy (LE), Near field communication (NFC) on non-iOS devices and biometric validation such as thumb print readers, as well as QR code reader software designed for mobile devices, offer new, secure ways for mobile phones to connect to access control systems. These control systems provide computer security and can also be used for controlling access to secure buildings.[58]

10.5.6 Secure operating systems

Main article: Security-focused operating system

One use of the term "computer security" refers to technology that is used to implement secure operating systems. Much of this technology is based on science developed in the 1980s and used to produce what may be some of the most impenetrable operating systems ever. Though still valid, the technology is in limited use today, primarily because it imposes some changes to system management and also because it is not widely understood. Such ultra-strong secure operating systems are based on operating system kernel technology that can guarantee that certain security policies are absolutely enforced in an operating environment. An example of such a Computer security policy is the Bell-LaPadula model. The strategy is based on a coupling of special microprocessor hardware features, often involving the memory management unit, to a special correctly implemented operating system kernel. This forms the foundation for a secure operating system which, if certain critical parts are designed and implemented correctly, can ensure the absolute impossibility of penetration by hostile elements. This capability is enabled because the configuration not only imposes a security policy, but in theory completely protects itself from corruption. Ordinary operating systems, on the other hand, lack the features that assure this maximal level of security. The design methodology to produce such secure systems is precise, deterministic and logical.

Systems designed with such methodology represent the state of the art of computer security although products using such security are not widely known. In sharp contrast to most kinds of software, they meet specifications with verifiable certainty comparable to specifications for size, weight and power. Secure operating systems designed this way are used primarily to protect national security information, military secrets, and the data of international financial institutions. These are very powerful security tools and very few secure operating systems have been certified at the highest level (Orange Book A-1) to operate over the range of "Top Secret" to "unclassified" (including Honeywell SCOMP, USAF SACDIN, NSA Blacker and Boeing MLS LAN). The assurance of security depends not only on the soundness of the design strategy, but also on the assurance of correctness of the implementation, and therefore there are degrees of security strength defined for COMPUSEC. The Common Criteria quantifies security strength of products in terms of two components, security functionality and assurance level (such as EAL levels), and these are specified in a Protection Profile for requirements and a Security Target for product descriptions. None of these ultra-high assurance secure general purpose operating systems have been produced for decades or certified under Common Criteria.

In USA parlance, the term High Assurance usually suggests the system has the right security functions that are implemented robustly enough to protect DoD and DoE classified information. Medium assurance suggests it can protect less valuable information, such as income tax information. Secure operating systems designed to meet medium robustness levels of security functionality and assurance have seen wider use within both government and commercial markets. Medium robust systems may provide the same security functions as high assurance secure operating systems but do so at a lower assurance level (such as Common Criteria levels EAL4 or EAL5). Lower levels mean we can be less certain that the security functions are implemented flawlessly, and therefore less dependable. These systems are found in use on web servers, guards, database servers, and management hosts and are used not only to protect the data stored on these systems but also to provide a high level of protection for network connections and routing services.

10.5.7   Secure coding

Main article: Secure coding

If the operating environment is not based on a secure operating system capable of maintaining a domain for its own execution, and capable of protecting application code from malicious subversion, and capable of protecting the system from subverted code, then high degrees of security are understandably not possible. While such secure operating systems are possible and have been implemented, most commercial systems fall in a 'low security' category because they rely on features not supported by secure operating systems (like portability, and others). In low security operating environments, applications must be relied on to participate in their own protection. There are 'best effort' secure coding practices that can be followed to make an application more resistant to malicious subversion.

In commercial environments, the majority of software subversion vulnerabilities result from a few known kinds of coding defects. Common software defects include buffer overflows, format string vulnerabilities, integer overflow, and code/command injection. These defects can be used to cause the target system to execute putative data. However, the "data" contain executable instructions, allowing the attacker to gain control of the processor.

Some common languages such as C and C++ are vulnerable to all of these defects (see Seacord, *"Secure Coding in C and C++"*).[59] Other languages, such as Java, are more resistant to some of these defects, but are still prone to code/command injection and other software defects which facilitate subversion.

Another bad coding practice occurs when an object is deleted during normal operation yet the program neglects to update any of the associated memory pointers, potentially causing system instability when that location is referenced again. This is called dangling pointer, and the first known exploit for this particular problem was presented in July 2007. Before this publication the problem was known but considered to be academic and not practically exploitable.[60]

Unfortunately, there is no theoretical model of "secure coding" practices, nor is one practically achievable, insofar as the code (ideally, read-only) and data (generally read/write) generally tends to have some form of defect.

10.5.8   Capabilities and access control lists

Main articles: Access control list and Capability (computers)

Within computer systems, two of many security models capable of enforcing privilege separation are access control lists (ACLs) and capability-based security. Using ACLs to confine programs has been proven to be insecure in many situations, such as if the host computer can be tricked into indirectly allowing restricted file access, an issue known as the confused deputy problem. It has also been shown that the promise of ACLs of giving access to an object to only one person can never be guaranteed in practice. Both of these problems are resolved by capabilities. This does not mean practical flaws exist in all ACL-based systems, but only that the designers of certain utilities must take responsibility to ensure that they do not introduce flaws.

Capabilities have been mostly restricted to research operating systems, while commercial OSs still use ACLs. Capabilities can, however, also be implemented at the language level, leading to a style of programming that is essentially a refinement of standard object-oriented design. An open source project in the area is the E language.

The most secure computers are those not connected to the Internet and shielded from any interference. In the real world, the most secure systems are operating systems where security is not an add-on.

10.5.9 Response to breaches

Responding forcefully to attempted security breaches (in the manner that one would for attempted physical security breaches) is often very difficult for a variety of reasons:

- Identifying attackers is difficult, as they are often in a different jurisdiction to the systems they attempt to breach, and operate through proxies, temporary anonymous dial-up accounts, wireless connections, and other anonymising procedures which make back-tracing difficult and are often located in yet another jurisdiction. If they successfully breach security, they are often able to delete logs to cover their tracks.

- The sheer number of attempted attacks is so large that organisations cannot spend time pursuing each attacker (a typical home user with a permanent (e.g., cable modem) connection will be attacked at least several times per day, so more attractive targets could be presumed to see many more). Note however, that most of the sheer bulk of these attacks are made by automated vulnerability scanners and computer worms.

- Law enforcement officers are often unfamiliar with information technology, and so lack the skills and interest in pursuing attackers. There are also budgetary constraints. It has been argued that the high cost of technology, such as DNA testing, and improved forensics mean less money for other kinds of law enforcement, so the overall rate of criminals not getting dealt with goes up as the cost of the technology increases. In addition, the identification of attackers across a network may require logs from various points in the network and in many countries, the release of these records to law enforcement (with the exception of being voluntarily surrendered by a network administrator or a system administrator) requires a search warrant and, depending on the circumstances, the legal proceedings required can be drawn out to the point where the records are either regularly destroyed, or the information is no longer relevant.

10.6   Notable computer security attacks and breaches

Some illustrative examples of different types of computer security breaches are given below.

10.6.1   Robert Morris and the first computer worm

Main article: Morris worm

In 1988, only 60,000 computers were connected to the Internet, and most were mainframes, minicomputers and professional workstations. On November 2, 1988, many started to slow down, because they were running a malicious code that demanded processor time and that spread itself to other computers - the first internet "computer worm".[61] The software was traced back to 23-year-old Cornell University graduate student Robert Tappan Morris, Jr. who said 'he wanted to count how many machines were connected to the Internet'.[61]

10.6.2   Rome Laboratory

In 1994, over a hundred intrusions were made by unidentified crackers into the Rome Laboratory, the US Air Force's main command and research facility. Using trojan horses, hackers were able to obtain unrestricted access to Rome's networking systems and remove traces of their activities. The intruders were able to obtain classified files, such as air tasking order systems data and furthermore able to penetrate connected networks of National Aeronautics and Space Administration's Goddard Space Flight Center, Wright-Patterson Air Force Base, some Defense contractors, and other private sector organizations, by posing as a trusted Rome center user.[62]

10.6.3   TJX loses 45.7m customer credit card details

In early 2007, American apparel and home goods company TJX announced that it was the victim of an unauthorized computer systems intrusion[63] and that the hackers had accessed a system that stored data on credit card, debit card, check, and merchandise return transactions.[64]

10.6.4   Stuxnet attack

The computer worm known as Stuxnet reportedly ruined almost one-fifth of Iran's nuclear centrifuges[65] by disrupting industrial programmable logic controllers (PLCs) in a targeted attack generally believed to have been launched by Israel and the United States[66][67][68][69] although neither has publicly acknowledged this.

10.6.5   Global surveillance disclosures

Main article: Global surveillance disclosures (2013–present)

In early 2013, massive breaches of computer security by the NSA were revealed, including deliberately inserting a backdoor in a NIST standard for encryption[70] and tapping the links between Google's data centres.[71] These were disclosed by NSA contractor Edward Snowden.[72]

10.6.6   Target and Home Depot breaches

In 2013 and 2014, a Russian/Ukrainian hacking ring known as "Rescator" broke into Target Corporation computers in 2013, stealing roughly 40 million credit cards,[73] and then Home Depot computers in 2014, stealing between 53 and 56 million credit card numbers.[74] Warnings were delivered at both corporations, but ignored; physical security breaches using self checkout machines are believed to have played a large role. "The malware utilized is absolutely unsophisticated and uninteresting," says Jim Walter, director of threat intelligence operations at security technology company McAfee - meaning that the heists could have easily been stopped by existing antivirus software had administrators responded to the warnings. The size of the thefts has resulted in major attention from state and Federal United States authorities and the investigation is ongoing.

10.7   Legal issues and global regulation

Conflict of laws in cyberspace has become a major cause of concern for computer security community. Some of the main challenges and complaints about the antivirus industry are the lack of global web regulations, a global base of common rules to judge, and eventually punish, cyber crimes and cyber criminals. There is no global cyber law and cybersecurity treaty that can be invoked for enforcing global cybersecurity issues.

International legal issues of cyber attacks are complicated in nature. Even if an antivirus firm locates the cyber criminal behind the creation of a particular virus or piece of malware or form of cyber attack, often the local authorities cannot take action due to lack of laws under which to prosecute.[75][76] Authorship attribution for cyber crimes and cyber attacks is a major problem for all law enforcement agencies.

"[Computer viruses] switch from one country to another, from one jurisdiction to another — moving around the world, using the fact that we don't have the capability to globally police operations like this. So the Internet is as if someone [had] given free plane tickets to all the online criminals of the world."[75] Use of dynamic DNS, fast flux and bullet proof servers have added own complexities to this situation.

10.8   Government

The role of the government is to make regulations to force companies and organizations to protect their systems, infrastructure and information from any cyber attacks, but also to protect its own national infrastructure such as the national power-grid.

The question of whether the government should intervene or not in the regulation of the cyberspace is a very polemical one. Indeed, for as long as it has existed and by definition, the cyberspace is a virtual space free of any government intervention. Where everyone agree that an improvement on cybersecurity is more than vital, is the government the best actor to solve this issue? Many government officials and experts think that the government should step in and that there is a crucial need for regulation, mainly due to the failure of the private sector to solve efficiently the cybersecurity problem. R. Clarke said during a panel discussion at the RSA Security Conference in San Francisco, he believes that the "industry only responds when you threaten regulation. If industry doesn't respond (to the threat), you have to follow through."[77] On the other hand, executives from the private sector agree that improvements are necessary but think that the government intervention would affect their ability to innovate efficiently.

10.9 Actions and teams in the US

10.9.1 Legislation

The 1986 18 U.S.C. § 1030, more commonly known as the Computer Fraud and Abuse Act is the key legislation. It prohibits unauthorized access or damage of "protected computers" as defined in 18 U.S.C. § 1030(e)(2).

Although various other measures have been proposed, such as the "Cybersecurity Act of 2010 - S. 773" in 2009, the "International Cybercrime Reporting and Cooperation Act - H.R.4962"[78] and "Protecting Cyberspace as a National Asset Act of 2010 - S.3480"[79] in 2010 - none of these has succeeded.

Executive order 13636 *Improving Critical Infrastructure Cybersecurity* was signed February 12, 2013.

10.9.2 Agencies

Homeland Security

The Department of Homeland Security has a dedicated division responsible for the response system, risk management program and requirements for cybersecurity in the United States called the National Cyber Security Division.[80][81] The division is home to US-CERT operations and the National Cyber Alert System.[81] The National Cybersecurity and Communications Integration Center brings together government organizations responsible for protecting computer networks and networked infrastructure.[82]

FBI

The third priority of the Federal Bureau of Investigation (FBI) is to: *"Protect the United States against cyber-based attacks and high-technology crimes"*,[83] and they, along with the National White Collar Crime Center (NW3C), and the Bureau of Justice Assistance (BJA) are part of the multi-agency task force, The Internet Crime Complaint Center, also known as IC3.[84]

In addition to its own specific duties, the FBI participates alongside non-profit organizations such as InfraGard.[85][86]

Department of Justice

In the criminal division of the United States Department of Justice operates a section called the Computer Crime and Intellectual Property Section. The CCIPS is in charge of investigating computer crime and intellectual property crime and is specialized in the search and seizure of digital evidence in computers and networks.[87]

USCYBERCOM

The United States Cyber Command, also known as US-CYBERCOM, is tasked with the defense of specified Department of Defense information networks and *"ensure US/Allied freedom of action in cyberspace and deny the same to our adversaries."*[88] It has no role in the protection of civilian networks.[89][90]

10.9.3 FCC

The U.S. Federal Communications Commission's role in cybersecurity is to strengthen the protection of critical communications infrastructure, to assist in maintaining the reliability of networks during disasters, to aid in swift recovery after, and to ensure that first responders have access to effective communications services.[91]

10.9.4 Computer Emergency Readiness Team

Computer Emergency Response Team is a name given to expert groups that handle computer security incidents. In the US, two distinct organization exist, although they do work closely together.

- US-CERT: part of the National Cyber Security Division of the United States Department of Homeland Security.[92]

- CERT/CC: created by the Defense Advanced Research Projects Agency (DARPA) and run by the Software Engineering Institute (SEI).

10.10 International actions

Many different teams and organisations exist, including:

- The Forum of Incident Response and Security Teams (FIRST) is the global association of CSIRTs.[93] The US-CERT, AT&T, Apple, Cisco, McAfee, Microsoft are all members of this international team.[94]

- The Council of Europe helps protect societies worldwide from the threat of cybercrime through the Convention on Cybercrime.[95]

- The purpose of the Messaging Anti-Abuse Working Group (MAAWG) is to bring the messaging industry together to work collaboratively and to successfully address the various forms of messaging abuse, such as spam, viruses, denial-of-service attacks and other messaging exploitations.[96] France Telecom, Facebook, AT&T, Apple, Cisco, Sprint are some of the members of the MAAWG.[96]

- ENISA : The European Network and Information Security Agency (ENISA) is an agency of the European Union with the objective to improve network and information security in the European Union.

10.10.1   Germany

Berlin starts National Cyber Defense Initiative

On June 16, 2011, the German Minister for Home Affairs, officially opened the new German NCAZ (National Center for Cyber Defense) Nationales Cyber-Abwehrzentrum, which is located in Bonn. The NCAZ closely cooperates with BSI (Federal Office for Information Security) Bundesamt für Sicherheit in der Informationstechnik, BKA (Federal Police Organisation) Bundeskriminalamt (Deutschland), BND (Federal Intelligence Service) Bundesnachrichtendienst, MAD (Military Intelligence Service) Amt für den Militärischen Abschirmdienst and other national organisations in Germany taking care of national security aspects. According to the Minister the primary task of the new organisation founded on February 23, 2011, is to detect and prevent attacks against the national infrastructure and mentioned incidents like Stuxnet.

10.10.2   South Korea

Following cyberattacks in the first half of 2013, when government, news-media, television station, and bank websites were compromised, the national government committed to the training of 5,000 new cybersecurity experts by 2017. The South Korean government blamed its northern counterpart for these attacks, as well as incidents that occurred in 2009, 2011,[97] and 2012, but Pyongyang denies the accusations.[98]

10.10.3   India

Some provisions for cybersecurity have been incorporated into rules framed under the Information Technology Act 2000.

The National Cyber Security Policy 2013 is a policy framework by Department of Electronics and Information Technology (DeitY) which aims to protect the public and private infrastructure from cyber attacks, and safeguard "information, such as personal information (of web users), financial and banking information and sovereign data".

The Indian Companies Act 2013 has also introduced cyber law and cyber security obligations on the part of Indian directors.

10.10.4   Canada

On October 3, 2010, Public Safety Canada unveiled Canada's Cyber Security Strategy, following a Speech from the Throne commitment to boost the security of Canadian cyberspace.[99][100] The aim of the strategy is to strengthen Canada's "cyber systems and critical infrastructure sectors, support economic growth and protect Canadians as they connect to each other and to the world."[101] Three main pillars define the strategy: securing government systems, partnering to secure vital cyber systems outside the federal government, and helping Canadians to be secure online.[101] The strategy involves multiple departments and agencies across the Government of Canada.[102] The Cyber Incident Management Framework for Canada outlines these responsibilities, and provides a plan for coordinated response between government and other partners in the event of a cyber incident.[103] The Action Plan 2010-2015 for Canada's Cyber Security Strategy outlines the ongoing implementation of the strategy.[104]

Public Safety Canada's Canadian Cyber Incident Response Centre (CCIRC) is responsible for mitigating and responding to threats to Canada's critical infrastructure and cyber systems. The CCIRC provides support to mitigate cyber threats, technical support to respond and recover from targeted cyber attacks, and provides online tools for members of Canada's critical infrastructure sectors.[105] The CCIRC posts regular cyber security bulletins on the Public Safety Canada website.[106] The CCIRC also operates an online reporting tool where individuals and organizations can report a cyber incident.[107] Canada's Cyber Security Strategy is part of a larger, integrated approach to critical infrastructure protection, and functions as a counterpart document to the National Strategy and Action Plan for Critical Infrastructure.[102]

On September 27, 2010, Public Safety Canada partnered with STOP.THINK.CONNECT, a coalition of nonprofit, private sector, and government organizations dedicated to informing the general public on how to protect themselves online.[108] On February 4, 2014, the Government of Canada launched the Cyber Security Cooperation Program.[109] The program is a $1.5 million five-year initiative aimed at improving Canada's cyber systems through grants and contributions to projects in support of this objective.[110] Public Safety Canada aims to begin an evaluation of Canada's Cyber Security Strategy in early 2015.[102] Public Safety Canada administers and routinely updates the GetCyberSafe portal for Canadian citizens, and carries out Cyber Security Awareness Month during October.[111]

10.11 National teams

Here are the main computer emergency response teams around the world. Every country have their own team to protect network security. February 27, 2014, the Chinese network security and information technology leadership team is established. The leadership team will focus on national security and long-term development, co-ordination of major issues related to network security and information technology economic, political, cultural, social, and military and other fields of research to develop network security and information technology strategy, planning and major macroeconomic policy promote national network security and information technology law, and constantly enhance security capabilities.

10.11.1 Europe

CSIRTs in Europe collaborate in the TERENA task force TF-CSIRT. TERENA's Trusted Introducer service provides an accreditation and certification scheme for CSIRTs in Europe. A full list of known CSIRTs in Europe is available from the Trusted Introducer website.

10.11.2 Other countries

- CERT Brazil, member of FIRST (Forum for Incident Response and Security Teams)

- CARNet CERT, Croatia, member of FIRST

- AE CERT, United Arab Emirates

- SingCERT, Singapore

- CERT-LEXSI, France, Canada, Singapore

10.12 Modern warfare

Main article: Cyberwarfare

Cybersecurity is becoming increasingly important as more information and technology is being made available on cyberspace. There is growing concern among governments that cyberspace will become the next theatre of warfare. As Mark Clayton from the *Christian Science Monitor* described in an article titled "The New Cyber Arms Race":

> In the future, wars will not just be fought by soldiers with guns or with planes that drop bombs. They will also be fought with the click of a mouse a half a world away that unleashes carefully weaponized computer programs that disrupt or destroy critical industries like utilities, transportation, communications, and energy. Such attacks could also disable military networks that control the movement of troops, the path of jet fighters, the command and control of warships.[112]

This has led to new terms such as *cyberwarfare* and *cyberterrorism*. More and more critical infrastructure is being controlled via computer programs that, while increasing efficiency, exposes new vulnerabilities. The test will be to see if governments and corporations that control critical systems such as energy, communications and other information will be able to prevent attacks before they occur. As Jay Cross, the chief scientist of the Internet Time Group, remarked, "Connectedness begets vulnerability."[112]

10.13 The cyber security job market

Cyber Security is a fast-growing[113] field of IT concerned with reducing organizations' risk of hack or data breach. Commercial, government and non-governmental organizations all employ cybersecurity professionals. However, the use of the term "cybersecurity" is more prevalent in government job descriptions.[114]

Typical cybersecurity job titles and descriptions include:[115]

Security Analyst Analyzes and assesses vulnerabilities in the infrastructure (software, hardware, networks), investigates available tools and countermeasures to remedy the detected vulnerabilities, and recommends solutions and best practices. Analyzes and assesses damage to the data/infrastructure as a result of security incidents, examines available recovery tools and processes, and recommends solutions. Tests for compliance with security policies and procedures. May assist in the creation, implementation, and/or management of security solutions.

Security Engineer

Performs security monitoring, security and data/logs analysis, and forensic analysis, to detect security incidents, and mounts incident response. Investigates and utilizes new technologies and processes to enhance security capabilities and implement improvements. May also review code or perform other security engineering methodologies.

Security Architect

Designs a security system or major components of a security system, and may head a security design team building a new security system.

Security Administrator

Installs and manages organization-wide security systems. May also take on some of the tasks of a security analyst in smaller organizations.

Chief Information Security Officer (CISO)

A high-level management position responsible for the entire information security division/staff. The position may include hands-on technical work.

Chief Security Officer (CSO)

A high-level management position responsible for the entire security division/staff. A newer position now deemed needed as security risks grow.

Security Consultant/Specialist/Intelligence

Broad titles that encompass any one or all of the other roles/titles, tasked with protecting computers, networks, software, data, and/or information systems against viruses, worms, spyware, malware, intrusion detection, unauthorized access, denial-of-service attacks, and an ever increasing list of attacks by hackers acting as individuals or as part of organized crime or foreign governments.

Student programs are also available to people interested in beginning a career in cybersecurity.[116][117] Meanwhile, a flexible and effective option for information security professionals of all experience levels to keep studying is online security training, including webcasts.[118][119][120]

10.14   Terminology

The following terms used with regards to engineering secure systems are explained below.

- Access authorization restricts access to a computer to group of users through the use of authentication systems. These systems can protect either the whole computer – such as through an interactive login screen – or individual services, such as an FTP server. There are many methods for identifying and authenticating users, such as passwords, identification cards, and, more recently, smart cards and biometric systems.

- Anti-virus software consists of computer programs that attempt to identify, thwart and eliminate computer viruses and other malicious software (malware).

- Applications with known security flaws should not be run. Either leave it turned off until it can be patched or otherwise fixed, or delete it and replace it with some other application. Publicly known flaws are the main entry used by worms to automatically break into a system and then spread to other systems connected to it. The security website Secunia provides a search tool for unpatched known flaws in popular products.

- Authentication techniques can be used to ensure that communication end-points are who they say they are.

- Automated theorem proving and other verification tools can enable critical algorithms and code used in secure systems to be mathematically proven to meet their specifications.

- Backups are a way of securing information; they are another copy of all the important computer files kept in another location. These files are kept on hard disks, CD-Rs, CD-RWs, tapes and more recently on the cloud. Suggested locations for backups are a fireproof, waterproof, and heat proof safe, or in a separate, off-site location than that in which the original files are contained. Some individuals and companies also keep their backups in safe deposit boxes inside bank vaults. There is also a fourth option, which involves using one of the file hosting services that backs up files over the Internet for both business and individuals, known as the cloud.

  - Backups are also important for reasons other than security. Natural disasters, such as earthquakes, hurricanes, or tornadoes, may strike the building where the computer is located. The building can be on fire, or an explosion may occur. There needs to be a recent backup at an alternate secure location, in case of such kind of disaster. Further, it is recommended that the alternate location be placed where the same disaster would not affect both locations. Examples of alternate disaster recovery sites being compromised by the same disaster that affected the primary site include having had a primary site in World Trade Center I and the recovery site in 7 World Trade Center, both of which were destroyed in the 9/11 attack, and having one's primary site and recovery site in the same coastal region, which leads to both being vulnerable to hurricane damage (for example, primary site in New Orleans and recovery site in Jefferson Parish, both of which were hit by Hurricane Katrina in 2005). The backup media should be moved between the geographic sites in a secure manner, in order to prevent them from being stolen.

- Capability and access control list techniques can be used to ensure privilege separation and mandatory access control. This section discusses their use.

- Chain of trust techniques can be used to attempt to ensure that all software loaded has been certified as authentic by the system's designers.

- Confidentiality is the nondisclosure of information except to another authorized person.[121]

- Cryptographic techniques can be used to defend data in transit between systems, reducing the probability that data exchanged between systems can be intercepted or modified.

- Cyberwarfare is an Internet-based conflict that involves politically motivated attacks on information and information systems. Such attacks can, for example, disable official websites and networks, disrupt or disable essential services, steal or alter classified data, and criple financial systems.

- Data integrity is the accuracy and consistency of stored data, indicated by an absence of any alteration in data between two updates of a data record.[122]

This is secret stuff, PSE do not...

➡ **5a0 (k$hQ% ...**

➡ **This is secret stuff, PSE do not...**

Cryptographic techniques involve transforming information, scrambling it so it becomes unreadable during transmission. The intended recipient can unscramble the message; ideally, eavesdroppers cannot.

- Encryption is used to protect the message from the eyes of others. Cryptographically secure ciphers are designed to make any practical attempt of breaking infeasible. Symmetric-key ciphers are suitable for bulk encryption using shared keys, and public-key encryption using digital certificates can provide a practical solution for the problem of securely communicating when no key is shared in advance.

- Endpoint security software helps networks to prevent exfiltration (data theft) and virus infection at network entry points made vulnerable by the prevalence of potentially infected portable computing devices, such as laptops and mobile devices, and external storage devices, such as USB drives.[123]

- Firewalls are an important method for control and security on the Internet and other networks. A network firewall can be a communications processor, typically a router, or a dedicated server, along with firewall software. A firewall serves as a gatekeeper system that protects a company's intranets and other computer networks from intrusion by providing a filter and safe transfer point for access to and from the Internet and other networks. It screens all network traffic for proper passwords or other security codes and only allows authorized transmission in and out of the network. Firewalls can deter, but not completely prevent, unauthorized access (hacking) into computer networks; they can also provide some protection from online intrusion.

- Honey pots are computers that are either intentionally or unintentionally left vulnerable to attack by crackers. They can be used to catch crackers or fix vulnerabilities.

- Intrusion-detection systems can scan a network for people that are on the network but who should not be there or are doing things that they should not be doing, for example trying a lot of passwords to gain access to the network.

- A microkernel is the near-minimum amount of software that can provide the mechanisms to implement an operating system. It is used solely to provide very low-level, very precisely defined machine code upon which an operating system can be developed. A simple example is the early '90s GEMSOS (Gemini Computers), which provided extremely low-level machine code, such as "segment" management, atop which an operating system could be built. The theory (in the case of "segments") was that—rather than have the operating system itself worry about mandatory access separation by means of military-style labeling—it is safer if a low-level, independently scrutinized module can be charged **solely** with the management of individually labeled segments, be they memory "segments" or file system "segments" or executable text "segments." If software below the visibility of the operating system is (as in this case) charged with labeling, there is no theoretically viable means for a clever hacker to subvert the labeling scheme, since the operating system *per se* does **not** provide mechanisms for interfering with labeling: the operating system is, essentially, a client (an "application," arguably) atop the microkernel and, as such, subject to its restrictions.

- Pinging The ping application can be used by potential crackers to find if an IP address is reachable. If a cracker finds a computer, they can try a port scan to detect and attack services on that computer.

- Social engineering awareness keeps employees aware of the dangers of social engineering and/or having a policy in place to prevent social engineering can reduce successful breaches of the network and servers.

10.15 Scholars

- Salvatore J. Stolfo
- Ross J. Anderson
- Annie Anton
- Adam Back
- Daniel J. Bernstein
- Matt Blaze
- Stefan Brands
- L. Jean Camp
- Lance Cottrell
- Lorrie Cranor
- Dorothy E. Denning
- Peter J. Denning
- Cynthia Dwork
- Deborah Estrin
- Joan Feigenbaum
- Ian Goldberg
- Shafi Goldwasser
- Lawrence A. Gordon
- Peter Gutmann

- Paul Kocher
- Monica S. Lam
- Butler Lampson
- Brian LaMacchia
- Carl Landwehr
- Kevin Mitnick
- Peter G. Neumann
- Susan Nycum
- Roger R. Schell
- Bruce Schneier
- Dawn Song
- Gene Spafford
- Joseph Steinberg
- Willis Ware
- Moti Yung

10.16 See also

- Attack tree
- CAPTCHA
- CERT
- CertiVox
- Cloud computing security
- Comparison of antivirus software
- Computer insecurity
- Computer security model
- Content security
- Countermeasure (computer)
- Cyber security standards
- Dancing pigs
- Data loss prevention products
- Data security
- Differentiated security
- Disk encryption

- Exploit (computer security)

- Fault tolerance

- Human-computer interaction (security)

- Identity Based Security

- Identity management

- Identity theft

- Information Leak Prevention

- Information Security Awareness

- Internet privacy

- ISO/IEC 15408

- IT risk

- List of Computer Security Certifications

- Mobile security

- Network security

- Network Security Toolkit

- Next-Generation Firewall

- Open security

- OWASP

- Penetration test

- Physical information security

- Presumed security

- Privacy software

- Proactive Cyber Defence

- Risk cybernetics

- Sandbox (computer security)

- Separation of protection and security

- Software Defined Perimeter

- Cyber Insurance

10.17 Further reading

- Chwan-Hwa (John) Wu and J. David Irwin, *Introduction to Computer Networks and Cybersecurity* (Boca Raton: CRC Press, 2013), ISBN 978-1466572133.

- Newton Lee, *Counterterrorism and Cybersecurity: Total Information Awareness (Second Edition)* (Switzerland: Springer International Publishing, 2015), ISBN 978-3-319-17243-9.

- P. W. Singer and Allan Friedman, *Cybersecurity and Cyberwar: What Everyone Needs to Know* (Oxford: Oxford University Press, 2014), ISBN 978-0199918119.

- Peter Kim, *The Hacker Playbook: Practical Guide To Penetration Testing* (Seattle: CreateSpace Independent Publishing Platform, 2014), ISBN 978-1494932633.

10.18 References

[1] Gasser, Morrie (1988). *Building a Secure Computer System* (PDF). Van Nostrand Reinhold. p. 3. ISBN 0-442-23022-2. Retrieved 6 September 2015.

[2] "Definition of computer security". *Encyclopedia*. Ziff Davis, PCMag. Retrieved 6 September 2015.

[3] Rouse, Margaret. "Social engineering definition". TechTarget. Retrieved 6 September 2015.

[4] "Reliance spells end of road for ICT amateurs", May 07, 2013, The Australian

[5] http://www.evolllution.com/opinions/cybersecurity-understanding-online-threat/

[6] Gallagher, Sean (May 14, 2014). "Photos of an NSA "upgrade" factory show Cisco router getting implant". Ars Technica. Retrieved August 3, 2014.

[7] Arcos Sergio. "Social Engineering" (PDF).

[8] J. C. Willemssen, "FAA Computer Security". GAO/T-AIMD-00-330. Presented at Committee on Science, House of Representatives, 2000.

[9] Pagliery, Jose. "Hackers attacked the U.S. energy grid 79 times this year". *CNN Money*. Cable News Network. Retrieved 16 April 2015.

[10] P. G. Neumann, "Computer Security in Aviation," presented at International Conference on Aviation Safety and Security in the 21st Century, White House Commission on Safety and Security, 1997.

[11] J. Zellan, Aviation Security. Hauppauge, NY: Nova Science, 2003, pp. 65–70.

[12] http://www.securityweek.com/
air-traffic-control-systems-vulnerabilities-could-make-unfriendly-skies-bit-less

[13] http://www.npr.org/blogs/
alltechconsidered/2014/08/04/337794061/
hacker-says-he-can-break-into-airplane-systems-using-in-flight-wi-fi

[14] http://www.reuters.com/article/2014/08/04/
us-cybersecurity-hackers-airplanes-idUSKBN0G40WQ20140804

[15] http://www.npr.org/blogs/
alltechconsidered/2014/08/06/338334508/
is-your-watch-or-thermostat-a-spy-cyber-security-firms-are-on-it

[16] Melvin Backman (18 September 2014). "Home Depot: 56 million cards exposed in breach". CNNMoney.

[17] "Staples: Breach may have affected 1.16 million customers' cards". Fortune.com. December 19, 2014. Retrieved 2014-12-21.

[18] "Target security breach affects up to 40M cards". *Associated Press via Milwaukee Journal Sentinel*. 19 December 2013. Retrieved 21 December 2013.

[19] Melvin Backman (18 September 2014). "Home Depot: 56 million cards exposed in breach". CNNMoney.

[20] "Staples: Breach may have affected 1.16 million customers' cards". Fortune.com. December 19, 2014. Retrieved 2014-12-21.

[21] Bright, Peter (February 15, 2011). "Anonymous speaks: the inside story of the HBGary hack". Arstechnica.com. Retrieved March 29, 2011.

[22] Anderson, Nate (February 9, 2011). "How one man tracked down Anonymous—and paid a heavy price". Arstechnica.com. Retrieved March 29, 2011.

[23] Palilery, Jose (December 24, 2014). "What caused Sony hack: What we know now". CNN Money. Retrieved January 4, 2015.

[24] James Cook (December 16, 2014). "Sony Hackers Have Over 100 Terabytes Of Documents. Only Released 200 Gigabytes So Far". *Business Insider*. Retrieved December 18, 2014.

[25] http://www.vox.com/2015/1/18/7629603/
car-hacking-dangers

[26] http://www.autosec.org/pubs/cars-usenixsec2011.pdf

[27] http://www.markey.senate.gov/imo/media/doc/
2015-02-06_MarkeyReport-Tracking_Hacking_
CarSecurity%202.pdf

[28] "Internet strikes back: Anonymous' Operation Megaupload explained". *RT*. January 20, 2012. Archived from the original on May 5, 2013. Retrieved May 5, 2013.

[29] "Gary McKinnon profile: Autistic 'hacker' who started writing computer programs at 14". *The Daily Telegraph* (London). 23 January 2009.

[30] "Gary McKinnon extradition ruling due by 16 October". BBC News. September 6, 2012. Retrieved September 25, 2012.

[31] Law Lords Department (30 July 2008). "House of Lords - Mckinnon V Government of The United States of America and Another". Publications.parliament.uk. Retrieved 30 January 2010. 15. ... alleged to total over $700,000

[32] "NSA Accessed Mexican President's Email", October 20, 2013, Jens Glüsing, Laura Poitras, Marcel Rosenbach and Holger Stark, spiegel.de

[33] Sanders, Sam (4 June 2015). "Massive Data Breach Puts 4 Million Federal Employees' Records At Risk". *NPR*. Retrieved 5 June 2015.

[34] Liptak, Kevin (4 June 2015). "U.S. government hacked; feds think China is the culprit". *CNN*. Retrieved 5 June 2015.

[35] Sean Gallagher. "Encryption "would not have helped" at OPM, says DHS official".

[36] Cashell, B., Jackson, W. D., Jickling, M., & Webel, B. (2004). The Economic Impact of Cyber-Attacks. Congressional Research Service, Government and Finance Division. Washington DC: The Library of Congress.

[37] Gordon, Lawrence; Loeb, Martin (November 2002). "The Economics of Information Security Investment". *ACM Transactions on Information and System Security* **5** (4): 438–457. doi:10.1145/581271.581274.

[38] RFC 2828 Internet Security Glossary

[39] CNSS Instruction No. 4009 dated 26 April 2010

[40] InfosecToday Glossary

[41] "Firms lose more to electronic than physical theft". Reuters.

[42] Harrison, J. (2003). "Formal verification at Intel". pp. 45–54. doi:10.1109/LICS.2003.1210044.

[43] Formal verification of a real-time hardware design. Portal.acm.org (1983-06-27). Retrieved on April 30, 2011.

[44] "Abstract Formal Specification of the seL4/ARMv6 API" (PDF). Retrieved May 19, 2015.

[45] Christoph Baumann, Bernhard Beckert, Holger Blasum, and Thorsten Bormer Ingredients of Operating System Correctness? Lessons Learned in the Formal Verification of PikeOS

[46] "Getting it Right" by Jack Ganssle

[47] Definitions: IT Security Architecture. SecurityArchitecture.org, Jan, 2006

[48] Jannsen, Cory. "Security Architecture". *Techopedia*. Janalta Interactive Inc. Retrieved 9 October 2014.

[49] *The Hacker in Your Hardware: The Next Security Threat* August 4, 2010 Scientific American

[50] Waksman, Adam; Sethumadhavan, Simha (2010), "Tamper Evident Microprocessors" (PDF), *Proceedings of the IEEE Symposium on Security and Privacy* (Oakland, California)

[51] "Sentinel HASP HL". E-Spin. Retrieved 2014-03-20.

[52] "Token-based authentication". SafeNet.com. Retrieved 2014-03-20.

[53] "Lock and protect your Windows PC". TheWindowsClub.com. Retrieved 2014-03-20.

[54] James Greene (2012). "Intel Trusted Execution Technology: White Paper" (PDF). Intel Corporation. Retrieved 2013-12-18.

[55] "SafeNet ProtectDrive 8.4". SCMagazine.com. 2008-10-04. Retrieved 2014-03-20.

[56] "Secure Hard Drives: Lock Down Your Data". PC-Mag.com. 2009-05-11.

[57] "Top 10 vulnerabilities inside the network". Network World. 2010-11-08. Retrieved 2014-03-20.

[58] "Forget IDs, use your phone as credentials". Fox Business Network. 2013-11-04. Retrieved 2014-03-20.

[59] "Secure Coding in C and C++, Second Edition". Cert.org. Retrieved 2013-09-25.

[60] New hacking technique exploits common programming error. SearchSecurity.com, July 2007

[61] Jonathan Zittrain, 'The Future of The Internet', Penguin Books, 2008

[62] Information Security. United States Department of Defense, 1986

[63] "THE TJX COMPANIES, INC. VICTIMIZED BY COMPUTER SYSTEMS INTRUSION; PROVIDES INFORMATION TO HELP PROTECT CUSTOMERS" (Press release). The TJX Companies, Inc. 2007-01-17. Retrieved 2009-12-12.

[64] Largest Customer Info Breach Grows. MyFox Twin Cities, 29 March 2007.

[65] "The Stuxnet Attack On Iran's Nuclear Plant Was 'Far More Dangerous' Than Previously Thought". Business Insider. 20 November 2013.

[66] Reals, Tucker (24 September 2010). "Stuxnet Worm a U.S. Cyber-Attack on Iran Nukes?". CBS News.

[67] Kim Zetter (17 February 2011). "Cyberwar Issues Likely to Be Addressed Only After a Catastrophe". Wired. Retrieved 18 February 2011.

[68] Chris Carroll (18 October 2011). "Cone of silence surrounds U.S. cyberwarfare". Stars and Stripes. Retrieved 30 October 2011.

[69] John Bumgarner (27 April 2010). "Computers as Weapons of War" (PDF). IO Journal. Retrieved 30 October 2011.

[70] "Can You Trust NIST?".

[71] "New Snowden Leak: NSA Tapped Google, Yahoo Data Centers", Oct 31, 2013, Lorenzo Franceschi-Bicchierai, mashable.com

[72] Seipel, Hubert. "Transcript: ARD interview with Edward Snowden". *La Foundation Courage*. Retrieved 11 June 2014.

[73] "Missed Alarms and 40 Million Stolen Credit Card Numbers: How Target Blew It"

[74] "Home Depot says 53 million emails stolen"

[75] "Mikko Hypponen: Fighting viruses, defending the net". TED.

[76] "Mikko Hypponen - Behind Enemy Lines". Hack In The Box Security Conference.

[77] Kirby, Carrie (June 24, 2011). "Former White House aide backs some Net regulation / Clarke says government, industry deserve 'F' in cybersecurity". *The San Francisco Chronicle*.

[78] "Text of H.R.4962 as Introduced in House: International Cybercrime Reporting and Cooperation Act - U.S. Congress". OpenCongress. Retrieved 2013-09-25.

[79] Archived July 4, 2015 at the Wayback Machine

[80] "National Cyber Security Division". U.S. Department of Homeland Security. Retrieved June 14, 2008.

[81] "FAQ: Cyber Security R&D Center". U.S. Department of Homeland Security S&T Directorate. Retrieved June 14, 2008.

[82] AFP-JiJi, "U.S. boots up cybersecurity center", October 31, 2009.

[83] "Federal Bureau of Investigation - Priorities". Federal Bureau of Investigation.

[84] Internet Crime Complaint Center

[85] "Infragard, Official Site". *Infragard*. Retrieved 10 September 2010.

[86] "Robert S. Mueller, III -- InfraGard Interview at the 2005 InfraGard Conference". *Infragard (Official Site) -- "Media Room"*. Retrieved 9 December 2009.

[87] "CCIPS".

[88] U.S. Department of Defense, Cyber Command Fact Sheet, May 21, 2010 http://www.stratcom.mil/factsheets/Cyber_Command/

[89] "Speech:". Defense.gov. Retrieved 2010-07-10.

[90] Shachtman, Noah. "Military's Cyber Commander Swears: "No Role" in Civilian Networks", The Brookings Institution, 23 September 2010.

[91] "FCC Cybersecurity". FCC.

[92] Verton, Dan (January 28, 2004). "DHS launches national cyber alert system". *Computerworld* (IDG). Retrieved 2008-06-15.

[93] "FIRST website".

[94] "First members".

[95] "European council".

[96] "MAAWG".

[97] "South Korea seeks global support in cyber attack probe". *BBC Monitoring Asia Pacific*. 7 March 2011.

[98] Kwanwoo Jun (23 September 2013). "Seoul Puts a Price on Cyberdefense". *Wall Street Journal*. Dow Jones & Company, Inc. Retrieved 24 September 2013.

[99] "Government of Canada Launches Canada's Cyber Security Strategy". *Market Wired*. 3 October 2010. Retrieved 1 November 2014.

[100] "Canada's Cyber Security Strategy".

[101] "Canada's Cyber Security Strategy". *Public Safety Canada*. Government of Canada. Retrieved 1 November 2014.

[102] "Action Plan 2010-2015 for Canada's Cyber Security Strategy". *Public Safety Canada*. Government of Canada. Retrieved 3 November 2014.

[103] "Cyber Incident Management Framework For Canada". *Public Safety Canada*. Government of Canada. Retrieved 3 November 2014.

[104] "Action Plan 2010-2015 for Canada's Cyber Security Strategy". *Public Safety Canada*. Government of Canada. Retrieved 1 November 2014.

[105] "Canadian Cyber Incident Response Centre". *Public Safety Canada*. Retrieved 1 November 2014.

[106] "Cyber Security Bulletins". *Public Safety Canada*. Retrieved 1 November 2014.

[107] "Report a Cyber Security Incident". *Public Safety Canada*. Government of Canada. Retrieved 3 November 2014.

[108] "Government of Canada Launches Cyber Security Awareness Month With New Public Awareness Partnership". *Market Wired* (Government of Canada). 27 September 2012. Retrieved 3 November 2014.

[109] "Cyber Security Cooperation Program". *Public Safety Canada*. Retrieved 1 November 2014.

[110] "Cyber Security Cooperation Program". *Public Safety Canada*.

[111] "GetCyberSafe". *Get Cyber Safe*. Government of Canada. Retrieved 3 November 2014.

[112] Clayton, Mark. "The new cyber arms race". *The Christian Science Monitor*. Retrieved 16 April 2015.

[113] "The Growth of Cybersecurity Jobs". Mar 2014. Retrieved 24 April 2014.

[114] de Silva, Richard (11 Oct 2011). "Government vs. Commerce: The Cyber Security Industry and You (Part One)". Defence IQ. Retrieved 24 Apr 2014.

[115] "Department of Computer Science". Retrieved April 30, 2013.

[116] "(Information for) Students". NICCS (US National Initiative for Cybercareers and Studies). Retrieved 24 April 2014.

[117] "Current Job Opportunities at DHS". U.S. Department of Homeland Security. Retrieved 2013-05-05.

[118] "Cybersecurity Training & Exercises". U.S. Department of Homeland Security. Retrieved 2015-01-09.

[119] "Cyber Security Awareness Free Training and Webcasts". MS-ISAC (Multi-State Information Sharing & Analysis Center. Retrieved 9 January 2015.

[120] "Security Training Courses". LearnQuest. Retrieved 2015-01-09.

[121] "Confidentiality". Retrieved 2011-10-31.

[122] "Data Integrity". Retrieved 2011-10-31.

[123] "Endpoint Security". Retrieved 2014-03-15.

10.19   External links

- Computer security at DMOZ

Chapter 11

CONDOR secure cell phone

Qualcomm built several prototype secure CDMA phones for NSA under a contract project called "Condor". The NSA insisted on hardware encryption, which Qualcomm originally implemented using Fortezza PC cards, but later it became apparent that what the NSA really wanted was developed as the STU-III.

11.1   See also

- Secure Communications Interoperability Protocol

- Sectéra Secure Module for Motorola GSM cell phones

Chapter 12

Crypto phone

Crypto phones are mobile telephones that provide security against eavesdropping and electronic surveillance.

The interception of telecommunications has become a major industry. Most of the world's intelligence agencies and many private organisations intercept telephone communications to obtain military, economic and political information. The price of simple mobile phone surveillance devices has become so low that many individuals can afford to use them.[1][2] Advances in technology have made it difficult to determine who is intercepting and recording private communications.

Crypto phones can protect calls from interception by using algorithms to encrypt the signals. The phones have a cryptographic chip that handles encryption and decryption. Two algorithms are programmed into the chip: A key-exchange algorithm for the key agreement protocol and a symmetric-key algorithm for voice encryption.

12.1 Prevention

For the system to work, both users must have crypto phones logged into crypto mode. As with other phones, the signal is encrypted by GSM but it is also encrypted by the cryptographic chip. When the IMSI-catcher performs a man-in-the-middle attack and disables the GSM encryption, the crypto phone encryption remains intact. Therefore, while the signal is still being intercepted, it can no longer be decoded and fake SMS messages can't be sent as the IMSI-catcher does not have the correct code.

12.2 Authentication

At the beginning of the call, both users get the same session key by using the hash function. Then the session key becomes a confirm code. The confirm code could be 3 letters or 4 numbers, depending on the phone's manufacturer. In the crypto mode, the user reads the confirm code over the encrypted line to his communication partner and verifies the confirm code his partner reads back. If there is a discrepancy in the confirm code, a man-in-the-middle attack has been detected.

12.3 Key Erase

The "session code" that has been established is used only for that specific call. At termination, all the parameters are wiped from memory, and there is no way to reconstruct the code. Intercepted and stored encrypted material can be kept for later analysis, but there is no way to break the code except, possibly, by the time consuming trial-and-error method.

12.4 See also

- Secure telephone
- Secure voice
- PGPfone
- ZRTP

12.5 References

[1] Hackers crack open mobile network BBC News Technology

[2] van den Broek, Fabian. Eavesdropping on GSM: state-of-affairs, Radboud University, Nijmegen, Institute for Computing and Information Sciences (iCIS).

12.6 External links

Chapter 13

CSipSimple

CSipSimple is a Voice over Internet Protocol (VoIP) application for Google Android operating system using the Session Initiation Protocol (SIP).[3][4] It is open source and free software released under the GNU General Public License.

13.1 Details

It relies on the PJSIP SIP stack and get features provided by this SIP stack.[5]

The key features of this software are:

- Multi-codec support: Speex (narrow-band/wideband), G.711 (u-law/a-law), GSM, iLBC, G.729 (support dropped with r2180, need to buy a licensed g729 pluggin), G.722, AMR (narrow-band), iSAC, SILK (narrow-band/wide-band/ultra wide-band) (support dropped in 2014)

- A plug-in adds support for Codec2, G.726, G.722.1 and Opus

- A plug-in adds video calling with VP8, H264 and H263-1998 codecs

- Multi-account support: up to 10 accounts can be activated at the same time

- Can use native audio driver

- NAT traversal using STUN, TURN and ICE

- Integration with Android operating system with filters and rewriting rules

- Security and encryption with SRTP, SIP over TLS 1.0 and ZRTP

- SIP SIMPLE messaging

- An API for third party applications is available [6]

- Packet loss concealment (PLC) using PJSIP[7]

- Support for IPv6 - If the hardware, Android version, ISP and all other parts of the connections involved can handle IPv6, then Csipsimple can be used to make direct end-to-end ipv6-to-ipv6 calls.

13.2 Reviews

As of 2011, reviews are favourable.[8][9]

13.3 See also

- Comparison of VoIP software
- List of SIP software
- Mobile VoIP

13.4 References

[1] https://code.google.com/p/csipsimple/source/diff?spec=svn2448&r=2448&format=side&path=/trunk/CSipSimple/AndroidManifest.xml

[2] "CSipSimple nightlies".

[3] "CSipSimple". *Google Code*. Google. Retrieved 22 August 2011.

[4] "CSipSimple "OpenSource SIP", for Android". *IPComms*. IP Communications. Retrieved 22 August 2011.

[5] "SIP and Media Features". *pjsip.org*. 12 December 2007.

[6] "Javadoc of CSipSimple API". r3gis3r. 8 April 2012.

[7] code.google.com

[8] Michael (7 January 2011). "OneSuite on Android Using CSipSimple". *Perk Up*. OneSuite Blog.

[9] "CSIP Simple: Mobile SIP Client for Android". *OnSIP*.
    Junction Networks. 5 May 2011.

13.5   External links

- Official website

Chapter 14

Data breach

A **data breach** is the intentional or unintentional release of secure information to an untrusted environment. Other terms for this phenomenon include **unintentional information disclosure**, **data leak** and also **data spill**. Incidents range from concerted attack by black hats with the backing of organized crime or national governments to careless disposal of used computer equipment or data storage media.

Definition: "A data breach is a security incident in which sensitive, protected or confidential data is copied, transmitted, viewed, stolen or used by an individual unauthorized to do so."[1] Data breaches may involve financial information such as credit card or bank details, personal health information (PHI), Personally identifiable information (PII), trade secrets of corporations or intellectual property.

According to the nonprofit consumer organization Privacy Rights Clearinghouse, a total of 227,052,199 individual records containing sensitive personal information were involved in security breaches in the United States between January 2005 and May 2008, excluding incidents where sensitive data was apparently not actually exposed.[2]

Many jurisdictions have passed data breach notification laws, requiring a company that has been subject to a data breach to inform customers and take other steps to remediate possible injuries.

14.1 Definition

This may include incidents such as theft or loss of digital media such as computer tapes, hard drives, or laptop computers containing such media upon which such information is stored unencrypted, posting such information on the world wide web or on a computer otherwise accessible from the Internet without proper information security precautions, transfer of such information to a system which is not completely open but is not appropriately or formally accredited for security at the approved level, such as unencrypted e-mail, or transfer of such information to the information systems of a possibly hostile agency, such as a competing corporation or a foreign nation, where it may be exposed to more intensive decryption techniques.[3]

ISO/IEC 27040 defines a data breach as: *compromise of security that leads to the accidental or unlawful destruction, loss, alteration, unauthorized disclosure of, or access to protected data transmitted, stored or otherwise processed.*

14.2 Trusted environment

The notion of a trusted environment is somewhat fluid. The departure of a trusted staff member with access to sensitive information can become a data breach if the staff member retains access to the data subsequent to termination of the trust relationship. In distributed systems, this can also occur with a breakdown in a web of trust.

14.3 Data privacy

Most such incidents publicized in the media involve private information on individuals, *i.e.* social security numbers, *etc.*. Loss of corporate information such as trade secrets, sensitive corporate information, details of contracts, *etc.* or of government information is frequently unreported, as there is no compelling reason to do so in the absence of potential damage to private citizens, and the publicity around such an event may be more damaging than the loss of the data itself.

14.4 Insider versus external threats

Those working inside an organization are a major cause of data breaches. Estimates of breaches caused by accidental "human factor" errors range from 37% by Ponemon Institute[4] to 14% by the Verizon 2013 Data Breach Investigations Report.[5] The external threat category includes hackers and state-sponsored actors. Professional associations for IT asset managers[6] work aggressively with IT professionals to educate them on best risk-reduction practices for both internal and external threats to IT assets, software and information.

14.5 Medical data breach

Main article: Medical data breach

Some celebrities have found themselves to be the victims of inappropriate medical record access breaches, albeit more so on an individual basis, not part of a typically much larger breach.[7] Given the series of medical data breaches and the lack of public trust, some countries have enacted laws requiring safeguards to be put in place to protect the security and confidentiality of medical information as it is shared electronically and to give patients some important rights to monitor their medical records and receive notification for loss and unauthorized acquisition of health information. The United States and the EU have imposed mandatory medical data breach notifications.[8]

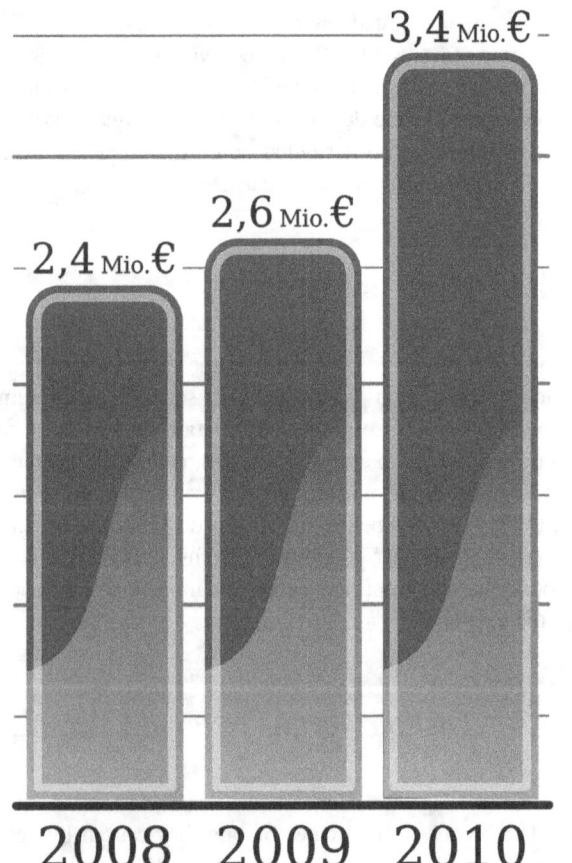

Average cost of data breaches in Germany[9]

14.6 Consequences

Although such incidents pose the risk of identity theft or other serious consequences, in most cases there is no lasting damage; either the breach in security is remedied before the information is accessed by unscrupulous people, or the thief is only interested in the hardware stolen, not the data it contains. Nevertheless, when such incidents become publicly known, it is customary for the offending party to attempt to mitigate damages by providing to the victims subscription to a credit reporting agency, for instance, new credit cards, or other instruments. In the case of Target, the 2013 breach cost Target a significant drop in profit, which dove an estimated 40 percent in the 4th quarter of the year.[10]

14.7 Major incidents

Notable incidents include:

14.7.1 2015

- In July 2015, adult website Ashley Madison suffered a data breach when a hacker group stole information on its 37 million users. The hackers threatened to reveal user names and specifics if Ashley Madison and a fellow site, EstablishedMen.com, did not shut down permanently.[11]

- In February 2015, Anthem suffered a data breach of nearly 80 million records, including personal information such as names, Social Security numbers, dates of birth, and other sensitive details.[12]

- In June 2015, The Office of Personnel Management of the U.S. government suffered a data breach in which the records of 4 million current and former federal employees of the United States were hacked and stolen.

14.7.2 2014

- In August 2014, nearly 200 photographs of celebrities were posted to the image board website 4chan. An investigation by Apple found that the images were obtained "by a very targeted attack on user names, passwords and security questions".[13]

- In September 2014, Home Depot suffered a data breach of 56 million credit card numbers.[14]

- In October 2014, Staples suffered a data breach of 1.16 million customer payment cards.[15]

- In November 2014 and for weeks after, Sony Pictures Entertainment suffered a data breach involving personal information about Sony Pictures employees and their families, e-mails between employees, information about executive salaries at the company, copies of (previously) unreleased Sony films, and other information. The hackers involved claim to have taken over 100 terabytes of data from Sony.[16]

14.7.3   2013

- In October 2013, Adobe Systems revealed that their corporate data base was hacked and some 130 million user records were stolen. According to Adobe, "For more than a year, Adobe's authentication system has cryptographically hashed customer passwords using the SHA-256 algorithm, including salting the passwords and iterating the hash more than 1,000 times. This system was not the subject of the attack we publicly disclosed on October 3, 2013. The authentication system involved in the attack was a backup system and was designated to be decommissioned. The system involved in the attack used Triple DES encryption to protect all password information stored."[17]
  Further information: Adobe Systems § Source code and customer data breach

- In late November to early December 2013, Target Corporation announced that data from around 40 million credit and debit cards was stolen. It is the second largest credit and debit card breach after the TJX Companies data breach where almost 46 million cards were affected.[18]

- In 2013, Edward Snowden published a series of secret documents that revealed widespread spying by the United States National Security Agency and similar agencies in other countries.

14.7.4   2012

- In the Summer of 2012, Wired.com Senior Writer Mat Honan claims that "hackers destroyed my entire digital life in the span of an hour" by hacking his Apple, Twitter, and Gmail passwords in order to gain access to his Twitter handle and in the process, claims the hackers wiped out every one of his devices, deleting all of his messages and documents, including every picture he had ever taken of his 18-month-old daughter.[19] The exploit was achieved with a combination of information provided to the hackers by Amazon's tech support

through social engineering, and the password recovery system of Apple which used this information.[20] Related to his experience, Mat Honan wrote a piece outlining why passwords cannot keep users safe.[21]

- In October 2012, a law enforcement agency contacted the South Carolina Department of Revenue (DoR) with evidence that Personally Identifiable Information (PII) of three individuals had been stolen.[22] It was later reported that an estimated 3.6 million Social Security numbers were compromised along with 387,000 credit card records.[23]

14.7.5   2011

- In April 2011, Sony experienced a data breach within their PlayStation Network. It is estimated that the information of 77 million users was compromised.

- In March 2011, RSA suffered a breach of their SecurID token system seed-key warehouse, where the seed keys for their 2-Factor authentication system were stolen, allowing the attackers to replicate the hardware tokens used for secure access in corporate and government environments.

- In June 2011, Citigroup disclosed a data breach within their credit card operation, affecting approximately 210,000 or 1% of their customers' accounts.[24][25]

- Throughout the year 2010, Chelsea Manning (then known as Bradley Manning) released large volumes of secret military data to the public.

14.7.6   2009

- In December 2009 a RockYou! password database was breached containing 32 million user names and plaintext passwords, further compromising the use of weak passwords for any purpose.

- In May 2009 the United Kingdom parliamentary expenses scandal was revealed by The Daily Telegraph. A hard disk containing scanned receipts of UK Members of Parliament and Peers in the House of Lords was offered to various UK newspapers in late April, with The Daily Telegraph finally acquiring it. They published details in installments from 8 May onwards. Although it was intended by Parliament that the data was to be published, this was to be in redacted form, with details the individual members considered "sensitive" blanked out. The newspaper published unredacted scans which showed details of the claims, many of which appeared to be in breach of the rules and suggested widespread abuse of the generous expenses system. The resulting media storm led to the resignation of the Speaker of the House of Commons and the prosecution and imprisonment of several MPs and Lords for fraud. The expenses system was overhauled and tightened up, being put more on a par with private industry schemes. The Metropolitan Police Service continues to investigate possible frauds, and the Crown Prosecution Service is considering further prosecutions. Several MPs and Lords apologised and made whole, partial or no restitution, and retained their seats. Others who had been shamed in the media did not offer themselves for re-election at the United Kingdom general election, 2010. Although numbering less than 1,500 individuals, the affair received the largest global media coverage of any data breach (as at February 2012).

- In January 2009 Heartland Payment Systems announced that it had been "the victim of a security breach within its processing system", possibly part of a "global cyber fraud operation".[26] The intrusion has been called the largest criminal breach of card data ever, with estimates of up to 100 million cards from more than 650 financial services companies compromised.[27]

14.7.7   2008

- In January 2008, GE Money, a division of General Electric, disclosed that a magnetic tape containing 150,000 social security numbers and in-store credit card information from 650,000 retail customers is known to be missing from an Iron Mountain Incorporated storage facility. J.C. Penney is among 230 retailers affected.[28]

- Horizon Blue Cross and Blue Shield of New Jersey, January, 300,000 members [2]

- Lifeblood, February, 321,000 blood donors [2]

- British National Party membership list leak,[29]

- In Early 2008, Countrywide Financial (since acquired by Bank of America) allegedly fell victim to a data breach when, according to news reports and court documents, employee Rene L. Rebollo Jr. stole and sold up to 2.5 million customers' personal information including social security numbers.[30][31] According to the legal complaint: "Beginning in 2008 - coincidentally after they sold their mortgage portfolios under wrongful and fraudulent 'securitization pools,' and coincidentally after their mortgage portfolio went into massive default as a result thereof - Countrywide

learned that the financial information of potentially millions of customers had been stolen by certain Countrywide agents, employees or other individuals."[32] In July 2010, Bank of America settled more than 30 related class-action lawsuits by offering free credit monitoring, identity theft insurance and reimbursement for losses to as many as 17 million consumers impacted by the alleged data breach. The settlement was estimated at $56.5 million not including court costs.[33]

14.7.8   2007

- D. A. Davidson & Co. 192,000 clients' names, customer account and social security numbers, addresses and dates of birth[34]

- The 2007 loss of Ohio and Connecticut state data by Accenture

- TJ Maxx, data for 45 million credit and debit accounts[35]

- 2007 UK child benefit data scandal

- CGI Group, August, 283,000 retirees from New York City [2]

- The Gap, September, 800,000 job applicants [2]

- Memorial Blood Center, December, 268,000 blood donors [2]

- Davidson County Election Commission, December, 337,000 voters [2]

14.7.9   2006

- AOL search data scandal (sometimes referred to as a "Data *Valdez*",[36][37][38] due to its size)

- Department of Veterans Affairs, May, 28,600,000 veterans, reserves, and active duty military personnel,[2][39]

- Ernst & Young, May, 234,000 customers of Hotels.com (after a similar loss of data on 38,000 employees of Ernst & Young clients in February) [2]

- Boeing, December, 382,000 employees (after similar losses of data on 3,600 employees in April and 161,000 employees in November, 2005) [2]

14.7.10   2005

- Ameriprise Financial, stolen laptop, December 24, 260,000 customer records [2]

14.8 See also

- Full disclosure (computer security)

14.9 References

[1] U.S. DEPARTMENT OF HEALTH AND HUMAN SERVICES Administration for Children and Families. Information Memorandum. Retrieved 2015-09-01. Available: http://www.acf.hhs.gov/sites/default/files/cb/im1504.pdf

[2] "Chronology of Data Breaches", Privacy Rights Clearinghouse

[3] *When we discuss incidents occurring on NSSs, are we using commonly defined terms?*, "Frequently Asked Questions on Incidents and Spills", National Archives Information Security Oversight Office

[4] Risk of Insider Fraud: Second Annual Study. Ponemon.org (2013-02-28). Retrieved on 2014-06-10.

[5] Verizon Data Breach Investigations Report | Verizon Enterprise Solutions. VerizonEnterprise.com. Retrieved on 2014-06-10.

[6] Welcome to IAITAM. Iaitam.org. Retrieved on 2014-06-10.

[7] Ornstein, Charles (2008-03-15). "Hospital to punish snooping on Spears". *Los Angeles Times*. Retrieved 2013-07-26.

[8] Kierkegaard, P. (2012) Medical data breaches: Notification delayed is notification denied, Computer Law & Security Report , 28 (2), p.163–183.

[9] "2010 Annual Study: German Cost of a Data Breach" (PDF). Ponemon Institute. February 2011. Retrieved 2011-10-12.

[10] http://www.nytimes.com/2014/02/27/business/target-reports-on-fourth-quarter-earnings.html?_r=0

[11] "Online Cheating Site AshleyMadison Hacked". krebsonsecurity.com. 2015-07-15. Retrieved 2015-07-20.

[12] "Data breach at health insurer Anthem could impact millions". 15 February 2015.

[13] "Apple Media Advisory: Update to Celebrity Photo Investigation". *Business Wire* (StreetInsider.com). September 2, 2014. Retrieved 2014-09-05.

[14] Melvin Backman (18 September 2014). "Home Depot: 56 million cards exposed in breach". CNNMoney.

[15] "Staples: Breach may have affected 1.16 million customers' cards". Fortune.com. December 19, 2014. Retrieved 2014-12-21.

[16] James Cook (December 16, 2014). "Sony Hackers Have Over 100 Terabytes Of Documents. Only Released 200 Gigabytes So Far". *Business Insider*. Retrieved December 18, 2014.

[17] Goodin, Dan. (2013-11-01) How an epic blunder by Adobe could strengthen hand of password crackers. Ars Technica. Retrieved on 2014-06-10.

[18] "Target security breach affects up to 40M cards". *Associated Press via Milwaukee Journal Sentinel*. 19 December 2013. Retrieved 21 December 2013.

[19] Honan, Mat (2012-11-15). "Kill the Password: Why a String of Characters Can't Protect Us Anymore". *Wired.com* (Condé Nast). Retrieved 2013-01-17.

[20] Honan, Mat (August 6, 2012). "How Apple and Amazon Security Flaws Led to My Epic Hacking". *Wired.com*. Retrieved 26 Jan 2013.

[21] "Protecting the Individual from Data Breach". *The National Law Review*. Raymond Law Group. 2014-01-14. Retrieved 2013-01-17.

[22] "Public Incident Response Report" (PDF). State of South Carolina. 2012-11-12. Retrieved 2014-10-10.

[23] "South Carolina: The mother of all data breaches". The Post and Courier. 2012-11-03. Retrieved 2014-10-10.

[24] Greenberg, Andy (9 June 2011). "Citibank Reveals One Percent Of Credit Card Accounts Exposed In Hacker Intrusion". *Forbes*. Retrieved 2014-09-05.

[25] Making Business a Little Less Risky: The Convergence of Data, Identity, and Regulatory Risks. LessRiskyBiz.blogspot.com (2011-06-13). Retrieved on 2014-06-10.

[26] Heartland Payment Systems Uncovers Malicious Software In Its Processing System

[27] Lessons from the Data Breach at Heartland, *MSNBC*, July 7, 2009

[28] GE Money Backup Tape With 650,000 Records Missing At Iron Mountain - Iron Mountain

[29] BNP activists' details published - BBC News

[30] "Bank of America settles Countrywide data theft suits"

[31] "Countrywide Sued For Data Breach, Class Action Suit Seeks $20 Million in Damages", *Bank Info Security*, April 9, 2010

[32] "Countrywide Sold Private Info, Class Claims", *Courthouse News*, April 05, 2010

[33] "The Convergence of Data, Identity, and Regulatory Risks", Making Business a Little Less Risky Blog

[34] Manning, Jeff (2010-04-13). "D.A. Davidson fined over computer security after data breach". *The Oregonian*. Retrieved 2013-07-26.

[35] "T.J. Maxx data theft worse than first reported". msnbc.com.
      2007-03-29. Retrieved 2009-02-16.

[36] *data Valdez* Doubletongued dictionary

[37] *AOL's Massive Data Leak*, Electronic Frontier Foundation

[38] *data Valdez*, Net Lingo

[39] "Active-duty troop information part of stolen VA data",
      Network World, June 6, 2006

14.10   External links

- "Data Loss Database" is a research project aimed at documenting known and reported data loss incidents world-wide.

- "Most Recent Data Breaches", TeamSHATTER.com, updated regularly

- "A Chronology of Data Breaches", Privacy Rights Clearinghouse, updated twice a week

- "Breaches Affecting 500 or More Individuals", Breaches reported to the U.S. Department of Health and Human Services by (HIPAA-covered) entities

Chapter 15

Defense Red Switch Network

A Multi Line Phone (MLP-1A), made by Electrospace Systems, which was part of the Defense Red Switch Network since 1983. The phone has the four extra MLPP buttons and 48 programmable buttons for access to both secure and nonsecure lines.[1]

The **Defense Red Switch Network** (DRSN) is a dedicated telephone network which provides global secure communication services for the command and control structure of the United States Armed Forces. The network is maintained by the Defense Information Systems Agency (DISA) and is secured for communications up to the level of Top Secret SCI.

The DRSN provides multilevel secure voice and voice-conferencing capabilities to the National Command Authority (NCA, being the President and the Secretary of Defense of the United States), the Joint Chiefs of Staff, the National Military Command Center (NMCC), Combatant Commanders and their command centers, warfighters, other DoD agencies, government departments, and NATO allies.

Department of Defense and federal government agencies can get access to the network with approval of the Joint Staff.[2] Upon approval by the Joint Staff, DISA will work with the customer and the appropriate military department to arrange the service.[3]

The Defense Red Switch Network consists of four major subsystems: the Switching Subsystem, the Transmission Subsystem, the Timing and Synchronization Subsystem, and the Network Management Subsystem. The Switching Subsystem uses both RED and BLACK switches to provide an integrated RED/BLACK service. End users are provided with a single telephone instrument with which they can access both secure and nonsecure networks.

The DRSN carried around 15,000 calls per day prior to September 11, 2001. DRSN usage subsequently peaked at 45,000 calls per day and by mid-2003 was running at around 25,000 calls per day. In that period the Defense Red Switch Network was expanded to support 18 additional US Federal Homeland Defense initiatives.[4]

Nowadays, this network is also called the **Multilevel Secure Voice** service. It's the core of the Global Secure Voice System (GSVS) during peacetime, crisis and time of conventional war, by hosting national-level conferencing and connectivity requirements and providing interoperability with both tactical and strategic communication networks.[5]

15.1 See also

- Defense Switched Network (DSN)

- Automatic Secure Voice Communications Network (AUTOSEVOCOM)

15.2 References

[1] Detailed pictures of the Electrospace MLP-1A

[2] Details are in the Chairman of the Joint Chiefs of Staff Instruction (CJCSI) 6215.01 (September 23, 2001).

[3] DISA Europe: Voice Services

[4] Janes Defense: Description of the Defense Red Switch Net-
       work (DRSN)

[5] DISA: Multilevel Secure Voice service

15.3   External links

- GlobalSecurity.org: Defense Red Switch Network
  (DRSN)

Chapter 16

Fishbowl (secure phone)

Fishbowl is a mobile phone architecture developed by the U.S. National Security Agency (NSA) to provide a secure Voice over IP (VoIP) capability using commercial grade products that can be approved to communicate classified information. It is the first phase of NSA's Enterprise Mobility Architecture. According to a presentation at the 2012 RSA Conference by Margaret Salter, a Technical Director in the Information Assurance Directorate, "The plan was to buy commercial components, layer them together and get a secure solution. It uses solely commercial infrastructure to protect classified data." Government employees were reportedly testing 100 of the phones as of the announcement.[1]

The initial version was implemented using Google's Android operating system, modified to ensure central control of the phone's configuration at all times. To minimize the chance of compromise, the phones use two layers of encryption protocols, IPsec and Secure Real-time Transport Protocol (SRTP), and employ NSA's Suite B encryption and authentication algorithms.

The phones are locked down in many ways. While they use commercial wireless channels, all communications must be sent through an enterprise-managed server. No direct voice calls are allowed, except for 9-1-1 emergency calls. Only NSA approved applications from the NSA enterprise app store can be installed. NSA has published a 100-page overview specification for the Mobility Capability Package.[2] In tandem with the Capability Package there are a series of Protection Profiles. [3] These Protection Profiles list out the requirements a commercial product must meet to be used in the mobile phone architecture.

[2] http://www.nsa.gov/ia/_files/Mobility_Capability_Pkg_Vers_2_1.pdf

[3] https://www.niap-ccevs.org/pp/index.cfm?&CFID=18170027&CFTOKEN=640af50d25b9baa2-3B6BCE4B-C0B3-B411-7F8F85AB279753D6

16.1   References

[1] http://www.scmagazine.com.au/News/292189,
nsa-builds-android-phone-for-top-secret-calls.aspx
NSA builds Android phone for top secret calls

Chapter 17

Greynet

For an alternate use of the term, see network telescope

Within the context of corporate and organizational networks, a **greynet** (or Grayware) is an elusive networked computer application that is downloaded and installed on end user systems without express permission from network administrators and often without awareness or cognition that it is deeply embedded in the organization's network fabric. These applications may be of some marginal use to the user, but inevitably consume system and network resources. In addition, greynet applications often open the door for end use systems to become compromised by additional applications, security risks and malware.

NOTE: Unfortunately, this computer application shares the same name as the Grey Literature Network Service (GreyNet) http://www.greynet.org Fortunately, however, this is the only thing they share in common.

17.1 Examples

- Public instant messaging (AOL Instant Messenger, Windows Live Messenger, Yahoo! Messenger)

- Web conferencing (webcam, VoIP telephony)

- Peer-to-peer (P2P) file sharing clients

- Distributed computing such as SETI@home

- Adware "utilities"

- Commercial spyware

- Keystroke logging

17.2 The dynamics of greynet growth

As computer workstations have become connected to the Internet, a variety of programs have proliferated that offer the ability to extend communications, gather and deliver information, and to serve the needs of marketing concerns. Among the first to emerge were instant messaging clients such as ICQ, AOL Instant Messenger and MSN Messenger. Developments in technology have added video capability through webcam units, all of which have worked together to take advantage of available bandwidth in single, small network, and corporate environments.

The growth of greynets takes advantage of software and hardware developments. Informal networks are now appearing that provide a variety of streaming media and content that is supplied or modified by end users. An emerging category is "podcasting", in which users generate content for widespread download on portable MP3 players.

17.3 Problems with greynet programs

The problem with greynet programs is fourfold. First, greynet programs create network security risks by causing broad vectors for malware dissemination. Second, they create privacy issues for the network by opening large holes for information leakage. Third, greynet programs create compliance issues for a computer network by creating an invisible parallel communications network. Fourth, they create issues on local machines through the consumption of local system resources and possible operating system or program stability concerns. All of these things increase network and IT administration time and costs.

Added to this in the corporate work environment is the loss of meaningful production time due to non-work related distractions through these greynet applications. Individual network environment policies may vary from non-existent to a full lockdown of end user system privileges. See the "Risks and Liabilities" section of instant messaging for a more detailed overview of threats, risks, and solutions to those problems for the most prevalent of the greynet programs, public IM.

Dealing with the security aspects of greynets has led to the emergence of specific administrative software packages that monitor and control traffic, as well as the enhancement of security suites and adware clients.

17.4 Security and monitoring

Among the first and most prevalent of the specific administrative software packages were products that secure networks against threats borne by IM and P2P networks. These products were first introduced in 2002, and now protect 10% to 15% of U.S. corporations.

17.5 References

- Dostart, Kate (April 12, 2007). "Instant messaging threats become more sophisticated". SearchVoIP.com.

- Kerner, Sean Michael (November 10, 2006). "Greynets Getting Greyer". internetnews.com.

- Joyce, Erin (August 2, 2005). "Spyware Skyrockets on Greynet Fuel". internetnews.com.

Chapter 18

The Guardian Project (software)

This article is about the open-source software initiative. For the comic, see The Guardian Project (comic).

The Guardian Project is a global collective of software developers, designers, advocates, activists and trainers who develop open source mobile security software and operating system enhancements.[1] They also create customized mobile devices to help individuals communicate more freely and protect themselves from intrusion and monitoring. The effort specifically focuses on users who live or work in high-risk situations, and who often face constant surveillance and intrusion attempts into their mobile devices and communication streams.

18.1 History

Founder Nathan Freitas speaking at the Unlike Us conference in 2013[2]

The Guardian Project was founded by Nathan Freitas in 2009 in Brooklyn, NY.[3][4][5] Since it was founded, the Guardian Project has developed more than a dozen mobile applications for Android and iOS with over two million downloads and hundreds of thousands of active users. It has also partnered with prominent open source software projects, activists groups, NGOs, commercial partners and news organizations to support their mobile security software capabilities.

In November 2014, "ChatSecure + Orbot" received a top score on the Electronic Frontier Foundation's secure messaging scorecard, along with Cryptocat, TextSecure, "Signal / RedPhone", Silent Phone, and Silent Text.[6] "Jitsi + Ostel" scored 6 out of 7 points on the Electronic Frontier Foundation's secure messaging scorecard. They lost a point because there has not been a recent independent code audit.[6]

18.2 Funding

The Guardian Project has received funding from Google, UC Berkeley with the MacArthur Foundation, Avaaz, Internews, Open Technology Fund, WITNESS, the Knight Foundation, Benetech, and Free Press Unlimited.[7]

Through work on partner projects like The Tor Project, Commotion mesh and StoryMaker, they have received indirect funding from both the US State Department through the Bureau of Democracy, Human Rights and Labor Internet Freedom program, and the Dutch Ministry of Foreign Affairs through HIVOS.

18.3 Active projects

18.3.1 Orbot

Main article: Orbot

Orbot brings the capabilities of Tor to Android. Tor uses Onion Routing to provide access to network services that may be blocked, censored or monitored, while also protecting the identity of the user requesting those resources.[8]

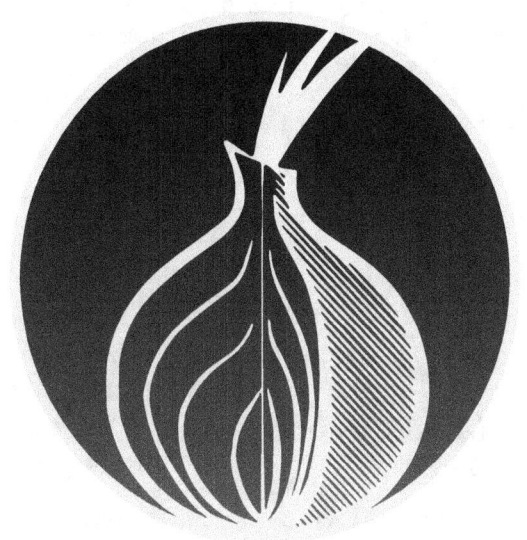

Orbot

18.3.2 Orfox

On 30 June 2015, The Guardian Project announced a stable alpha of Orfox, the mobile counterpart of the Tor Browser. Orfox is built from *Fennec* (Firefox for Android) code and the Tor Browser code repository, and is given security hardening patches by the Tor Browser development team. Some of the Orfox build work is based on the [Fennec] F-Droid project.[9]

In Orfox, the project removed the WebRTC component, Chromecast connectivity, app permissions to access the camera, microphone, contacts (address book), location data (GPS et al.), and NFC.[9][10] Orfox is to supersede the Orweb browser project.[9]

18.3.3 Orweb

Orweb is a privacy enhanced web browser that supports proxies. When used with Orbot, Orweb protects against network analysis, blocks cookies, keeps no local browsing history, and disables Flash to keep the user safe.[8]

18.3.4 ChatSecure

ChatSecure, formerly Gibberbot,[11] is a full featured instant messaging application integrated with the Off-the-Record encrypted chat protocol. The app is built on Google's open-source Talk app and modified to support the Jabber XMPP protocol.[8]

18.3.5 ObscuraCam

A secure camera app that can obscure, encrypt or destroy pixels within an image. This project is in partnership with WITNESS, a human rights video advocacy and training organization.[8]

18.3.6 Ostel

A tool for having end-to-end encrypted VoIP calls.[12] This is a public testbed of the Open Secure Telephony Network (OSTN) project, with the goal of promoting the use of free, open protocols, standards and software, to power end-to-end secure voice communications on mobile devices, as well as with desktop computers.[8]

18.4 Additional projects

Third party projects that are supported, developed on, and recommended by The Guardian Project:[8]

18.4.1 K-9 and OpenKeychain

Main articles: K-9 Mail and OpenKeychain

K-9 Mail is an open-source app based on Android's built-in Email app. The project is focused on making it easy to manage multiple accounts and large volumes of email, as well supporting OpenPGP encryption using OpenKeychain.[8]

18.4.2 CSipSimple

Main article: CSipSimple

CSipSimple is a free and open source SIP client for Android that provides end-to-end encryption using ZRTP. This app is compatible with an Ostel account for making secure VoIP calls on Android.[8]

18.4.3 TextSecure

Main article: TextSecure

TextSecure, developed by Open Whisper Systems, provides a robust encrypted instant messaging and text messaging solution on Android.[8] It is intended to be used in place of the standard text messaging application.[13] TextSecure users can exchange encrypted messages, media and attachments.

The application uses end-to-end encryption to secure all messages that are sent to other TextSecure users.[14][15][16]

18.4.4 Linphone

Main article: Linphone

Linphone is a free and open source SIP client for Apple iOS, Android, and desktop that provides end-to-end encryption using ZRTP. This app is compatible with an Ostel account for making secure voice and video calls.

18.4.5 Osmand

Main article: Osmand

Osmand is the most private map app because it works completely offline. The map data is downloaded from OpenStreetMap. As of version 2.0, it can also understand many different location sharing URLs, as well as download map data via Tor by setting the proxy.

18.5 Distribution

The Guardian Project offers downloads of its apps from Google Play, Amazon Appstore, directly from their website, and through an F-Droid compatible repository.[8][17] Direct downloads are signed and can be verified with the developer's key.[18]

18.6 See also

- Freedom of information
- Internet privacy

18.7 References

[1] Thomas Lowenthal (19 April 2011). "For paranoid Androids, Guardian Project offers smartphone security". ArsTechnica.com. Retrieved 15 May 2013.

[2] http://networkcultures.org/unlikeus/2013/03/25/nathan-freitas-checking-in-for-the-greater-good/

[3] Nathan Freitas Tweet on Twitter

[4] Nathan Freitas (20 March 2009). "Nathan Freitas on Guardian". YouTube.com. Retrieved 15 May 2013.

[5] NANCY SCOLA (31 March 2011). "The Guardian Project: Building Mobile Security for a Dangerous World". TechPresident.com. Retrieved 15 May 2013.

[6] "Secure Messaging Scorecard. Which apps and tools actually keep your messages safe?". Electronic Frontier Foundation. 2014-11-04.

[7] https://guardianproject.info/home/partners/

[8] The Guardian Project. "Secure Mobile Apps". GuardianProject.info. Retrieved 15 July 2015.

[9] n8fr8 (2015-06-30). "Orfox: Aspiring to bring Tor Browser to Android". The Guardian Project. Retrieved 2015-07-11.

[10] Long, Jacob (2015-07-01). "Orfox Is The Guardian Project's Latest App For Bringing The Tor Browser Experience To Android, First Alpha Release Is Available". *Android Police*. Illogical Robot LLC. Retrieved 2015-07-21.

[11] Nathan Freitas (24 October 2013). "ChatSecure v12 Provides Comprehensive Mobile Security and a Whole New Look". GuardianProject.info. Retrieved 24 October 2013.

[12] "Ostel OSTN". Retrieved 6 September 2014.

[13] "TextSecure / README.md". Retrieved 26 February 2014.

[14] Molly Wood (19 February 2014). "Privacy Please: Tools to Shield Your Smartphone". The New York Times. Retrieved 26 February 2014.

[15] Moxie Marlinspike (24 February 2014). "The New TextSecure: Privacy Beyond SMS". Open WhisperSystems. Retrieved 26 February 2014.

[16] Martin Brinkmann (24 February 2014). "TextSecure is an open source messaging app with strong security features". Ghacks Technology News. Retrieved 26 February 2014.

[17] Hans-Christoph Steiner (30 June 2014). "New Official Guardian Project app repo for FDroid!". Retrieved 27 July 2014.

[18] "Signing Keys". The Guardian Project. Retrieved 19 September 2014.

18.8 External links

- Website of The Guardian Project
- Blog of The Guardian Project
- Guardian Project on Twitter
- The Guardian Project on GitHub
- Official F-Droid repository (Open in F-Droid app)

Chapter 19

HTTPS

This article is about the secure protocol. For the obsolete and little-used Secure Hypertext Transfer Protocol (S-HTTP), see Secure Hypertext Transfer Protocol.
For more information about Wikipedia and HTTPS, see Wikipedia:Secure server

HTTPS (also called **HTTP over TLS**,[1][2] **HTTP over SSL**,[3] and **HTTP Secure**[4][5]) is a protocol for secure communication over a computer network which is widely used on the Internet. HTTPS consists of communication over Hypertext Transfer Protocol (HTTP) within a connection encrypted by Transport Layer Security or its predecessor, Secure Sockets Layer. The main motivation for HTTPS is authentication of the visited website and to protect the privacy and integrity of the exchanged data.

In its popular deployment on the internet, HTTPS provides authentication of the website and associated web server with which one is communicating, which protects against man-in-the-middle attacks. Additionally, it provides bidirectional encryption of communications between a client and server, which protects against eavesdropping and tampering with and/or forging the contents of the communication.[6] In practice, this provides a reasonable guarantee that one is communicating with precisely the website that one intended to communicate with (as opposed to an impostor), as well as ensuring that the contents of communications between the user and site cannot be read or forged by any third party.

Historically, HTTPS connections were primarily used for payment transactions on the World Wide Web, e-mail and for sensitive transactions in corporate information systems. In the late 2000s and early 2010s, HTTPS began to see widespread use for protecting page authenticity on all types of websites, securing accounts and keeping user communications, identity and web browsing private.

19.1 Overview

For more details on this topic, see Transport Layer Security.
The *HTTPS* uniform resource identifier (URI) scheme has

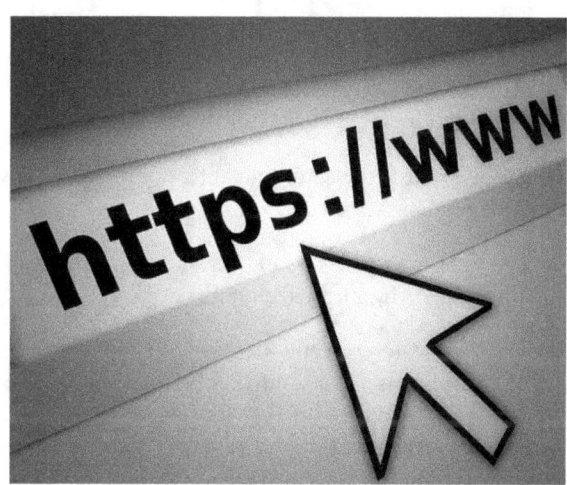

Logo of the networking protocol https and the www letters

identical syntax to the standard HTTP scheme, aside from its scheme token. However, HTTPS signals the browser to use an added encryption layer of SSL/TLS to protect the traffic. SSL is especially suited for HTTP since it can provide some protection even if only one side of the communication is authenticated. This is the case with HTTP transactions over the Internet, where typically only the server is authenticated (by the client examining the server's certificate).

HTTPS creates a secure channel over an insecure network. This ensures reasonable protection from eavesdroppers and man-in-the-middle attacks, provided that adequate cipher suites are used and that the server certificate is verified and trusted.

Because HTTPS piggybacks HTTP entirely on top of TLS, the entirety of the underlying HTTP protocol can be encrypted. This includes the request URL (which particular web page was requested), query parameters, headers, and cookies (which often contain identity information about the user). However, because host (website) addresses and port numbers are necessarily part of the underlying TCP/IP protocols, HTTPS cannot protect their disclosure. In practice this means that even on a correctly configured web server, eavesdroppers can infer the IP address and port number of the web server (sometimes even the domain name e.g. www.example.org, but not the rest of the URL) that one is communicating with as well as the amount (data transferred) and duration (length of session) of the communication, though not the content of the communication.[6]

Web browsers know how to trust HTTPS websites based on certificate authorities that come pre-installed in their software. Certificate authorities (such as Symantec, Comodo, GoDaddy and GlobalSign) are in this way being trusted by web browser creators to provide valid certificates. Therefore, a user should trust an HTTPS connection to a website if and only if all of the following are true:

- The user trusts that the browser software correctly implements HTTPS with correctly pre-installed certificate authorities.

- The user trusts the certificate authority to vouch only for legitimate websites.

- The website provides a valid certificate, which means it was signed by a trusted authority.

- The certificate correctly identifies the website (e.g., when the browser visits "https://example.com", the received certificate is properly for "example.com" and not some other entity).

- The user trusts that the protocol's encryption layer (TLS/SSL) is sufficiently secure against eavesdroppers.

HTTPS is especially important over insecure networks (such as public WiFi access points), as anyone on the same local network can packet sniff and discover sensitive information not protected by HTTPS. Additionally, many free to use and even paid for WLAN networks engage in packet injection in order to serve their own ads on webpages. However, this can be exploited maliciously in many ways, such as injecting malware onto webpages and stealing users' private information.[7]

HTTPS is also very important for connections over the Tor anonymity network, as malicious Tor nodes can damage or alter the contents passing through them in an insecure fashion and inject malware into the connection. This is one reason why the Electronic Frontier Foundation and the Tor project started the development of HTTPS Everywhere,[6] which is included in the Tor Browser Bundle.[8]

As more information is revealed about global mass surveillance and hackers stealing personal information, the use of HTTPS security on all websites is becoming increasingly important regardless of the type of Internet connection being used.[9][10] While metadata about individual pages that a user visits is not sensitive, when combined together, they can reveal a lot about the user and compromise the user's privacy.[2][11][12]

Deploying HTTPS also allows the use of SPDY, a networking protocol designed to reduce page load times and latency.

It is recommended to use HTTP Strict Transport Security (HSTS) with HTTPS to protect users from man-in-the-middle attacks, especially SSL stripping.[12][13]

HTTPS should not be confused with the little-used Secure HTTP (S-HTTP) specified in RFC 2660.

19.1.1   Usage in websites

As of October 12, 2014, 32.8% of the Internet's 151,509 most popular websites have a secure implementation of HTTPS.[14]

19.1.2   Browser integration

Most browsers display a warning if they receive an invalid certificate. Older browsers, when connecting to a site with an invalid certificate, would present the user with a dialog box asking if they wanted to continue. Newer browsers display a warning across the entire window. Newer browsers also prominently display the site's security information in the address bar. Extended validation certificates turn the address bar green in newer browsers. Most browsers also display a warning to the user when visiting a site that contains a mixture of encrypted and unencrypted content.

Firefox uses HTTPS for Google searches as of version 14,[15] to "shield our users from network infrastructure that may be gathering data about the users or modifying/censoring their search results".[16]

The Electronic Frontier Foundation, opining that "In an ideal world, every web request could be defaulted to HTTPS", has provided an add-on called HTTPS Everywhere for Mozilla Firefox that enables HTTPS by default for hundreds of frequently used websites. A beta version of this plugin is also available for Google Chrome and Chromium.[17][18]

19.2 Security

The security of HTTPS is therefore that of the underlying TLS, which uses long-term public and secret keys to exchange a short term session key to encrypt the data flow between client and server. X.509 certificates are used to guarantee one is talking to the partner with whom one wants to talk. As a consequence, certificate authorities and a public key infrastructure are necessary to verify the relation between the owner of a certificate and the certificate, as well as to generate, sign, and administer the validity of certificates. While this can be more beneficial than verifying the identities via a web of trust, the 2013 mass surveillance disclosures made it more widely known that certificate authorities are a weak point from a security standpoint, allowing man-in-the-middle attacks.[19][20] Another important property in this context is perfect forward secrecy (PFS), so the short-term session key cannot be derived from the long-term asymmetric secret key; however, PFS is not widely adopted.[21]

A site must be completely hosted over HTTPS, without having part of its contents loaded over HTTP - for example, having scripts loaded insecurely - or the user will be vulnerable to some attacks and surveillance. Also having only a certain page that contains sensitive information (such as a log-in page) of a website loaded over HTTPS, while having the rest of the website loaded over plain HTTP, will expose the user to attacks. On a site that has sensitive information somewhere on it, every time that site is accessed with HTTP instead of HTTPS, the user and the session will get exposed. Similarly, cookies on a site served through HTTPS have to have the secure attribute enabled.[12]

19.3 Technical

19.3.1 Difference from HTTP

HTTPS URLs begin with "https://" and use port 443 by default, whereas HTTP URLs begin with "http://" and use port 80 by default.

HTTP is not encrypted and is vulnerable to man-in-the-middle and eavesdropping attacks, which can let attackers gain access to website accounts and sensitive information, and modify webpages to inject malware or advertisements. HTTPS is designed to withstand such attacks and is considered secure against them (with the exception of older, deprecated versions of SSL).

19.3.2 Network layers

HTTP operates at the highest layer of the TCP/IP model, the Application layer; as does the SSL security protocol (operating as a lower sublayer of the same layer), which encrypts an HTTP message prior to transmission and decrypts a message upon arrival. Strictly speaking, HTTPS is not a separate protocol, but refers to use of ordinary HTTP over an encrypted SSL/TLS connection.

Everything in the HTTPS message is encrypted, including the headers, and the request/response load. With the exception of the possible CCA cryptographic attack described in the limitations section below, the attacker can only know that a connection is taking place between the two parties and their domain names and IP addresses.

19.3.3 Server setup

To prepare a web server to accept HTTPS connections, the administrator must create a public key certificate for the web server. This certificate must be signed by a trusted certificate authority for the web browser to accept it without warning. The authority certifies that the certificate holder is the operator of the web server that presents it. Web browsers are generally distributed with a list of signing certificates of major certificate authorities so that they can verify certificates signed by them.

Acquiring certificates

Authoritatively signed certificates may be free[22][23] or cost between 8 USD[24] and 70 USD[25] per year (in 2012–2014).

Organizations may also run their own certificate authority, particularly if they are responsible for setting up browsers to access their own sites (for example, sites on a company intranet, or major universities). They can easily add copies of their own signing certificate to the trusted certificates distributed with the browser.

There also exists a peer-to-peer certificate authority, CACert. However, it is not included in the trusted root certificates of many popular browsers (e.g. Firefox, Chrome, Internet Explorer), which may cause warning messages to be displayed to end users.

An upcoming certificate authority, Let's Encrypt, is to be launched by the end of 2015[26] and will provide free and automated SSL/TLS certificates to websites.[27] According to the Electronic Frontier Foundation, "Let's Encrypt" will make switching from HTTP to HTTPS "as easy as issuing one command, or clicking one button."[28]

Use as access control

The system can also be used for client authentication in order to limit access to a web server to authorized users. To do this, the site administrator typically creates a certificate for each user, a certificate that is loaded into his/her browser. Normally, that contains the name and e-mail address of the authorized user and is automatically checked by the server on each reconnect to verify the user's identity, potentially without even entering a password.

In case of compromised secret (private) key

An important property in this context is perfect forward secrecy (PFS). Possessing one of the long term asymmetric secret keys used to establish an HTTPS session should not make it easier to derive the short term session key to then decrypt the conversation, even at a later time. Diffie–Hellman key exchange (DHE) and Elliptic curve Diffie–Hellman key exchange (ECDHE) are in 2013 the only ones known to have that property. Only 30% of Firefox, Opera, and Chromium Browser sessions use it, and nearly 0% of Apple's Safari and Microsoft Internet Explorer sessions.[21] Among the larger internet providers, only Google supports PFS since 2011 (State of September 2013).

A certificate may be revoked before it expires, for example because the secrecy of the private key has been compromised. Newer versions of popular browsers such as Firefox,[29] Opera,[30] and Internet Explorer on Windows Vista[31] implement the Online Certificate Status Protocol (OCSP) to verify that this is not the case. The browser sends the certificate's serial number to the certificate authority or its delegate via OCSP and the authority responds, telling the browser whether or not the certificate is still valid.[32]

19.3.4 Limitations

SSL comes in two options, simple and mutual. The mutual version is more secure, but requires the user to install a personal browser certificate into their web browser in order to authenticate themselves.

Whatever strategy is used (simple or mutual), the level of protection strongly depends on the correctness of the implementation of the web browser and the server software and the actual cryptographic algorithms supported.

SSL does not prevent the entire site from being indexed using a web crawler, and in some cases the URI of the encrypted resource can be inferred by knowing only the intercepted request/response size.[33] This allows an attacker to have access to the plaintext (the publicly available static content), and the encrypted text (the encrypted version of the static content), permitting a cryptographic attack.

Because SSL operates below HTTP and has no knowledge of higher-level protocols, SSL servers can only strictly present one certificate for a particular IP/port combination.[34] This means that, in most cases, it is not feasible to use name-based virtual hosting with HTTPS. A solution called Server Name Indication (SNI) exists, which sends the hostname to the server before encrypting the connection, although many older browsers do not support this extension. Support for SNI is available since Firefox 2, Opera 8, Safari 2.1, Google Chrome 6, and Internet Explorer 7 on Windows Vista.[35][36][37]

From an architectural point of view:

1. An SSL/TLS connection is managed by the first front machine that initiates the SSL connection. If, for any reasons (routing, traffic optimization, etc.), this front machine is not the application server and it has to decipher data, solutions have to be found to propagate user authentication information or certificate to the application server, which needs to know who is going to be connected.

2. For SSL with mutual authentication, the SSL/TLS session is managed by the first server that initiates the connection. In situations where encryption has to be propagated along chained servers, session timeOut management becomes extremely tricky to implement.

3. With mutual SSL/TLS, security is maximal, but on the client-side, there is no way to properly end the SSL connection and disconnect the user except by waiting for the SSL server session to expire or closing all related client applications.

A sophisticated type of man-in-the-middle attack called SSL stripping was presented at the Blackhat Conference 2009. This type of attack defeats the security provided by HTTPS by changing the https: link into an http: link, taking advantage of the fact that few Internet users actually type "https" into their browser interface: they get to a secure site by clicking on a link, and thus are fooled into thinking that they are using HTTPS when in fact they are using HTTP. The attacker then communicates in clear with the client.[38] This prompted the development of a countermeasure in HTTP called HTTP Strict Transport Security.

In May 2010, a research paper by researchers from Microsoft Research and Indiana University discovered that detailed sensitive user data can be inferred from side channels such as packet sizes. More specifically, the researchers found that an eavesdropper can infer the illnesses/medications/surgeries of the user, his/her family income and investment secrets, despite HTTPS protection in several high-profile, top-of-the-line web applications in healthcare, taxation, investment and web search.[39]

19.4 History

Netscape Communications created HTTPS in 1994 for its Netscape Navigator web browser.[40] Originally, HTTPS was used with the SSL protocol. As SSL evolved into Transport Layer Security (TLS), the current version of HTTPS was formally specified by RFC 2818 in May 2000.

19.5 See also

- AAA protocol
- Bullrun (decryption program) — a secret anti-encryption program run by the U.S. National Security Agency
- Computer security
- curl-loader
- HTTPsec
- Moxie Marlinspike
- Opportunistic encryption
- Stunnel

19.6 References

[1] Network Working Group (May 2000). "HTTP Over TLS". The Internet Engineering Task Force. Retrieved February 27, 2015.

[2] "HTTPS as a ranking signal". *Google Webmaster Central Blog*. Google Inc. August 6, 2014. Retrieved February 27, 2015. You can make your site secure with HTTPS (Hypertext Transfer Protocol Secure) [...]

[3] "Enabling HTTP Over SSL". Adobe Systems Incorporated. Retrieved February 27, 2015.

[4] "Secure your site with HTTPS". *Google Support*. Google, Inc. Retrieved February 27, 2015.

[5] "What is HTTPS?". Comodo CA Limited. Retrieved February 27, 2015. Hyper Text Transfer Protocol Secure (HTTPS) is the secure version of HTTP [...]

[6] "HTTPS Everywhere FAQ". Retrieved 3 May 2012.

[7] "Hotel Wifi JavaScript Injection". Retrieved 24 July 2012.

[8] The Tor Project, Inc. "Tor". *torproject.org*.

[9] Konigsburg, Eitan; Pant, Rajiv; Kvochko, Elena (November 13, 2014). "Embracing HTTPS". The New York Times. Retrieved February 27, 2015.

[10] Gallagher, Kevin (September 12, 2014). "Fifteen Months After the NSA Revelations, Why Aren't More News Organizations Using HTTPS?". Freedom of the Press Foundation. Retrieved February 27, 2015.

[11] Grigorik, Ilya; Far, Pierre (June 26, 2014). "Google I/O 2014 - HTTPS Everywhere". Google Developers. Retrieved February 27, 2015.

[12] "How to Deploy HTTPS Correctly". Retrieved 13 June 2012.

[13] "HTTP Strict Transport Security". *Mozilla Developer Network*.

[14] "SSL Pulse". Trustworthy Internet Movement. Retrieved 2014-10-12.

[15] "Firefox 14.0.1 Release Notes". Retrieved 24 July 2012.

[16] "Firefox Rolling Out HTTPS Google search". Retrieved 24 July 2012.

[17] Peter Eckersley: Encrypt the Web with the HTTPS Everywhere Firefox Extension EFF blog, 17 June 2010

[18] HTTPS Everywhere EFF projects

[19] Law Enforcement Appliance Subverts SSL, Wired, 2010-04-03.

[20] New Research Suggests That Governments May Fake SSL Certificates, EFF, 2010-03-24.

[21] SSL: Intercepted today, decrypted tomorrow, Netcraft, 2013-06-25.

[22] "Free SSL Certificates from a Free Certificate Authority". sslshopper.com. Retrieved 2009-10-24.

[23] Justin Fielding (2006-07-16). "Secure Outlook Web Access with (free) SSL: Part 1". TechRepublic. Retrieved 2009-10-24.

[24] "Namecheap.com SSL Services". namecheap. Retrieved 30 Jan 2012.

[25] "Secure Site Pro with SSL Certificate". Retrieved 23 Aug 2014.

[26] "Launch schedule". *Let's Encrypt*. Retrieved 21 September 2015.

[27] Kerner, Sean Michael (November 18, 2014). "Let's Encrypt Effort Aims to Improve Internet Security". *eWeek.com*. Quinstreet Enterprise. Retrieved February 27, 2015.

[28] Eckersley, Peter (November 18, 2014). "Launching in 2015: A Certificate Authority to Encrypt the Entire Web". Electronic Frontier Foundation. Retrieved February 27, 2015.

[29] "Mozilla Firefox Privacy Policy". Mozilla Foundation. 27 April 2009. Retrieved 13 May 2009.

[30] "Opera 8 launched on FTP". Softpedia. 19 April 2005. Retrieved 13 May 2009.

[31] Lawrence, Eric (31 January 2006). "HTTPS Security Improvements in Internet Explorer 7". MSDN. Retrieved 13 May 2009.

[32] Myers, M; Ankney, R; Malpani, A; Galperin, S; Adams, C (June 1999). "Online Certificate Status Protocol – OCSP". Internet Engineering Task Force. Retrieved 13 May 2009.

[33] Pusep, Stanislaw (31 July 2008). "The Pirate Bay un-SSL". Retrieved 6 March 2009.

[34] "SSL/TLS Strong Encryption: FAQ". *apache.org*.

[35] Lawrence, Eric (22 October 2005). "Upcoming HTTPS Improvements in Internet Explorer 7 Beta 2". Microsoft. Retrieved 12 May 2009.

[36] "Server Name Indication (SNI)". *inside aebrahim's head*.

[37] Pierre, Julien. "Browser support for TLS server name indication" (2001-12-19). *Bugzilla*. Mozilla Foundation. Retrieved 2010-12-15.

[38] "sslstrip". Retrieved 2011-11-26.

[39] Shuo Chen, Rui Wang, XiaoFeng Wang, and Kehuan Zhang (May 2010). "Side-Channel Leaks in Web Applications: a Reality Today, a Challenge Tomorrow" (PDF). IEEE Symposium on Security & Privacy 2010.

[40] Walls, Colin (2005). *Embedded software*. Newnes. p. 344. ISBN 0-7506-7954-9.

19.7   External links

- RFC 2818: HTTP Over TLS

- Hacking HTTPS by man-in-the-middle attack

- RFC 5246: The Transport Layer Security Protocol 1.2

- SSL 3.0 Specification (IETF)

Chapter 20

HTTPS Everywhere

HTTPS Everywhere is a free and open source web browser extension for Google Chrome, Mozilla Firefox and Opera, a collaboration by The Tor Project and the Electronic Frontier Foundation (EFF).[3] It automatically makes websites use the more secure HTTPS connection instead of HTTP, if they support it.[4]

20.1 Development

HTTPS Everywhere was inspired by Google's increased use of HTTPS,[5] and is designed to make HTTPS automatically used whenever possible.[6] The code in part is based on NoScript's HTTP Strict Transport Security implementation, but HTTPS Everywhere is intended to be simpler to use than NoScript.[7] The EFF provides information for users on how to add HTTPS rulesets to HTTPS Everywhere,[8] and information on which websites support HTTPS.[9]

20.1.1 Platform support

A public beta of HTTPS Everywhere for Firefox was released in 2010,[10] and version 1.0 was released in 2011.[11] A beta for Google Chrome was released in February 2012.[12] In 2014, a version was released for Android phones.[13]

20.2 SSL Observatory

The SSL Observatory is a feature in HTTPS Everywhere introduced in version 2.0.1[12] which analyzes public key certificates to determine if certificate authorities have been compromised,[14] and if the user is vulnerable to man-in-the-middle attacks.[15] The ICANN Security and Stability Advisory Committee (SSAC) notes that the dataset used by the SSL Observatory often treats intermediate authorities as different entities, thus inflating the number of certifi-

cate authorities. The SSAC criticizes SSL Observatory for potentially significantly undercounting internal name certificates, and notes that it uses a data set from 2010.[16]

20.3 Reception

Two studies have recommended building in HTTPS Everywhere functionality into Android browsers.[17][18] In 2012, Eric Phetteplace described it as "perhaps the best response to Firesheep-style attacks available for any platform".[19] In 2011, Vincent Toubiana and Vincent Verdot pointed out some drawbacks of the HTTPS Everywhere plugin, including that the list of services which support HTTPS needs maintaining, and that some services are redirected to HTTPS even though they are not yet available in HTTPS, not allowing the user of the extension to get to the service.[20]

20.4 See also

- Transport Layer Security

- Privacy Badger, also created by the EFF

- Switzerland (software), also created by the EFF

20.5 References

[1] "Changelog.txt". EFF. 2015-08-25. Retrieved 2015-08-29.

[2] HTTPS Everywhere Development Electronic Frontier Foundation

[3] "HTTPS Everywhere | Electronic Frontier Foundation". Eff.org. Retrieved 2014-04-14.

[4] "HTTPS Everywhere reaches 2.0, comes to Chrome as beta - The H Open: News and Features". H-online.com. 2012-02-29. Retrieved 2014-04-14.

[5] "Automatic web encryption (almost) everywhere - The H Open Source: News and Features". H-online.com. 2010-06-18. Archived from the original on 2010-06-23. Retrieved 2014-04-15.

[6] Kate Murphy: New hacking tools pose bigger threats to Wi-Fi users. The New York Times, February 17, 2011.

[7] "HTTPS Everywhere | Electronic Frontier Foundation". Eff.org. Retrieved 2014-06-04.

[8] "HTTPS Everywhere Rulesets | Electronic Frontier Foundation". Eff.org. 2014-01-24. Retrieved 2014-05-19.

[9] "HTTPS Everywhere Atlas". *eff.org*. Retrieved 2014-05-24..

[10] Mills, Elinor (2010-06-18). "Firefox add-on encrypts sessions with Facebook, Twitter". CNET. Retrieved 2014-04-14.

[11] Scott Gilbertson (2011-08-05). "Firefox Security Tool HTTPS Everywhere Hits 1.0 | Webmonkey". WIRED. Retrieved 2014-04-14.

[12] "HTTPS Everywhere & the Decentralized SSL Observatory | Electronic Frontier Foundation". Eff.org. 2012-02-29. Retrieved 2014-06-04.

[13] Brian, Matt (2014-01-27). "Browsing on your Android phone just got safer, thanks to the EFF". Engadget.com. Retrieved 2014-04-14.

[14] Lemos, Robert (2011-09-21). "EFF builds system to warn of certificate breaches | Encryption". InfoWorld. Retrieved 2014-04-14.

[15] Vaughan, Steven J. (2012-02-28). "New 'HTTPS Everywhere' Web browser extension released". ZDNet. Retrieved 2014-04-14.

[16] "1 SSAC Advisory on Internal Name Certificates" (PDF). ICANN Security and Stability Advisory Committee (SSAC). 15 March 2013.

[17] Fahl, Sascha et al. "Why Eve and Mallory love Android: An analysis of Android SSL (in)security" (PDF). *Proceedings of the 2012 ACM conference on Computer and communications security* (ACM, 2012).

[18] Davis, B.; Chen, H. (2013). "Retro *Skeleton*". *Proceeding of the 11th annual international conference on Mobile systems, applications, and services - Mobi* Sys *'13*. p. 181. doi:10.1145/2462456.2464462. ISBN 9781450316729.

[19] Kern, M. Kathleen, and Eric Phetteplace. "Hardening the browser." Reference & User Services Quarterly 51.3 (2012): 210-214. http://eprints.rclis.org/16837/

[20] Toubiana, Vincent; Verdot, Vincent (2011). "Show Me Your Cookie And I Will Tell You Who You Are". arXiv:1108.5864 [cs.CR].

Chapter 21

Infinit

Infinit is a file transfer service operated by Infinit International Inc. Infinit has released four products, two for desktop and two for mobile. The desktop products are natives clients running on OS X (10.7+) and Windows. On mobile, Infinit has also released native apps, one for Android and another one for iOS.[1] A native Linux client is expected to be released later in 2015.[2]

The products enable users both to send files to each other and to a user's own devices through a simple drag/drop process.[3] The data is transferred by connecting directly to the recipient in a peer-to-peer configuration, as opposed to the server-centric model used by cloud services. The service has been reported to be faster than the competitive solutions, including Apple's local file transfer protocol AirDrop.[4] In addition, the files are encrypted before being transferred and are never stored in the cloud (on a third-party machine). Files are encrypted before leaving the sender's computer so that third-parties cannot view the files; only the recipient can decrypt them.[5]

21.1 Company

Infinit has headquarters in 25, rue Titon, 75011, Paris, France and 1407 Broadway, New York City, United States.

It was reported the company had a $0.5 million investment from Alive Ideas[6] and other business angels and after 4 months in stealth mode, the company released its first product, its file transfer app for Mac.[7]

Infinit was a member of the second season of Le Camping, a Paris-based startup accelerator, participating in a 6-month program from September 2011 to March 2012.[8] Infinit was selected in July 2013 by Agoranov, a research-based startup incubator in Paris.[9]

Infinit has been mentioned in some startup competitions such as Qualcomm QPrize 2012[10] and in March 2014 it was one of the 13 companies selected (among over 1000 applicants) by Techstars, a mentorship-driven startup incubator, to join its New York City program.[11] By May 2014

the company raised an additional $1.8 million investment from Alven Capital Partners and 360 Capital Partners.[12]

As of November 2014, over 50 petabytes of data have been transfer using Infinit.[13]

21.2 Technology

The company maintains and extends the Infinit technology originally developed by one of its founders, Julien Quintard, during his research as a PhD student at the University of Cambridge.[14]

21.3 References

[1] "Infinit and beyond: The ultimate file-transfer app just launched on mobile". *Venture Beat*. March 2015.

[2] "Infinit and beyond: This could become the ultimate file-sharing desktop app for creatives". *The Next Web*. June 2014.

[3] "Infinit's New Mobile Apps Might Be The Best Way To Transfer Those Pesky HD Videos". *TechCrunch*. March 2015.

[4] "Infinit Makes File Transfers Between Macs Fast and Painless".

[5] "Infinit launches a new, fast file-transfer app for Mac OS".

[6] "Alive Ideas". Alive Ideas web site. Archived from the original on 23 July 2013. Retrieved 10 August 2013.

[7] "Infinit: A fast, unlimited file-transfer desktop app for creatives".

[8] "Season 2 Le Camping". *Web site*. Retrieved 30 May 2013.

[9] "12 Projets Incubée par Agoranov Rècompensès au Concours National la Cèation D'Énteprises Innovantes 2013". *Agoranov web site*. 16 July 2013. Retrieved 10 August 2013. (French)

[10] "Winner of Qualcomm Venture QPrize 2012". *Wired*. October 2012.

[11] "13 New Techstars Companies for Spring Session in NYC". March 2014.

[12] "Infinit Raises $1.8 Million To Become The Definitive File Sharing App". *Techcrunch*. May 2014.

[13] "How to safely share files without the cloud". *Fox Business*. November 2014.

[14] Julien Quintard (12 June 2012). "Towards a worldwide storage infrastructure". Retrieved 10 August 2013.

21.4   External links

- Official website

Chapter 22

Inter-protocol exploitation

Inter-protocol exploitation is a class of security vulnerabilities that takes advantage of interactions between two communication protocols,[1] for example the protocols used in the Internet. It is commonly discussed in the context of the Hypertext Transfer Protocol (HTTP).[2] This attack uses the potential of the two different protocols meaningfully communicating commands and data.

It was popularized in 2007 and publicly described in research[3] of the same year. The general class of attacks that it refers to has been known since at least 1994 (see the Security Considerations section of RFC 1738).

Internet protocol implementations allow for the possibility of encapsulating exploit code to compromise a remote program which uses a different protocol. Inter-protocol exploitation can utilize inter-protocol communication to establish the preconditions for launching an inter-protocol exploit. For example, this process could negotiate the initial authentication communication for a vulnerability in password parsing. Inter-protocol exploitation is where one protocol attacks a service running a different protocol. This is a legacy problem because the specifications of the protocols did not take into consideration an attack of this type.

22.1 Technical details

The two protocols involved in the vulnerability are termed the carrier and target. The carrier encapsulates the commands and/or data. The target protocol is used for communication to the intended victim service. Inter-protocol communication will be successful if the carrier protocol can encapsulate the commands and/or data sufficiently to meaningfully communicate to the target service.

Two preconditions need to be met for successful communication across protocols: encapsulation and error tolerance. The carrier protocol must encapsulate the data and commands in a manner that the target protocol can understand. It is highly likely that the resulting data stream with induce parsing errors in the target protocol.

The target protocol be must be sufficiently forgiving of errors. During the inter-protocol connection it is likely that a percentage of the communication will be invalid and cause errors. To meet this precondition, the target protocol implementation must continue processing despite these errors.

22.2 Current implications

One of the major points of concern is the potential for this attack vector to reach through firewalls and DMZs. Inter-protocol exploits can be transmitted over HTTP and launched from web browsers on an internal subnet. An important point is the web browser is not exploited though any conventional means.

22.3 Example

JavaScript delivered over HTTP and communicating over the IRC protocol.

```
var form = document.createElement('form');
form.setAttribute('method', 'post');
form.setAttribute('action', 'http://irc.example.net:6667');
form.setAttribute('enctype', 'multipart/form-data');
var textarea = document.createElement('textarea');
textarea.innerText = "USER A B C D \nNICK
turtle\nJOIN #hack\nPRIVMSG #hackers: I like
turtles\n"; form.appendChild(textarea); document.body.appendChild(form); form.submit();
```

Known examples of the vulnerability were also demonstrated on files constructed to be valid HTML code and BMP image at the same time.[4][5][6]

22.4 References

[1] "Inter-protocol Communication" (PDF). 2006-08. Check

65

date values in: |date= (help)

[2] "HTML Form Protocol Attack".

[3] "Inter-protocol Exploitation". 2007-03-05.

[4] "Marco Ramilli's Blog: Hacking through images". *marco-ramilli.blogspot.co.uk*. Retrieved 2015-05-13.

[5] Buccafurri, F.; Caminiti, G.; Lax, G. (August 2008). "Signing the document content is not enough: A new attack to digital signature". pp. 520–525. doi:10.1109/ICADIWT.2008.4664402. Retrieved 2015-05-13.

[6] "http://www.softcomputing.net/jias/buccafurri.pdf" (PDF). *www.softcomputing.net*. Retrieved 2015-05-13.

22.5   External links

- http://www.theregister.co.uk/2007/06/27/wade_alcorn_metasploit_interview/

Chapter 23

Internet Security Research Group

The **Internet Security Research Group** (**ISRG**) is a California public-benefit corporation which focuses on Internet security. [2][3]

Let's Encrypt—its first major initiative—aims to make Secure Sockets Layer/Transport Layer Security (SSL/TLS) certificates available for free in an automated fashion.

Josh Aas, of Mozilla, serves as the group's executive director and board chair.[4][1] The board also contains individuals from Akamai, Cisco, University of Michigan, Mozilla, Stanford Law School, CoreOS, and the Electronic Frontier Foundation.[1]

23.1 References

[1] About Let's Encrypt

[2] Let's Encrypt: Delivering SSL/TLS Everywhere

[3] EFF, Mozilla back new certificate authority that will offer free SSL certificates

[4] Privacy push means free encryption for websites

23.2 External links

- Seth Schoen's Libre Planet 2015 lecture on Let's Encrypt

Chapter 24

Key ring file

A **key ring** is a file which contains multiple public keys of Certificate Authority (CA).

A key ring is a file which is necessary for Secure Sockets Layer (SSL) connection over the web. It is securely stored on the server which hosts the website. It contains the public/private key pair for the particular website. It also contains the public/private key pairs from various certificate authorities and the trusted root certificate for the various certification authorities. It also contains intermediate certificates from the intermediate certificate authorities.

An entity or website administrator has to send a certificate signing request (CSR) to the CA. The CA then returns a signed certificate to the entity. This certificate received from the CA has to be stored in the key ring.

24.1 References

- Joseph Steinberg, Tim Speed, *SSL VPN: Understanding, Evaluating, and Planning Secure, Web-based Remote Access*, Packt Publishing Ltd, 2005 ISBN 1847190014.

24.2 External links

- Creating a Lotus Notes Domino Server Key Ring File

Chapter 25

Let's Encrypt

Let's Encrypt is an upcoming certificate authority to be launched on November 16, 2015[1] that will provide free X.509 certificates for Transport Layer Security encryption (TLS) via an automated process designed to eliminate the current complex process of manual creation, validation, signing, installation and renewal of certificates for secure websites.[2][3]

25.1 Overview

The project aims to make encrypted connections in the World Wide Web the default case. By getting rid of payment, web server configuration, validation emails and dealing with expired certificates it is meant to significantly lower the complexity of setting up and maintaining TLS encryption.[4] On a Linux web server, execution of only two commands is said to be sufficient in order to set up HTTPS encryption, acquire and install certificates within 20 to 30 seconds.[5][6]

To that end, the inclusion of a software package into the official Debian software repositories is being worked on.[7] Current initiatives of big browser vendors to deprecate unencrypted HTTP are counting on the availability of Let's Encrypt.[8][9] The project is acknowledged to have the potential to accomplish encrypted connections as the default case for the entire web.[10]

So-called domain validation certificates are being issued. Organization validation and Extended Validation Certificates will not be offered.[11]

Being as transparent as possible is hoped to both protect their own trustworthiness and guarding against attacks and manipulation attempts. For that purpose they regularly publish transparency reports,[12] publicly log all ACME transactions, and use open standards and free software as much as possible. [5]

The name of the certificate authority software "Boulder" is a hint at a product of the fictional Acme Corporation from the animated cartoon series around Wile E. Coyote and The Road Runner.

There is currently no concrete plans to support wildcard certificates, though it has not been ruled out either. The reason given for the lack of support is that the ease of getting non-wildcard Let's Encrypt certificates issued makes wildcard certificates unnecessary,[13] though some users have opined that there are still use cases where wildcard certificates are easier to use or even technically necessary.[14]

25.2 Involved parties

Let's Encrypt is a service provided by the Internet Security Research Group (ISRG), a public benefit organization. Major sponsors are the Electronic Frontier Foundation (EFF), the Mozilla Foundation, Akamai, and Cisco Systems. Other partners include the certificate authority IdenTrust, the University of Michigan (U-M), the Stanford Law School, the Linux Foundation[15] as well as Stephen Kent from Raytheon/BBN Technologies and Alex Polvi from CoreOS.[5]

25.3 Technology

Let's Encrypt owns a RSA root certificate that is stored on a Hardware security module and doesn't get used directly. It is meant to be replaced by an ECDSA certificate later. It will be used to sign two intermediate certificates which are cross-signed by the certificate authority IdenTrust.[16] One of these will be used to sign issued certificates, the other as backup in case of problems with the first one. Because the IdenTrust certificate is preinstalled in major web browsers, Let's Encrypt certificates can normally be validated and are accepted out of the box right from the start. [17] In the long run, it is projected to get Let's Encrypt certificates preinstalled into applications directly.

25.3.1   Protocol

The challenge–response protocol used to automate enrolling with this new certificate authority is called Automated Certificate Management Environment (ACME). It involves various requests to the web server on the domain that is covered by the certificate. Based on whether the resulting responses match the expectations, control of the enrollee over the domain is assured (domain validation). In order to do that, the ACME client software sets up a special TLS server on the server system that gets queried by the ACME certificate authority server with special requests using Server Name Indication (Domain Validation using Server Name Indication, DVSNI). This process is only accepted for the first certificate being issued for any given domain (trust on first use, TOFU). Afterwards, the alternative way of validation via an existing certificate is used. Therefore, if control over an existing certificate is lost, a certificate has to be acquired from a third party in order to be able to obtain another Let's Encrypt certificate.

The validation processes are run multiple times over separate network paths. Checking DNS entries is provisioned to be done from multiple geographically diverse locations to make DNS spoofing attacks harder to do.

ACME interactions are based on exchanging JSON documents over HTTPS connections.[18] A draft specification is available on GitHub,[19] and a version has been submitted to the Internet Engineering Task Force (IETF) as a proposal for an Internet standard.[20]

25.3.2   Software implementation

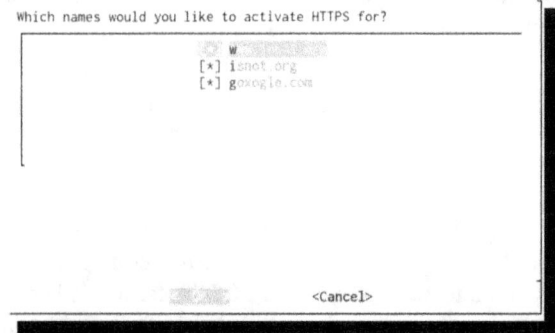

Domain selection dialogue

The certificate authority basically consists of a piece of software called Boulder, written in Go, that implements the server side of the ACME protocol. It is published as free software with source code under the terms of version 2 of the Mozilla Public License (MPL).[21] It provides a

RESTful API that can be accessed over a TLS-encrypted channel.

An Apache-licensed[22] Python certificate management program called letsencrypt gets installed on the client side (the web server of an enrollee). This is used to order the certificate, to conduct the domain validation process, to install the certificate, to configure the HTTPS encryption in the HTTP server, and later to regularly renew the certificate.[23][5] After installation and agreeing to the user license, executing a single command is enough to get a valid certificate installed. Additional options like OCSP stapling or HTTP Strict Transport Security (HSTS) can also be enabled.[18] Automatic setup initially only works with Apache and nginx.

25.4   History and schedule

Roots of the project lie in a project run by the Electronic Frontier Foundation in cooperation with the University of Michigan and an independent project by Mozilla that were combined into Let's Encrypt. In 2014, the mother organisation, the ISRG, was founded. The start of Let's Encrypt was announced on November 18, 2014.[24]

On January 28, 2015, the ACME protocol was officially submitted to the IETF for standardisation.[25] On April 9, 2015, the ISRG and the Linux Foundation declared their collaboration.[15] The root and intermediate certificates were generated in the beginning of June.[17] On June 16, 2015, the final launch schedule for the service was announced, with the first certificate expected to be issued sometime in the week of July 27, 2015, followed by a limited issuance period to test security and scalability. General availability of the service is expected to begin sometime in the week of September 14, 2015, provided everything goes as planned.[26] On August 7, 2015, the launch schedule was amended to provide more time for ensuring system security and stability, with the first certificate to be issued in the week of September 7, 2015 followed by general availability in the week of November 16, 2015.[1] The cross-signature from IdenTrust is planned to be available when Let's Encrypt opens for the public.[16]

On September 14, 2015, Let's Encrypt issued its first certificate, which was for the domain *helloworld.letsencrypt.org*. On the same day, ISRG submitted its root program applications to Mozilla, Microsoft, Google and Apple.[27]

25.5   Further reading

- Richard Barnes, Jacob Hoffman-Andrews, James Kasten (21 July 2015), IETF, ed., "Automatic Certificate Management Environment (ACME)" (in English), *Active Internet-Drafts*, https://tools.ietf.org/html/draft-barnes-acme (latest standardisation draft of the ACME specification)

25.6 References

[1] "Updated Let's Encrypt Launch Schedule". August 7, 2015.

[2] Kerner, Sean Michael (November 18, 2014). "Let's Encrypt Effort Aims to Improve Internet Security". *eWeek.com*. Quinstreet Enterprise. Retrieved February 27, 2015.

[3] Eckersley, Peter (November 18, 2014). "Launching in 2015: A Certificate Authority to Encrypt the Entire Web". Electronic Frontier Foundation. Retrieved February 27, 2015.

[4] Liam Tung (ZDNet), November 19, 2014: EFF, Mozilla to launch free one-click website encryption

[5] Fabian Scherschel (heise.de), November 19, 2014: Let's Encrypt: Mozilla und die EFF mischen den CA-Markt auf

[6] Rob Marvin (SD Times), November 19, 2014: EFF wants to make HTTPS the default protocol

[7] ITP: letsencrypt – Let's Encrypt client that can update Apache configurations

[8] Richard Barnes (Mozilla), April 30, 2015: Deprecating Non-Secure HTTP

[9] The Chromium Projects – Marking HTTP As Non-Secure

[10] Glyn Moody, November 25, 2014: The Coming War on Encryption, Tor, and VPNs – Time to stand up for your right to online privacy

[11] Steven J. Vaughan-Nichols (ZDNet), April 9, 2015: the web once and for all: The Let's Encrypt Project

[12] Zeljka Zorz (Help Net Security), July 6, 2015: Let's Encrypt CA releases transparency report before its first certificate

[13] https://community.letsencrypt.org/t/frequently-asked-questions-faq/26

[14] https://community.letsencrypt.org/t/please-support-wildcard-certificates/258/42

[15] Sean Michael Kerner (eweek.com), April 9, 2015: Let's Encrypt Becomes Linux Foundation Collaborative Project

[16] Reiko Kaps (heise.de), June 17, 2015: SSL-Zertifizierungsstelle Lets Encrypt will Mitte September 2015 öffnen

[17] Reiko Kaps (heise.de), June 5, 2015: Let's Encrypt: Meilenstein zu kostenlosen SSL-Zertifikaten für alle

[18] Chris Brook (Threatpost), November 18, 2014: EFF, Others Plan to Make Encrypting the Web Easier in 2015

[19] "Draft ACME specification".

[20] R. Barnes, P. Eckersley, S. Schoen, A. Halderman, J. Kasten (January 28, 2015). "Automatic Certificate Management Environment (ACME) draft-barnes-acme-01".

[21] https://github.com/letsencrypt/boulder/blob/master/LICENSE.txt

[22] https://github.com/letsencrypt/letsencrypt/blob/master/LICENSE.txt

[23] James Sanders (TechRepublic), November 25, 2014: Let's Encrypt initiative to provide free encryption certificates

[24] Joseph Tsidulko (2014-11-18). "Let's Encrypt, A Free And Automated Certificate Authority, Comes Out Of Stealth Mode". *crn.com*. Retrieved 2015-08-26.

[25] History for draft-barnes-acme

[26] Josh Aas (June 16, 2015). "Let's Encrypt Launch Schedule". *letsencrypt.org*. Let's Encrypt. Retrieved June 19, 2015.

[27] Michael Mimoso. "First Let's Encrypt Free Certificate Goes Live". Threatpost.com, Kaspersky Labs. Retrieved 2015-09-16.

25.7 External links

- Official website

- Code repositories hosted on GitHub

- Seth Schoen's Libre Planet 2015 lecture on Let's Encrypt

- technical introduction in a blog post by David Wong

- pde's talk on Let's Encrypt at CCCamp 2015

Chapter 26

Linoma Software

Linoma Software is a developer of managed file transfer and encryption solutions. Mid-sized companies, large enterprises and government entities use Linoma's solutions to protect sensitive data and comply with data security regulations such as PCI DSS, HIPAA/HITECH, SOX, GLBA and state privacy laws. Linoma's solutions run on a variety of platforms including Windows, Linux, UNIX, IBM i (iSeries), AIX, Solaris, HP-UX and Mac OS X.

26.1 History

Linoma Group, Inc. (the parent company of Linoma Software) was founded in 1994. The company was started in Lincoln, Nebraska by Robert and Christina Luebbe. Throughout most of the 1990s, the Linoma Group performed consulting and contract programming services for organizations in the Nebraska/Iowa area.

Linoma Software was formed in 1998 to address the needs of the IBM AS/400 platform (now known as IBM i) by developing productivity tools to help IT departments and end users. These tools were sold throughout the world and helped Linoma establish itself as an innovative software company.

In 2002, Linoma released Transfer Anywhere, which was a solution for automating and managing file transfers from the AS/400. Over the next 2–3 years, Linoma added encryption capabilities to Transfer Anywhere including support for Open PGP encryption, SFTP and FTPS. These encryption capabilities helped organizations protect sensitive data transmissions such as ACH payments, direct deposits, financial data, credit card authorizations, personally identifiable information (PII) and other confidential data.

Linoma expanded into other platforms when it completely redesigned Transfer Anywhere into an open OS solution with a graphical browser-based interface, renaming it GoAnywhere Director. Released in early 2008, GoAnywhere Director included comprehensive security controls, key management, trading partner wizards and detailed audit trails for compliance requirements.

In 2009, Linoma released GoAnywhere Services as a collection of secure file services including an FTP Server, FTPS Server, SFTP Server and HTTPS server.

GoAnywhere Director and Services were merged in 2015 to become GoAnywhere MFT. GoAnywhere MFT merged the workflow automation capabilities (adapted from GoAnywhere Director) with secure FTP server and collaboration features (adapted from GoAnywhere Services). This provides a unified browser-based interface, centralized logging and reporting. GoAnywhere MFT is in the Managed File Transfer software category of products, but can also be used for ETL functions.

GoAnywhere Gateway was released in 2010 as an enhanced reverse proxy to protect the DMZ and help organizations meet strict compliance requirements. GoAnywhere Gateway was enhanced in 2011 to provide forward proxy functions.

Linoma Software also performs encryption of data at rest on the IBM i (iSeries) platform with its Crypto Complete product. This product also includes key management, security controls and audit trails for PCI compliance.

As of 2015, Linoma Software serves more than 3,000 customers around the world.

26.2 Certifications

- VMware Ready

- Novell Ready for SUSE Linux Enterprise Server

- Works With Windows Server 2008 R2

- IBM Ready for Power Systems Software – IBM Power Systems

- IBM Ready for Systems with Linux - IBM Chiphopper

26.3 Associations

- Microsoft Partner

  - Silver Independent Software Vendor (ISV) competency.[1]
  - Silver Application Integration competency.[2]

- IBM Advanced Business Partner[3]

- VMware Elite Partner[4]

- Oracle Partner Network (OPN)[5]

- PCI Security Standards[6]

- COMMON[7]

- Better Business Bureau[8]

- Red Hat ISV Partner[9]

- OpenPGP Alliance[10]

- Apple Developer[11]

- Novell ISV Partner[12]

- GSA Advantage Schedule[13]

26.4 Current Software

26.4.1 GoAnywhere MFT

GoAnywhere MFT is a managed file transfer solution for the exchange of data between systems, employees, customers and trading partners. It provides a single point of control with security settings, detailed audit trails and reports. Data transfers are secured using protocols for FTP servers (FTPS, SFTP, and SCP) and Web servers (HTTPS and AS2). It supports popular encryption protocols and offers a NIST-certified FIPS 140-2 Validated Encryption module.

GoAnywhere MFT's interface and workflow features help to eliminate the need for custom programs/scripts, single-function tools and manual processes that were traditionally needed. This improves the quality of file transfers and helps organizations to comply with data security policies and regulations.

With integrated support for clustering, GoAnywhere MFT can process high volumes of file transfers for enterprises by load balancing processes across multiple systems. The clustering technology in GoAnywhere MFT also provides active-active automatic failover for disaster recovery.

A secure email module is also available that allows users to send messages and files as secure packages. Recipients receive an email with a unique link to each package that allows them to view or download the files via a secure HTTPS connection. There is no limit on file size or type, and each package can be subject to password protection as well as other security features.

26.4.2 GoDrive by GoAnywhere

GoDrive is an on-premise solution that provides Enterprise File Sync and Sharing (EFSS) services for employees and partners. GoDrive files and folders can be easily shared between users with advanced collaboration features including file revision tracking, commenting, trash bin, media viewing and synchronization with Windows and Mac devices.

GoDrive is an alternative to cloud-based file sharing services. It provides on-site file storage with localized control, end-to-end encryption, and detailed audit trails. If an end-user device is lost or stolen, its GoDrive data can be deactivated and wiped remotely.

26.4.3 GoAnywhere Gateway

GoAnywhere Gateway provides an additional layer of network security by masquerading server identities when exchanging data with trading partners. The application does not store user credentials or data in the DMZ / local network. When using a reverse proxy, inbound ports do not need to be opened into the private network, which is essential for compliance with PCI DSS,[14] HIPAA, HITECH, SOX, GLBA and state privacy laws. The current version is 2.0.1.

A reverse proxy is used by the application for the file-sharing services (for example, FTP/S, SFTP, HTTP/S servers) it front-ends in the DMZ. GoAnywhere Gateway's service broker binds file transfer requests to the appropriate service in the private network through a secure control channel.

GoAnywhere Gateway makes connections to external systems on behalf of users and applications in the private network. Routing outbound requests through a centralized point helps manage file transfers through a firewall. This method keeps inbound ports closed. The forward proxy hides the identities and locations of internal systems for security purposes.

26.4.4 GoAnywhere OpenPGP Studio

GoAnywhere OpenPGP Studio is a free desktop tool that protects sensitive files using the OpenPGP encryption standard. Documents can be encrypted, decrypted, signed and verified from a PC or workstation using this tool. An integrated key manager allows users to create, import, export and manage OpenPGP keys needed to encrypt and decrypt files. GoAnywhere OpenPGP Studio will run on almost any operating system including Windows, Linux, Mac OS X, Solaris and UNIX.

26.4.5 Crypto Complete

Crypto Complete is a program for the IBM i that protects sensitive data using strong encryption, tokenization, integrated key management, and auditing. This software encrypts database fields, can automatically encrypt IFS files.

The application also locates[15] sensitive information that should be encrypted using the FNDDBFLD utility, which is available at no cost to IBM i users.[16] The current version is 3.3.0.

The key management system is integrated within the Crypto Complete policy controls, encryption functions and auditing facilities. Along with the integrated security native to the IBM i, access to key maintenance/usage activities is controlled to help meet compliance requirements.

The backup encryption component encrypts the data written to tape devices. Crypto Complete encrypts the backups of any user data in IBM i libraries, objects, and IFS files.

The field encryption registry works with IBM's Field Procedures and remembers which fields in a database should be encrypted. This process can be automated whenever any data is added to the field. When the data is decrypted, the returned values are masked or displayed based on the authority of the user.

Tokenization is the process of replacing sensitive data with unique identification numbers (tokens) and storing the original data on a central server (typically in encrypted form). Tokenization can help thwart hackers and minimize the scope of compliance audits when it is stored in a single central location. Tokenization is used to protect sensitive data like credit card personal account numbers (PAN), bank account numbers, social security numbers, driver's license numbers and other personally identifiable information (PII).

26.4.6 Surveyor/400

A productivity suite for working with iSeries data, files, libraries, and objects. Surveyor/400 operates in a GUI frontend, but provides options for either IBM 5250 or "Command Line" emulation. The current version is 4.0.4.[17]

26.4.7 RPG Toolbox

RPG Toolbox was developed to help developers upgrade their older RPG and System/36 code to the new RPG IV or OS/400 standard. The program allows developers to save code "snippets" for re-use or testing. The current version is 4.06

26.5 Platforms

The GoAnywhere applications are VMware Ready and operate in a virtualized or static environment on the following operating systems.

- Linux
- Novell SUSE Linux Enterprise Server (SLES)
- Red Hat Enterprise Linux
- Unix
- Mac OS X
- Windows
- HP-UX
- Solaris
- IBM System p (AIX)
- IBM System i
- IBM System z

26.6 See also

- Comparison of FTP server software

26.7 Notes

[1] Microsoft Silver Independent Software Vendor (ISV)

[2] Microsoft Silver Application Integration

[3] IBM Advanced Business Partner

[4] VMware Elite Partner

[5] Oracle Partner Network (OPN)

[6] Participating Organization in the Payment Card Industry Security Standards Council (PCI SSC),

[7] Member of the COMMON User Group.

[8] A+ Better Business Bureau Member

[9] Beta tester and performance testing for current and upcoming Red Hat Enterprise Linux (RHEL) distributions.,

[10] OpenPGP standards group, started by Philip Zimmermann,

[11] Apple Developers actively participate in testing and building on the Mac OS X and Mac OS X Server platforms. Apple Developers also test the accessibility of their products on other Apple Web-based devices like the iPhone, iPad, and iPod Touch.

[12] Novell ISV partnership and SUSE Studio appliance participant.

[13] NAICS codes: 511210, 518210, 522320, 541512, 423430

[14] PCI DSS 1.3

[15] Find Database Fields - IBM Systems Magazine

[16] FNDDBFLD

[17] IBM Journal - 13 December 2010

26.8 External Reviews/Links

Sys-Con Media - GoAnywhere 3.0
Business Wire - GoAnywhere Services
Four Hundred Stuff - Crypto Complete 2.2
IBM Systems Magazine - Crypto Complete
IT Jungle - Surveyor/400 3.7
GoAnywhere

Chapter 27

Nautilus (secure telephone)

Nautilus is a program which allows two parties to securely communicate using modems or TCP/IP. It runs from a command line and is available for the Linux and Windows operating systems. The name was based upon Jules Verne's Nautilus and its ability to overcome a Clipper ship as a play on Clipper chip.

Nautilus is historically significant in the realm of secure communications because it was one of the first programs which were released as open source to the general public which used strong encryption. It was created as a response to the Clipper chip in which the US government planned to use a key escrow scheme on all products which used the chip. This would allow them to monitor "secure" communications. Once this program and another similar program PGPfone were available on the internet, the proverbial cat was "out of the bag" and it would have been nearly impossible to stop the use of strong encryption for telephone communications.

The project must move end of May 2014 due to the decision of Fraunhofer FOCUS to shut down the developer platform that hosted dozens of vital free software projects like mISDN, gpsd, etc.

27.1   External links

- new Nautilus homepage from May 1 2014 on

- "Can Nautilus Sink Clipper?" Article in **Wired**, Aug 1995

Chapter 28

Network Security Services

In computing, **Network Security Services** (NSS) comprises a set of libraries designed to support cross-platform development of security-enabled client and server applications with optional support for hardware TLS/SSL acceleration on the server side and hardware smart cards on the client side. NSS provides a complete open-source implementation of cryptographic libraries supporting Transport Layer Security (TLS) / Secure Sockets Layer (SSL) and S/MIME. Previously tri-licensed under the Mozilla Public License 1.1, the GNU General Public License, and the GNU Lesser General Public License, NSS upgraded to GPL-compatible MPL 2.0 with release 3.14.[2]

28.1 History

NSS originated from the libraries developed when Netscape invented the SSL security protocol.

28.1.1 FIPS 140 validation and NISCC testing

The NSS software crypto module has been validated five times (1997, 1999, 2002, 2007, and 2010) for conformance to FIPS 140 at Security Levels 1 and 2.[3] NSS was the first open source cryptographic library to receive FIPS 140 validation.[3] The NSS libraries passed the NISCC TLS/SSL and S/MIME test suites (1.6 million test cases of invalid input data).[3]

28.2 Applications that use NSS

AOL, Red Hat, Sun Microsystems/Oracle Corporation, Google and other companies and individual contributors have co-developed NSS. Mozilla provides the source code repository, bug tracking system, and infrastructure for mailing lists and discussion groups. They and others named below use NSS in a variety of products, including the following:

- Mozilla client products, including Firefox, Thunderbird, SeaMonkey, and Firefox for mobile (Fennec).[4]

- AOL Communicator and AOL Instant Messenger (AIM)

- Google Chrome and Chromium except for Android / OS X,[5] [6] and Opera

- Open source client applications such as Evolution, Pidgin, OpenOffice.org 2.0 and later, and Apache OpenOffice.

- Server products from Red Hat: Red Hat Directory Server, Red Hat Certificate System, and the mod nss SSL module for the Apache web server.

- Sun server products from the Sun Java Enterprise System, including Sun Java System Web Server, Sun Java System Directory Server, Sun Java System Portal Server, Sun Java System Messaging Server, and Sun Java System Application Server, open source version of Directory Server OpenDS.

- Libreswan IKE/IPsec requires NSS. It is a fork of Openswan which could optionally use NSS.

28.3 Architecture

NSS includes a framework to which developers and OEMs can contribute patches, such as assembly code, to optimize performance on their platforms. Mozilla has certified NSS 3.x on 18 platforms.[7][8] NSS makes use of Netscape Portable Runtime (NSPR), a platform-neutral open-source API for system functions designed to facilitate cross-platform development. Like NSS, NSPR has been used heavily in multiple products.

28.3.1   Software development kit

In addition to libraries and APIs, NSS provides security tools required for debugging, diagnostics, certificate and key management, cryptography-module management, and other development tasks. NSS comes with an extensive and growing set of documentation, including introductory material, API references, man pages for command-line tools, and sample code.

Programmers can utilize NSS as source and as shared (dynamic) libraries. Every NSS release is backward-compatible with previous releases, allowing NSS users to upgrade to new NSS shared libraries without recompiling or relinking their applications.

28.3.2   Interoperability and open standards

NSS supports a range of security standards, including the following:[9][1]

- TLS 1.0 (RFC 2246), 1.1 (RFC 4346), and 1.2 (RFC 5246). The Transport Layer Security (TLS) protocol from the IETF supersedes SSL v3.0 while remaining backward-compatible with SSL v3 implementations.

- SSL 2.0 and 3.0. The Secure Sockets Layer (SSL) protocol allows mutual authentication between a client and server and the establishment of an authenticated and encrypted connection.

- DTLS 1.0 (RFC 4347) and 1.2 (RFC 6347).

- DTLS-SRTP (RFC 5764).

- The following PKCS standards:

  - PKCS #1. RSA standard that governs implementation of public-key cryptography based on the RSA algorithm.

  - PKCS #3. RSA standard that governs implementation of Diffie–Hellman key agreement.

  - PKCS #5. RSA standard that governs password-based cryptography, for example to encrypt private keys for storage.

  - PKCS #7. RSA standard that governs the application of cryptography to data, for example digital signatures and digital envelopes.

  - PKCS #8. RSA standard that governs the storage and encryption of private keys.

  - PKCS #9. RSA standard that governs selected attribute types, including those used with PKCS #7, PKCS #8, and PKCS #10.

  - PKCS #10. RSA standard that governs the syntax for certificate requests.

  - PKCS #11. RSA standard that governs communication with cryptographic tokens (such as hardware accelerators and smart cards) and permits application independence from specific algorithms and implementations.

  - PKCS #12. RSA standard that governs the format used to store or transport private keys, certificates, and other secret material.

- Cryptographic Message Syntax, used in S/MIME (RFC 2311 and RFC 2633). IETF message specification (based on the popular Internet MIME standard) that provides a consistent way to send and receive signed and encrypted MIME data.

- X.509 v3. ITU standard that governs the format of certificates used for authentication in public-key cryptography.

- OCSP (RFC 2560). The Online Certificate Status Protocol (OCSP) governs real-time confirmation of certificate validity.

- PKIX Certificate and CRL Profile (RFC 3280). The first part of the four-part standard under development by the Public-Key Infrastructure (X.509) working group of the IETF (known as PKIX) for a public-key infrastructure for the Internet.

- RSA, DSA, ECDSA, Diffie–Hellman, EC Diffie–Hellman, AES, Triple DES, Camellia, IDEA, SEED, DES, RC2, RC4, SHA-1, SHA-256, SHA-384, SHA-512, MD2, MD5, HMAC: Common cryptographic algorithms used in public-key and symmetric-key cryptography.

- FIPS 186-2 pseudorandom number generator.

28.3.3   Hardware support

NSS supports the PKCS #11 interface for access to cryptographic hardware like SSL accelerators, HSM-s and smart cards. Since most hardware vendors such as SafeNet Inc. and Thales also support this interface, NSS-enabled applications can work with high-speed crypto hardware and use private keys residing on various smart cards, if vendors provide the necessary middleware. NSS version 3.13 and above support the Advanced Encryption Standard New Instructions (AES-NI).[10]

28.3.4 Java support

Network Security Services for Java (JSS) consists of a Java interface to NSS. It supports most of the security standards and encryption technologies supported by NSS. JSS also provides a pure Java interface for ASN.1 types and BER/DER encoding. The Mozilla CVS tree makes source code for a Java interface to NSS available.

28.4 See also

- Information security

- Comparison of TLS implementations

- mbed TLS

- OpenSSL

- wolfSSL

- MatrixSSL

28.5 References

[1] "NSS 3.20 release notes". Mozilla. 2015-08-19. Retrieved 2015-08-20.

[2] "NSS 3.14 release notes". *MDN*. Mozilla Developer Network. Retrieved 2015-09-01. The NSS license has changed to MPL 2.0. Previous releases were released under a MPL 1.1/GPL 2.0/LGPL 2.1 tri-license.

[3] "FIPS". Mozilla. 2012-02-01. Retrieved 2013-05-17.

[4] "Does Fennec use NSS?". *mozilla.dev.security.policy newsgroup*. April 2010. Retrieved 2013-05-17.

[5] "External: Chrome, NSS, and OpenSSL". 2014-01-26. Retrieved 2014-06-22.

[6] "The Chromium Project: BoringSSL". Retrieved 2015-03-06.

[7] "Network Security Services". Mozilla. 2013-05-16. Retrieved 2013-05-17.

[8] "NSS FAQ". Mozilla. 2013-05-16. Retrieved 2013-05-17.

[9] "Encryption Technologies Available in NSS 3.11". Mozilla. 2012-02-01. Retrieved 2013-05-17.

[10] "AES-NI enhancements to NSS on Sandy Bridge systems". 2012-05-02. Retrieved 2013 05-17.

28.6 External links

- Network Security Services

- Network Security Services

- JSS toolkit

- "Validated FIPS 140-1 and FIPS 140-2 Cryptographic Modules: 1997". NIST. 2013-04-02.

- "Validated FIPS 140-1 and FIPS 140-2 Cryptographic Modules: 2002". NIST. 2013-04-02.

- "Validated FIPS 140-1 and FIPS 140-2 Cryptographic Modules: 2010". NIST. 2013-05-17.

Chapter 29

Numbers relay page

The **Numbers Relay Page** (**NRP**) was conceived as an Internet alternative to serve the same function as Number Stations. The NRP are mostly in guest-book format and public, so that anyone can post and read the messages. The messages them-self are mostly encrypted with One-time pads, but other enciphering methods can also be used. The advantage of this system is that it is hard to determine for whom the message was meant, also, if the sender uses an anonymous Internet connection to send the message, then the sender and the receiver remain anonymous. One-time pads offer absolute security if used correctly, and are therefore best suitable for encrypted messages on NRP's.

29.1 Disadvantages

If both the sender and the receiver wish to remain completely anonymous, then computers with public access to the Internet need to be used. If someone sends or reads the messages from their private computers, their IP addresses could be traced by their ISP. Further improvement on the anonymity of the sender and receiver can be made by using software for anonymous web browsing, like the "TOR anonymity network", but one should alway be aware that any computer system can be compromised. By tracing the IP, the contents of the message wont be compromised if the encryption was done with One-time pads the old fashioned way, pen and paper. But the sender and the receiver could be traced.

29.2 External links

- More on NRP's

Some examples of NRP's

- Example 1
- Example 2

- example 3

Chapter 30

"ObserveIT"

ObserveIT is a User Activity Monitoring software company. ObserveIT provides insider threat security solutions to more than 1,200 customers in over 70 countries. The company's software is deployed primarily in the financial services, healthcare, manufacturing and retail markets.

30.1 History

Co-founders Gaby Friedlander and Avi Amos founded ObserveIT in Israel in 2006. ObserveIT initially focused on monitoring remote vendor monitoring, but now has expanded its focus to monitoring employees, privileged users and third parties in order to detect insider threats. The company grew until 2014, when Bain Capital invested $20 Million into ObserveIT, and opened up its U.S. headquarters in Boston, MA. Now under the leadership of CEO Paul Brady, ObserveIT is currently on its version 5.8.3 release.

30.2 ObserveIT Development Timeline

April 2006: ObserveIT Monitors the "Invisible Component" in IT Infrastructure – People [1]

April 2010: Lieberman Software and ObserveIT Partner to Monitor and Audit Privileged Access Activity within the Enterprise [2]

January 2012: Fujitsu Partners with ObserveIT to Bring Seamless Security Monitoring Applications to Enterprises in Hong Kong and Macau [3]

November 2013: ObserveIT unveils ObserveIT 5.6.8

April 2014: ObserveIT Appoints Paul Brady CEO [4]

September 2014: ObserveIT Releases v5.7 [5]

April 2015: ObserveIT Receives Three Awards in 2015 Security Industry's Global Excellence Awards [6]

May 2015: ObserveIT Launches CloudThreat Amazon Web Services Security Solution [7]

30.3 References

[1] "ObserveIT Monitors the Invisible Component in IT Infrastructure". *People*. 16 April 2006. Retrieved 20 July 2015.

[2] "Lieberman Software and ObserveIT Partner to Monitor and Audit Privileged Access Activity within the Enterprise". *Liebsoft*. 12 April 2010. Retrieved 20 July 2015.

[3] "Fujitsu Partners with ObserveIT to Bring Seamless Security Monitoring Applications to Enterprises in Hong Kong and Macau". *Fujitsu*. 16 January 2012. Retrieved 20 July 2015.

[4] "ObserveIT Appoints Paul Brady CEO". *BusinessWire*. 1 May 2014. Retrieved 20 July 2015.

[5] "ObserveIT Alerts and Analytics Help Companies Address User-Based Attacks in Real Time". *BusinessWire*. 15 September 2014. Retrieved 20 July 2015.

[6] "ObserveIT Receives Three Awards in 2015 Security Industry's Global Excellence Awards". *BusinessWire*. 22 April 2014. Retrieved 20 July 2015.

[7] "ObserveIT Launches CloudThreat Amazon Web Services Security Solution". *BusinessWire*. 20 May 2015. Retrieved 20 July 2015.

Chapter 31

Open Whisper Systems

This article is about an open-source software project. For the Twitter subsidiary, see Whisper Systems.

Open Whisper Systems is a nonprofit software group[3] that develops collaborative open source projects with a mission to "make private communication simple".[1] The group was established in 2013 and consists of a small team of dedicated grant-funded developers, as well as a large community of volunteer open source contributors.[1] Open Whisper Systems is funded by a combination of donations and grants, and all of its products are published as free and open-source software under the terms of the GNU General Public License (GPL) version 3.

31.1 History

31.1.1 Background

Security researcher Moxie Marlinspike and roboticist Stuart Anderson co-founded a startup company called Whisper Systems in 2010.[4][5] The company produced proprietary enterprise mobile security software. Among these were TextSecure and RedPhone.[6] They also developed a firewall and tools for encrypting other forms of data.[4]

In November 2011, Whisper Systems announced that it had been acquired by Twitter. The financial terms of the deal were not disclosed by either company.[7] The acquisition was done "primarily so that Mr. Marlinspike could help the then-startup improve its security".[8] Shortly after the acquisition, Whisper Systems' RedPhone service was made unavailable.[9] Some criticized the removal, arguing that the software was "specifically targeted [to help] people under repressive regimes" and that it left people like the Egyptians in "a dangerous position" during the events of the 2011 Egyptian revolution.[10]

Twitter released TextSecure as free and open-source software under the GPLv3 license in December 2011.[4][11][12][13] RedPhone was also released under the same license in July 2012.[14] Marlinspike later left Twitter and founded Open Whisper Systems[2] as a collaborative Open Source project for the continued development of TextSecure and RedPhone.[15]

31.1.2 Establishment

Open Whisper Systems' website was launched in January 2013.[15]

Toward the end of July 2014, Open Whisper Systems announced Flock, a private contact and calendar cloud sync, and plans to unify its RedPhone and TextSecure applications as Signal.[16] These announcements coincided with the initial release of Signal as a RedPhone counterpart for iOS. The developers said that their next steps would be to provide TextSecure instant messaging capabilities for iOS, unify the RedPhone and TextSecure applications on Android, and launch a web client.[17] Signal was the first iOS app to enable easy, strongly encrypted voice calls for free.[2][18]

On November 18, 2014, Open Whisper Systems announced a partnership with WhatsApp to provide end-to-end encryption by incorporating the protocol used in TextSecure into each WhatsApp client platform.[19] Open Whisper Systems asserted that they have already incorporated the protocol into the latest WhatsApp client for Android and that support for other clients, group/media messages, and key verification would be coming soon.[20] WhatsApp confirmed the partnership to reporters, but there was no announcement or documentation about the encryption feature on the official website, and further requests for comment were declined.[21]

In March 2015, Open Whisper Systems released Signal 2.0 with support for TextSecure private messaging on iOS.[22][23]

31.2 Funding

Open Whisper Systems is funded by a combination of donations and grants. The project has received financial support from, among others, the Freedom of the Press Foundation,[24] the Knight Foundation,[25] the Shuttleworth Foundation,[26] and the Open Technology Fund,[27] a U.S. government program that has also funded other privacy projects like the anonymity software Tor and the encrypted instant messaging application Cryptocat.

Open Whisper Systems uses a system called BitHub to distribute small donations appropriately among contributors. The system automatically pays a percentage of Bitcoin funds for every submission to one of Open Whisper Systems' GitHub repositories.[18][28]

31.3 Reception

Former NSA contractor Edward Snowden has endorsed Open Whisper Systems' applications on multiple occasions. In his keynote speech at SXSW in March 2014, he praised TextSecure and RedPhone for their ease-of-use.[29] During an interview with The New Yorker in October 2014, he recommended using "anything from Moxie Marlinspike and Open Whisper Systems".[30] During a remote appearance at an event hosted by Ryerson University and Canadian Journalists for Free Expression in March 2015, Snowden said that Signal is "very good" and that he knew the security model.[31] Asked about encrypted messaging apps during a Reddit AMA in May 2015, he recommended "Signal for iOS, Redphone/TextSecure for Android".[32][33]

In October 2014, the Electronic Frontier Foundation (EFF) included TextSecure, RedPhone, and Signal in their updated surveillance self-defense guide.[34] In November 2014, all three received top scores on the EFF's secure messaging scorecard, along with Cryptocat, Silent Phone, and Silent Text.[35] They received points for having communications encrypted in transit, having communications encrypted with keys the providers don't have access to (end-to-end encryption), making it possible for users to independently verify their correspondent's identities, having past communications secure if the keys are stolen (forward secrecy), having their code open to independent review (open source), having their security designs well-documented, and having recent independent security audits.[35]

On December 28, 2014, *Der Spiegel* published slides from an internal NSA presentation dating to June 2012 in which the NSA deemed RedPhone on its own as a "major threat" to its mission, and when used in conjunction with other privacy tools such as Cspace, Tor, Tails, and TrueCrypt was ranked as "catastrophic," leading to a "near-total loss/lack of insight to target communications, presence..."[36][37]

31.4 Projects

31.4.1 Active

- **AxolotlKit:** A free implementation of the Axolotl key management protocol. It is designed to be a drop-in library that can be easily integrated into existing projects. The implementation is written in Objective-C and is publisher under the GPLv2 license.[38]

- **BitHub:** A service that will automatically pay a percentage of Bitcoin funds for every submission to a GitHub repository.[39][40]

RedPhone

- **RedPhone:** A free and open-source encrypted voice calling application for Android. RedPhone integrates with the system dialer to make calls, but uses ZRTP to set up an end-to-end encrypted VoIP channel for the actual call. RedPhone was designed specifically for mobile devices, using audio codecs and buffer algorithms tuned to the characteristics of mobile networks, and uses push notifications to preserve the user's device's battery life while still remaining responsive.[41] RedPhone calls are compatible with Signal calls on iOS.[2] All calls are made over a Wi-Fi or data connection and are free of charge, including long distance and international.[18] All RedPhone calls to other RedPhone users and to Signal users are automatically end-to-end encrypted. The keys that are used to encrypt the user's communications are generated and stored at the endpoints (i.e. by users, not by servers). The encryption implements forward secrecy.[35] RedPhone and Signal have a built-in mechanism for verifying that no man-in-the-middle attack has occurred. During a call, the apps display two words (selected from the PGP word list) on the screen. If the words match on both ends of the call, the call is secure.[18][42] The software is published under the GPLv3 license.[41]

Signal

- **Signal:** A free and open-source encrypted voice calling and instant messaging application for iOS. Signal communications are compatible with RedPhone and TextSecure on Android. It uses end-to-end encryption with forward secrecy and deniable authentication to secure all communications to Signal, RedPhone, and TextSecure users.[23][35] Open Whisper Systems has set up dozens of servers to handle the encrypted calls in more than 10 countries around the world to minimize latency.[2] The software is published under the GPLv3 license.[43]

- **TextSecure:** A free and open-source encrypted messaging application for Android.[44][45] TextSecure can be used to send and receive SMS, MMS, and instant messages.[46] TextSecure instant messages are compatible with Signal messages on iOS. It uses end-to-end encryption with forward secrecy and deniable authentication to secure all instant messages to TextSecure and Signal users.[35][45][47][48] The software is published under the GPLv3 license.[44]

TextSecure

- **TextSecure-Server:** The software that handles message routing for the TextSecure data channel. Client-server communication is protected by TLS.[49] Communication is handled by a REST API and push messaging (both GCM and APN).[50] Support for WebSocket has been added.[51] The software is published under the AGPLv3 license.[50]

31.4.2 Discontinued

- **Flock:** A service that synced calendar and contact information on Android devices. Users had the ability to host their own server. Flock was shut down permanently on October 1, 2015. The developer cited technological choices that lead to high server costs as a reason for the EOL.[52] Flock's source code is still available on GitHub under the GPLv3 license.[53]

31.5   See also

- Freedom of speech

- The Guardian Project

- Internet privacy

- Secure communication

31.6 References

[1] Open Whisper Systems. "Open Whisper Systems". Retrieved 2015-05-01.

[2] Andy Greenberg (29 July 2014). "Your iPhone Can Finally Make Free, Encrypted Calls". Wired. Retrieved 18 January 2015.

[3] Franceschi-Bicchierai, Lorenzo (18 November 2014). "WhatsApp messages now have Snowden-approved encryption on Android". Mashable. Retrieved 23 January 2015.

[4] Garling, Caleb (2011-12-20). "Twitter Open Sources Its Android Moxie | Wired Enterprise". Wired.com. Retrieved 2011-12-21.

[5] "Company Overview of Whisper Systems Inc.". Bloomberg Businessweek. Retrieved 2014-03-04.

[6] Andy Greenberg (2010-05-25). "Android App Aims to Allow Wiretap-Proof Cell Phone Calls". Forbes. Retrieved 2014-02-28.

[7] Tom Cheredar (November 28, 2011). "Twitter acquires Android security startup Whisper Systems". VentureBeat. Retrieved 2011-12-21.

[8] Yadron, Danny (9 July 2015). "Moxie Marlinspike: The Coder Who Encrypted Your Texts". *The Wall Street Journal*. Retrieved 10 July 2015.

[9] Andy Greenberg (2011-11-28). "Twitter Acquires Moxie Marlinspike's Encryption Startup Whisper Systems". Forbes. Retrieved 2011-12-21.

[10] Garling, Caleb (2011-11-28). "Twitter Buys Some Middle East Moxie | Wired Enterprise". Wired.com. Retrieved 2011-12-21.

[11] Chris Aniszczyk (20 December 2011). "The Whispers Are True". *The Twitter Developer Blog*. Twitter. Archived from the original on 24 October 2014. Retrieved 22 January 2015.

[12] "TextSecure is now Open Source!". Whisper Systems. 20 December 2011. Archived from the original on 6 January 2012. Retrieved 22 January 2015.

[13] Pete Pachal (2011-12-20). "Twitter Takes TextSecure, Texting App for Dissidents, Open Source". Mashable. Retrieved 2014-03-01.

[14] "RedPhone is now Open Source!". Whisper Systems. 18 July 2012. Archived from the original on 31 July 2012. Retrieved 22 January 2015.

[15] "A New Home". Open Whisper Systems. 2013-01-21. Retrieved 2014-03-01.

[16] "Free, Worldwide, Encrypted Phone Calls for iPhone". Open Whisper Systems. 29 July 2014.

[17] Michael Mimoso (29 July 2014). "New Signal App Brings Encrypted Calling to iPhone". Threatpost.

[18] Jon Evans (29 July 2014). "Talk Private To Me: Free, Worldwide, Encrypted Voice Calls With Signal For iPhone". *TechCrunch*. AOL.

[19] Jon Evans (2014-11-18). "WhatsApp Partners With Open Whisper Systems To End-To-End Encrypt Billions Of Messages A Day". TechCrunch. Retrieved 2014-11-19.

[20] "Open Whisper Systems partners with WhatsApp to provide end-to-end encryption". Open Whisper Systems. November 18, 2014. Retrieved November 18, 2014.

[21] "Facebook's messaging service WhatsApp gets a security boost". Forbes. 18 Nov 2014. Retrieved 21 Nov 2014.

[22] Micah Lee (2015-03-02). "You Should Really Consider Installing Signal, an Encrypted Messaging App for iPhone". The Intercept. Retrieved 2015-03-03.

[23] Megan Geuss (2015-03-03). "Now you can easily send (free!) encrypted messages between Android, iOS". Ars Technica. Retrieved 2015-03-04.

[24] "Donate to Support Encryption Tools for Journalists". Freedom of the Press Foundation. Retrieved 19 May 2015.

[25] "TextSecure". Knight Foundation. Retrieved 5 January 2015.

[26] "Moxie Marlinspike". Shuttleworth Foundation. Retrieved 14 January 2015.

[27] "Open Whisper Systems". Open Technology Fund. Retrieved 19 May 2015.

[28] Marlinspike, Moxie (16 December 2013). "BitHub = Bitcoin + GitHub. An experiment in funding privacy OSS.". Open Whisper Systems. Retrieved 23 January 2015.

[29] Max Eddy (11 March 2014). "Snowden to SXSW: Here's How To Keep The NSA Out Of Your Stuff". PC Magazine: SecurityWatch. Retrieved 2014-03-16.

[30] "The Virtual Interview: Edward Snowden - The New Yorker Festival". *YouTube*. The New Yorker. Oct 11, 2014. Retrieved May 24, 2015.

[31] Dell Cameron (Mar 6, 2015). "Edward Snowden tells you what encrypted messaging apps you should use". The Daily Dot. Retrieved May 24, 2015.

[32] Alan Yuhas (May 21, 2015). "NSA surveillance powers on the brink as pressure mounts on Senate bill – as it happened". The Guardian. Retrieved May 24, 2015.

[33] Zack Beauchamp (May 21, 2015). "The 9 best moments from Edward Snowden's Reddit Q&A". Vox Media. Retrieved May 24, 2015.

[34] "Surveillance Self-Defense. Communicating with Others". Electronic Frontier Foundation. 2014-10-23.

[35] "Secure Messaging Scorecard. Which apps and tools actually keep your messages safe?". Electronic Frontier Foundation. 2014-11-04.

[36] SPIEGEL Staff (28 December 2014). "Prying Eyes: Inside the NSA's War on Internet Security". *Der Spiegel*. Retrieved 23 January 2015.

[37] "Presentation from the SIGDEV Conference 2012 explaining which encryption protocols and techniques can be attacked and which not" (PDF). *Der Spiegel*. 28 December 2014. Retrieved 23 January 2015.

[38] Open Whisper Systems. "AxolotlKit". *GitHub*. Retrieved 1 April 2015.

[39] Open Whisper Systems. "BitHub". *GitHub*. Retrieved 14 January 2015.

[40] Finley, Klint (17 December 2013). "Love Child of Bitcoin and GitHub Pays Cash for Code". *Wired*. Condé Nast. Retrieved 22 September 2015.

[41] Open Whisper Systems. "RedPhone". *GitHub*. Retrieved 14 January 2015.

[42] "Exactly how does Zfone and ZRTP protect against a man-in-the-middle (MiTM) attack?". The Zfone Project. Retrieved 25 January 2015.

[43] Open Whisper Systems. "Signal-iOS". *GitHub*. Retrieved 14 January 2015.

[44] Open Whisper Systems. "TextSecure". *GitHub*. Retrieved 26 February 2014.

[45] Molly Wood (19 February 2014). "Privacy Please: Tools to Shield Your Smartphone". The New York Times. Retrieved 26 February 2014.

[46] DJ Pangburn (3 March 2014). "TextSecure Is the Easiest Encryption App To Use (So Far)". Motherboard. Retrieved 14 March 2014.

[47] Moxie Marlinspike (24 February 2014). "The New TextSecure: Privacy Beyond SMS". Open Whisper Systems. Retrieved 26 February 2014.

[48] Martin Brinkmann (24 February 2014). "TextSecure is an open source messaging app with strong security features". Ghacks Technology News. Retrieved 26 February 2014.

[49] Frosch, Tilman; Mainka, Christian; Bader, Christoph; Bergsma, Florian; Schwenk, Jörg; Holz, Thorsten. "How Secure is TextSecure?" (PDF). Horst Görtz Institute for IT Security, Ruhr University Bochum. Retrieved 4 November 2014.

[50] Open Whisper Systems. "TextSecure-Server". *GitHub*. Retrieved 22 April 2015.

[51] Open Whisper Systems. "Why do I need Google Play installed to use TextSecure on Android?". Retrieved 2014-03-13.

[52] rhodey (16 July 2015). "RE: Flock shutting down". *GitHub Gist*. Retrieved 8 September 2015.

[53] Open Whisper Systems. "Flock". *GitHub*. Retrieved 4 October 2015.

31.7 External links

- Official website
- Open Whisper Systems on GitHub

Chapter 32

Petname

Petname systems are naming systems that claim to possess all three naming properties of Zooko's triangle - global, secure, and memorable.[1] Software that uses such a system can satisfy all three requirements. Such systems can be used to enhance security, such as preventing phishing attacks.[2]

32.1 Examples

CapDesk — a desktop environment.

32.1.1 Firefox extension

There is a Petname Tool extension available for Firefox that allows petnames to be assigned to secure websites.[3] Use of this extension can help prevent phishing attacks.[4] [5]

32.2 PetName Markup Language

The **PetName Markup Language** (PNML) is an XML proposal for using petname systems ubiquitously.[6]

PNML consists of two tags:

- <pn>pet-name-string</pn>

- <key>stringified-cryptographic-key</key>

32.3 References

[1] "An Introduction to Petname Systems".

[2] Sadek Ferdous, Audun Jøsang, Kuldeep Singh, Ravishankar Borgaonkar (2009). *Security Usability of Petname Systems*. Lecture Notes in Computer Science. Springer Science+Business Media. ISBN 9783642047657.

[3] Petname Tool extension

[4] Markus Jakobsson, Steven Myers (2006). *Phishing and Countermeasures: Understanding the Increasing Problem of Electronic Identity Theft*. Wiley-Interscience. ISBN 0471782459.

[5] "Ten Firefox extensions to keep your browsing private and secure".

[6] "The PetName Markup Language".

32.4 External links

- Petname Site

- An Introduction to Petname Systems

- The PetName Markup Language

Chapter 33

PGPfone

PGPfone was a secure voice telephony system developed by Philip Zimmermann in 1995. The PGPfone protocol had little in common with Zimmermann's popular PGP email encryption package, except for the use of the name. It used ephemeral Diffie-Hellman protocol to establish a session key, which was then used to encrypt the stream of voice packets. The two parties compared a short authentication string to detect a Man-in-the-middle attack, which is the most common method of wiretapping secure phones of this type. PGPfone could be used point-to-point (with two modems) over the public switched telephone network, or over the Internet as an early Voice over IP system.

In 1996, there were no protocol standards for Voice over IP. Ten years later, Zimmermann released the successor to PGPfone, Zfone and ZRTP, a newer and secure VoIP protocol based on modern VoIP standards. Zfone builds on the ideas of PGPfone.

According to the MIT PGPfone web page,[1] "MIT is no longer distributing PGPfone. Given that the software has not been maintained since 1997, we doubt it would run on most modern systems."

33.1 See also

- Zfone

- ZRTP

- Nautilus (secure telephone)

- PGP word list

- Secure telephone

33.2 References

[1] http://web.mit.edu/network/pgpfone/

33.3 External links

- PGPfone homepage on PGPi

- Old PGPfone homepage on MIT

Chapter 34

Phoner

34.1 Description

Phoner and **PhonerLite** are softphone applications for Windows operating systems available as freeware. Phoner is a multiprotocol telephony application supporting telephony via CAPI, TAPI and VoIP, while PhonerLite provides a specialized and optimized user interface for VoIP only. Beside the different user interface focus both programs share the same code base.

Both programs use the Session Initiation Protocol for VoIP call signalisation. Calls are supported via server-based infrastructure or direct IP to IP. Media streams are transmitted via the Real-time Transport Protocol which may be encrypted with the Secure Real-time Transport Protocol (SRTP) and the ZRTP security protocols. Phoner provides as well an interface for configuring and using all supplementary ISDN services provided via CAPI and thus needs an ISDN terminal adapter hardware installed in the computer.

Phoner and **PhonerLite** support IPv4 and IPv6 connections by using UDP, TCP and TLS.

34.2 Supported audio formats

- G.711 A-law: 64 kbit/s payload, 8 kHz sampling rate
- G.711 μ-law: 64 kbit/s payload, 8 kHz sampling rate
- G.722: 64 kbit/s payload, 16 kHz sampling rate
- G.726: 16, 24, 32 or 40 kbit/s payload, 8 kHz sampling rate
- GSM: 13 kbit/s payload, 8 kHz sampling rate
- iLBC: 13.3 or 15.2 kbit/s payload, 8 kHz sampling rate
- Speex narrow band: 15 kbit/s payload, 8 kHz sampling rate
- Speex wide band: 30 kbit/s payload, 16 kHz sampling rate
- Opus: 10-50 kbit/s, up to 48 kHz sampling rate

34.3 See also

- Comparison of VoIP software
- List of SIP software
- Opportunistic encryption

34.4 References

[1] Phoner history

[2] Release history on Phoner download page

[3] Changelog on PhonerLite download page

34.5 External links

- Phoner Main Website

Chapter 35

Presumed security

Presumed security is a principle in security engineering that a system is safe from attack due to an attacker assuming, on the basis of probability, that it is secure. Presumed security is the opposite of security through obscurity. A system relying on security through obscurity may have actual security vulnerabilities, but its owners or designers deliberately make the system more complex in the hope that attackers are unable to find a flaw. Conversely a system relying on presumed security makes no attempt to address its security flaws, which may be publicly known, but instead relies upon potential attackers simply assuming that the target is not worth attacking. The reasons for an attacker to make this assumption may range from personal risk (the attacker believes the system owners can easily identify, capture and prosecute them) to technological knowledge (the attacker believes the system owners have sufficient knowledge of security techniques to ensure no flaws exist, rendering an attack moot).

Although this approach to security is implicitly understood by security professionals, it is rarely discussed or documented. The phrase "presumed security" appears to have been first coined by the security commentary website Zero Flaws.[1] The article uses the Royal Military Academy Sandhurst as an example, focusing on the apparent lack of entry security and contrasting it against the presumed security a military installation will have. The article also details the flaws inherent in a trust seal such as the Verisign Secure Site seal, and explains why this presumed security approach is actually detrimental to an overall security posture.

35.1   References & notes

[1] Zero Flaws: Presumed Security

Chapter 36

Quicknet

Quicknet is an Ajax framework (using XMLHttpRequest in JavaScript) designed to develop web applications or websites that use passwords to identify correct users. Using this framework, no cleartext password would be sent over the network or stored in the server. Quicknet supports multi-language, JavaScript cooperative multitasking, AJAX call, session and password management, modular structure, XML content, and JavaScript animation. It uses PHP on the server side, and JavaScript on the client side.

36.1 System requirements

Server-side Quicknet should run on any server with Apache 2.2+, MySQL 5.1+ and PHP 5+ .

Client-side Quicknet should be compatible with Internet Explorer 7+, Firefox 3+, Opera 9+, Safari 3+ and Google Chrome 1+ .

36.2 Session and Password Management

Quicknet is an AJAX framework that aims to protect users' passwords with specially designed algorithm. This is achieved by using the same Cryptographic hash function in JavaScript code on the client-side, as well as PHP code on the server-side, to generate and compare hash results based on users' passwords and some random data. However, no cleartext password would be sent over the network or stored in the server. It is believed that it is impossible to steal a session or discover the user's original password, even if the data sent over the network and/or stored on the server is known.

36.3 Secure Data Transmission

Currently, Quicknet is possibly the only PHP AJAX framework that provides secure data transmission without SSL.

36.4 Multi-language

Currently, Quicknet is possibly the only PHP AJAX framework with built-in support for multi-language. Developers could easily add new language to build their own systems.

36.5 See also

- Ajax framework

36.6 External links

- Official website

Chapter 37

Red/black concept

For other uses, see Red-black.

The **red/black concept**, sometimes called the **red-black**

Red/black box

architecture[1] or **red/black engineering**,[2][3] refers to the careful segregation in cryptographic systems of signals that contain sensitive or classified plaintext information (**red signals**) from those that carry encrypted information, or ciphertext (**black signals**).

In NSA jargon, encryption devices are often called blackers, because they convert red signals to black. TEMPEST standards spelled out in NSTISSAM Tempest/2-95 specify shielding or a minimum physical distance between wires or equipment carrying or processing red and black signals.[4]

Different organizations have differing requirements for the separation of red and black fiber optic cables.

Red/black terminology is also applied to cryptographic keys. Black keys have themselves been encrypted with a "key encryption key" (KEK) and are therefore benign. Red keys are not encrypted and must be treated as highly sensitive material.[5]

37.1   See also

- Secure by design

- Computer security

- Security engineering

37.2   References

[1] David Kleidermacher. "Bringing Android to military communications devices". 2010.

[2] "MIL-HDBK-232A: Red/black engineering -- installation guidelines". 1988.

[3] "Cabling for Secure Government Networks".

[4] McConnell, J. M. (12 December 1995). "NSTISSAM TEMPEST/2-95". Archived from the original on 2007-04-08. Retrieved 2007-12-02.

[5] Clark, Tom (2003). *Designing Storage Area Networks*. Addison-Wesley Professional. ISBN 0-321-13650-0.

Chapter 38

RetroShare

RetroShare is free software for encrypted filesharing, serverless email, instant messaging, chatrooms, and BBS, based on a friend-to-friend network built on GPG (GNU Privacy Guard). It is not strictly a darknet since optionally, peers may communicate certificates and IP addresses from and to their friends.[3][4]

38.1 History

There has been an unofficial build for the single-board computer Raspberry Pi, named PiShare, since 2012.[5]

The web site *PRISM Break* has recommended RetroShare for anonymous file sharing since 2013.[6]

On November 4, 2014, RetroShare scored 6 out of 7 points on the Electronic Frontier Foundation's secure messaging scorecard. It lost a point because there has not been a recent independent code audit.[7]

38.2 Features

38.2.1 Authentication and connectivity

After initial installation, the user generates a pair of (GPG) cryptographic keys with RetroShare.

After authentication and exchanging an asymmetric key, SSH is used to establish a connection. End to end encryption is accomplished by using OpenSSL. Friends of friends cannot connect by default, but they can see each other, if the users allow it.

IPv6 support is planned for the RetroShare 0.6 release branch with a possible release in 2015.

38.2.2 File sharing

It is possible to share folders between friends. File transfer is carried on using a multi-hop swarming system (in-

spired by the "Turtle Hopping" feature from the Turtle F2F project, but implemented differently). In essence, data is only exchanged between friends, although it is possible that the ultimate source and destination of a given transfer are multiple friends apart. A search function performing anonymous multi-hop search is another source of finding files in the network.

Files are represented by their SHA-1 hash value, and HTTP-compliant file and links may be exported, copied, and pasted into/out of RetroShare to publish their virtual location into the RetroShare network.

38.2.3 Communication

The services that RetroShare offers for communication are :

- a private chat

- a private mailing system allow secure communication between known friends and distant friends.

- public and private multi-user chat lobbies.

- a forum system allowing both anonymous and authenticated forums which distributes posts from friends to friends.

- a channel system offers the possibility to auto-download files posted in a given channel to every subscribed peer, similar to rss feeds.

- a Posted links system, where links to important information can be shared.

- VoIP calls, and videocalls since version 0.6.0.

- tor support, for further anonymisation - since version 0.6.0

38.2.4 User interface

The core of the RetroShare software is based on an offline library, to which two executables are plugged :

- a command-line interface executable, that offers nearly no control.

- a graphical user interface written in Qt4, which is the one most users would use. In addition to functions quite common to other file sharing software, such as a search tab and visualization of transfers, RetroShare gives users the potential to manage their network by collecting optional information about neighbour friends and visualizing it as a trust matrix or as a dynamic network graph. The appearance can be changed by choosing one of several available style sheets.

38.2.5 Anonymity

The friend-to-friend structure of the RetroShare network makes it difficult to intrude and hardly possible to monitor from an external point of view. The degree of anonymity may be improved further by deactivating the DHT and IP/certificate exchange services, making the Retroshare network a real Darknet.

Friends of friends may not connect directly with each other; however, the possibility exists of anonymously sharing files with friends of friends, if enabled by the user. Search, access, and both uploading and downloading of these files is made by "routing" through a series of friends. This means that communications between the source of data (the uploader) and the destination of the data (the downloader) is indirect through mutual friends. Although the intermediary friends cannot determine the original source or ultimate destination, they can see their very next links in the communication chain (their friends). Since the data stream is encrypted, only original source and ultimate destination are able to see what data is transferred.

38.3 See also

- Anonymous P2P

- Private peer-to-peer – RetroShare is a private peer-to-peer software

- Comparison of file sharing applications

- I2P – a polyvalent software like RetroShare

- Gnunet – a file sharing software

- Tribler – an open source anonymous P2P decentralized BitTorrent client

- Tox (software) - Tox is a free and open-source, peer-to-peer, encrypted instant messaging and video calling software.

38.4 References

[1] Interview with RetroShare founder

[2] http://www.transifex.com/projects/p/retroshare/

[3] RetroShare: Anonymous, Decentralized and Uncensored File-Sharing is Booming

[4] Handbook of Peer-to-Peer Networking. Shen, X.; Yu, H.; Buford, J.; Akon, M. (Eds.)

[5] http://sourceforge.net/projects/pishare/

[6] https://prism-break.org/en/projects/

[7] "Secure Messaging Scorecard. Which apps and tools actually keep your messages safe?". Electronic Frontier Foundation. 2014-11-04.

38.5 External links

- Official website

Chapter 39

Secret broadcast

A **secret broadcast** is, simply put, a broadcast that is not for the consumption of the general public. The invention of the wireless was initially greeted as a boon by armies and navies. Units could now be coordinated by nearly instant communications. An adversary could glean valuable and sometimes decisive intelligence from intercepted radio signals:

- messages that were not encrypted or poorly encrypted could be read

- order of battle and future intentions could be deduced by traffic analysis

- individual units could be located using direction finding

In the 1920s the United States was able to track Japanese fleet exercises even through fog banks by monitoring their radio transmissions.

A doctrine was developed of having units in the field, particularly ships at sea, maintain radio silence except for urgent situations, such as reporting contact with enemy forces. Ships in formation reverted to pre-wireless methods, including semaphore and signal flags, with signal lamps used at night. Communication from headquarters were sent by one-way radio broadcasts.

39.1 "Personal messages" on propaganda stations

During WWII, the BBC would include "personal messages" in its broadcasts of news and entertainment to occupied-Europe. Often they were coded messages intended for secret agents. Leo Marks attributes this idea to Georges Bégué, an agent for the Special Operations Executive who felt their use could eliminate a lot of the two-way radio traffic that often compromised agents. Such messages were also used to authenticate agents to sources of assistance in the field. The agent would arrange to have the BBC broadcast any short phrase the other person chose.

39.2 Numbers stations

Main article: Numbers station

In the mid-twentieth century, the High Frequency radio bands were used by numerous stations sending seemingly random Morse code, usually in five-letter groups. As more advanced communications methods, such as teleprinter and satellite, took over, the number of such stations diminished, but another type appeared that transmitted spoken and also seemingly random number and letter groups, the latter usually using words from a radio alphabet such as ICAO/NATO alphabet.

Though there has been no official confirmation (beyond a 1998 article in *The Daily Telegraph* which quoted a spokesperson for the Department of Trade and Industry as saying, "These [numbers stations] are what you suppose they are. People shouldn't be mystified by them. They are not for, shall we say, public consumption."[1]) there is little doubt that most of these numbers stations are primarily used to send messages to spies and other clandestine agents (additional possible uses include communication with embassies when a crisis might dictate destruction of cryptographic equipment and as a backup to normal command systems in wartime). Other intended recipients of secret broadcasts have faster and easier-to-use equipment at their disposal. But number stations are ideal for spies in that they require no special equipment, beyond a short-wave receiver. Morse code skills, once a staple of spy training, are no longer required.

39.3 Problems with secret broadcast

An issue in the past has been the limited bandwidth of the broadcast. Morse code was typically sent at 25 words per minute. Teleprinters could operate at or above 60 words per minute. The military uses a message precedence system to prioritize critical traffic, but all too often, senior commanders insisted on high precedence for lengthy messages lacking real urgency.

39.4 See also

- Letter beacon

- Numbers station

- Pirate radio - Piracy in amateur and two-way radio

- Traffic flow security

39.5 References

[1] "Salon People Feature | Counting spies". Salon.com. 1999-09-16. Retrieved 2010-08-26.

Chapter 40

Secure Communications Interoperability Protocol

The **Secure Communications Interoperability Protocol** (**SCIP**) is a multinational standard for secure voice and data communication. SCIP derived from the US Government **Future Narrowband Digital Terminal** (**FNBDT**) project after the US offered to share details of FNBDT with other nations in 2003.[1] SCIP supports a number of different modes, including national and multinational modes which employ different cryptography. Many nations and industries develop SCIP devices to support the multinational and national modes of SCIP.

SCIP has to operate over the wide variety of communications systems, including commercial land line telephone, military radios, communication satellites, Voice over IP and the several different cellular telephone standards. Therefore it was designed to make no assumptions about the underlying channel other than a minimum bandwidth of 2400 Hz. It is similar to a dial-up modem in that once a connection is made, two SCIP phones first negotiate the parameters they need and then communicate in the best way possible.

US SCIP or FNBDT systems were used since 2001, beginning with the CONDOR secure cell phone. The standard is designed to cover wideband as well as narrowband voice and data security.

SCIP was designed by the Department of Defense Digital Voice Processor Consortium (DDVPC) in cooperation with the U.S. National Security Agency and is intended to solve problems with earlier NSA encryption systems for voice, including STU-III and Secure Terminal Equipment (STE) which made assumptions about the underlying communication systems that prevented interoperability with more modern wireless systems. STE sets can be upgraded to work with SCIP, but STU-III cannot. This has led to some resistance since various government agencies already own over 350,000 STU-III telephones at a cost of several thousand dollars each.

There are several components to the SCIP standard: key management, voice compression, encryption and a sig-nalling plan for voice, data and multimedia applications.

40.1 Key Management (120)

To set up a secure call, a new Traffic Encryption Key (**TEK**) must be negotiated. For Type 1 security (classified calls), the SCIP signalling plan uses an enhanced FIREFLY messaging system for key exchange. FIREFLY is an NSA key management system based on public key cryptography. At least one commercial grade implementation uses Diffie-Hellman key exchange.

STEs use security tokens to limit use of the secure voice capability to authorized users while other SCIP devices only require a PIN code, 7 digits for Type 1 security, 4 digits for unclassified.

40.2 Voice compression using Voice Coders (vocoders)

SCIP can work with a variety of vocoders. The standard requires, as a minimum, support for the mixed-excitation linear prediction (MELP) coder, an enhanced MELP algorithm known as MELPe, with additional preprocessing, analyzer and synthesizer capabilities for improved intelligibility and noise robustness. The old MELP and the new MELPe are interoperable and both operate at 2400 bit/s, sending a 54 bit data frame every 22.5 milliseconds but the MELPe has optional additional rates of 1200 bit/s and 600 bit/s.

2400 bit/s MELPe is the only mandatory voice coder required for SCIP. Other voice coders can be supported in terminals. These can be used if all terminals involved in the call support the same coder (agreed during the negotiation stage of call setup) and the network can support the required throughput. G.729D is the most widely supported non-mandatory voice coder in SCIP terminals as it offers a good compromise between higher voice quality without dramatically increasing the required throughput.

40.3   Encryption (SCIP 23x)

The security used by the multinational and national modes of SCIP is defined by the SCIP 23x family of documents. SCIP 231 defines AES based cryptography which can be used multinationally. SCIP 232 defines an alternate multi-national cryptographic solution. Several nations have defined, or are defining, their own national security modes for SCIP.

40.4   US National Mode (SCIP 230)

SCIP 230 defines the cryptography of the US national mode of SCIP. The rest of this section refers to SCIP 230. For security, SCIP uses a block cipher operating in counter mode. A new Traffic Encryption Key (**TEK**) is negotiated for each call. The block cipher is fed a 64-bit state vector (**SV**) as input. If the cipher's block size is longer than 64 bits, a fixed filler is added. The output from the block cipher is xored with the MELP data frames to create the cipher text that is then transmitted.

The low-order two bits of the state vector are reserved for applications where the data frame is longer than the block cipher output. The next 42 bits are the counter. Four bits are used to represent the transmission mode. This allows more than one mode, e.g. voice and data, to operate at the same time with the same TEK. The high-order 16 bits are a sender ID. This allows multiple senders on a single channel to all use the same TEK. Note that since overall SCIP encryption is effectively a stream cipher, it is essential that the same state vector value never be used twice for a given TEK. At MELP data rates, a 42-bit counter allows a call over three thousand years long before the encryption repeats.

For Type 1 security, SCIP uses BATON, a 128-bit block design. With this or other 128-bit ciphers, such as AES, SCIP specifies that two data frames are encrypted with each cipher output bloc, the first beginning at bit 1, the second at bit 57 (i.e. the next byte boundary). At least one commercial grade implementation uses the Triple DES cipher.

40.5   Signalling plan (210)

The SCIP signalling plan is common to all national and multinational modes of SCIP. SCIP has two mandatory types of transmission. The mandatory data service uses an ARQ protocol with forward error correction (FEC) to ensure reliable transmission. The receiving station acknowledges accurate receipt of data blocks and can ask for a block to be re-transmitted, if necessary. For voice, SCIP simply sends a stream of voice data frames (typically MELPe frames, but possibly G.729D or another codec if that has been negotiated between the terminals). To save power on voice calls, SCIP stops sending if there is no speech input. A synchronization block is sent roughly twice a second in place of a data frame. The low order 14 bits of the encryption counter are sent with every sync block. The 14 bits are enough to cover a fade out of more than six minutes. Part of the rest of the state vector are sent as well so that with receipt of three sync blocks, the entire state vector is recovered. This handles longer fades and allows a station with the proper TEK to join a multi station net and be synchronized within 1.5 seconds.

40.6   Availability

As of March 2011 a range of SCIP documents, including the SCIP-210 signalling standard, are publicly available from the IAD website.[2]

Prior to this, SCIP specifications were not widely diffused or easily accessible. This made the protocol for government use rather "opaque" outside governments or defense industries. No public implementation of the Type 1 security and transport protocols are available, precluding its security from being publicly verified.

40.7   See also

- Secure voice
- ZRTP
- MELP
- MELPe
- CVSD
- CELP
- LPC-10e
- FS1015
- FS1016
- ANDVT
- Secure Terminal Equipment

- L-3 Omni/Omni xi

- Sectéra secure voice family

40.8 Notes

[1] Introduction to FNBDT by NC3A discusses the prospects for FNBDT for NATO in 2003

[2] SCIP-related documents are made available through the Information Assurance Directorate web site. Documents can be retrieved by typing "SCIP" into the IAD SecurePhone document search web page

40.9 References

- *Securing the Wireless Environment (FNBDT)*, briefing available from http://wireless.securephone.net/

- *Secure Communications Interoperability Protocols, SCIP*, HFIA briefing available at http://www.hfindustry.com/Sept05/Sept2005_Presentations/HFIAbriefing.ppt

Chapter 41

Secure Electronic Transaction

Secure Electronic Transaction (SET) was a communications protocol standard for securing credit card transactions over insecure networks, specifically, the Internet. SET was not itself a payment system, but rather a set of security protocols and formats that enabled users to employ the existing credit card payment infrastructure on an open network in a secure fashion. However, it failed to gain attraction in the market. VISA now promotes the 3-D Secure scheme.

41.1 History and development

SET was developed by the **SET Consortium**, established in 1996 by VISA and MasterCard in cooperation with GTE, IBM, Microsoft, Netscape, SAIC, Terisa Systems, RSA, and VeriSign.[1] The consortium's goal was to combine the card associations' similar but incompatible protocols (STT from Visa/Microsoft and SEPP from MasterCard/IBM) into a single standard.

SET allowed parties to identify themselves to each other and exchange information securely. Binding of identities was based on X.509 certificates with several extensions.[2] SET used a cryptographic blinding algorithm that, in effect, would have let merchants substitute a certificate for a user's credit-card number. If SET were used, the merchant itself would never have had to know the credit-card numbers being sent from the buyer, which would have provided verified good payment but protected customers and credit companies from fraud.

SET was intended to become the de facto standard payment method on the Internet between the merchants, the buyers, and the credit-card companies.

41.2 Key features

To meet the business requirements, SET incorporates the following features:

- Confidentiality of information

- Integrity of data

- Cardholder account authentication

- Merchant authentication

41.3 Participants

A SET system includes the following participants:

- Cardholder

- Merchant

- Issuer

- Acquirer

- Payment gateway

- Certification authority

41.3.1 How it Works

Both cardholders and merchants must register with CA (certificate authority) first, before they can buy or sell on the Internet, which we will talk about later. Once registration is done, cardholder and merchant can start to do transactions, which involve 9 basic steps in this protocol, which is simplified.

1. Customer browses website and decides on what to purchase

2. Customer sends order and payment information, which includes 2 parts in one message:

a. Purchase Order – this part is for merchant b. Card Information – this part is for merchant's bank only.

1. Merchant forwards card information (part b) to their bank

2. Merchant's bank checks with Issuer for payment authorization

3. Issuer send authorization to Merchant's bank

4. Merchant's bank send authorization to merchant

5. Merchant completes the order and sends confirmation to the customer

6. Merchant captures the transaction from their bank

7. Issuer prints credit card bill (invoice) to customer

41.4   Dual signature

As described in (Stallings 2000):

> An important innovation introduced in SET is the *dual signature*. The purpose of the dual signature is to link two messages that are intended for two different recipients. In this case, the customer wants to send the order information (OI) to the merchant and the payment information (PI) to the bank. The merchant does not need to know the customer's credit-card number, and the bank does not need to know the details of the customer's order. The customer is afforded extra protection in terms of privacy by keeping these two items separate. However, the two items must be linked in a way that can be used to resolve disputes if necessary. The link is needed so that the customer can prove that this payment is intended for this order and not for some other goods or service.

The message digest (MD) of the OI and the PI are independently calculated by the customer. The dual signature is the encrypted MD (with the customer's secret key) of the concatenated MD's of PI and OI. The dual signature is sent to both the merchant and the bank. The protocol arranges for the merchant to see the MD of the PI without seeing the PI itself, and the bank sees the MD of the OI but not the OI itself. The dual signature can be verified using the MD of the OI or PI. It doesn't require the OI or PI itself. Its MD does not reveal the content of the OI or PI, and thus privacy is preserved.

41.5   Notes

[1] Merkow p.248

[2] SET Specification Book 2 p.214

41.6   References

Mark S. Merkow (2004). "Secure Electronic Transactions (SET)". In Hossein Bidgoli. *The Internet Encyclopedia.* John Wiley & Sons. pp. 247–260. ISBN 978-0-471-22203-3.

Stallings, William (Nov 1, 2000). "The SET Standard & E-Commerce". *Dr. Dobbs.*

SET Secure Electronic Transaction Specification (V1.0) Book 1 (PDF). Mastercard and Visa. May 1997.

SET Secure Electronic Transaction Specification (V1.0) Book 2 (PDF). Mastercard and Visa. May 1997.

SET Secure Electronic Transaction Specification (V1.0) Book 3 (PDF). Mastercard and Visa. May 1997.

External Interface Guide to SET Secure Electronic Transaction (PDF). Mastercard and Visa. September 1997.

SETco Main Page, SETco, archived from the original on 2002-09-28, retrieved 2013-11-07

Chapter 42

Secure Terminal Equipment

STE desk set. Note slot in front for Crypto PC Card.

Secure Terminal Equipment (**STE**) is the U.S. Government's current (as of 2008), encrypted telephone communications system for wired or "landline" communications. STE is designed to use ISDN telephone lines which offer higher speeds of up to 128 kbit/s and are all digital. The greater bandwidth allows higher quality voice and can also be utilized for data and fax transmission through a built-in RS-232 port. STE is intended to replace the older STU-III office system and the KY-68 tactical system. STE sets are backwards compatible with STU-III phones, but not with KY-68 sets.[1]

STE sets look like ordinary high-end office desk telephones and can place unsecured calls to anywhere on the public switched telephone network (PSTN), as well as secured calls on it via the phone's backwards compatible STU-III mode. There is a PC Card slot in the STE that allows a Fortezza Plus (KOV-14) Crypto Card or KSV-21 Enhanced Crypto Card to be inserted. When an NSA configured Crypto Card is present, secure calls can be placed to other STE phones. STE phones are "releasable" (unlike STU-III sets). All cryptographic algorithms are in the crypto card.

Newer STE sets can communicate with systems that use the

Secure Communications Interoperability Protocol (SCIP) (formerly Future Narrowband Digital Terminal (FNBDT)). There are upgrade kits available for older units.[2]

42.1 Models

- **Office** - The Office STE is the most widely used STE and provides voice and data access to ISDN (Integrated Services Digital Network) and PSTN (Public Switched Telephone Network) telecommunications systems.

- **Tactical** - The Tactical STE is similar to the Office STE but can also access the TRI-TAC (TRI Service TACtical) network and has a serial EIA-530A/EIA-232 BDI (Black Digital Interface) port.

- **Data** - The Data STE provides remote access for voice, fax, data and video-conferencing. This model has two serial EIA-530A/EIA-232 BDI ports and allows for data transfers to multiple destinations.

- **C2** - The C2 STE is similar to the Tactical STE but C2 has modified software for use with its Tactical Terminal Locking Handset mechanism.

- **STE-R** - Similar to the Data STE, the STE-R*emote* provides dial-in access to the Defense Red Switch Network (DRSN).

- **VoIP** - The STE now has Voice over Internet Protocol (VoIP) capability, available as an upgrade to the current models, or built into some new models.

As of 2007, a typical STE terminal cost about $3,100, not including the crypto card.[2]

42.2 References

[1] "Secure Telephone Unit Third Generation (STU-III) / Secure Terminal Equipment (STE)". Federation of American

Scientists. Retrieved September 2, 2010.

[2] "STE: Secure Terminal Equipment: Direct Sale Pricing". Archived from the original on 2006-10-15.

42.3 External links

- The NAVY INFOSEC WebSite on STU-III and STE

Chapter 43

Secure voice

Gretacoder 210 secure radio system.

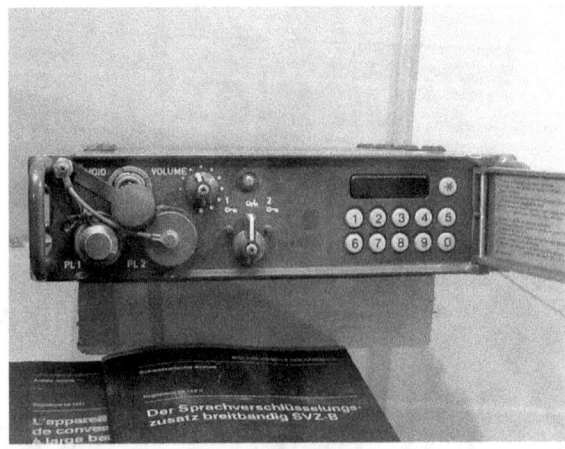

CVX-396 secure voice system, Crypto AG

Secure voice (alternatively **secure speech** or **ciphony**) is a term in cryptography for the encryption of voice communication over a range of communication types such as radio,

telephone or IP.

43.1 History

The implementation of voice encryption dates back to World War II when secure communication was paramount to the US armed forces. During that time, noise was simply added to a voice signal to prevent enemies from listening to the conversations. Noise was added by playing a record of noise in synch with the voice signal and when the voice signal reached the receiver, the noise signal was subtracted out, leaving the original voice signal. In order to subtract out the noise, the receiver need to have exactly the same noise signal and the noise records were only made in pairs; one for the transmitter and one for the receiver. Having only two copies of records made it impossible for the wrong receiver to decrypt the signal. To implement the system, the army contracted Bell Laboratories and they developed a system called SIGSALY. With SIGSALY, ten channels were used to sample the voice frequency spectrum from 250 Hz to 3 kHz and two channels were allocated to sample voice pitch and background hiss. In the time of SIGSALY, the transistor had not been developed and the digital sampling was done by circuits using the model 2051 Thyratron vacuum tube. Each SIGSALY terminal used 40 racks of equipment weighing 55 tons and filled a large room. This equipment included radio transmitters and receivers and large phonograph turntables. The voice was keyed to two 16-inch vinyl phonograph records that contained a Frequency Shift Keying (FSK) audio tone. The records were played on large precise turntables in synch with the voice transmission.

From the introduction of voice encryption to today, encryption techniques have evolved drastically. Digital technology has effectively replaced old analog methods of voice encryption and by using complex algorithms, voice encryption has become much more secure and efficient. One relatively modern voice encryption method is Sub-band coding. With Sub-band Coding, the voice signal is split into multiple frequency bands, using multiple bandpass filters that cover specific frequency ranges of interest. The output signals from the bandpass filters are then lowpass translated to reduce the bandwidth, which reduces the sampling rate. The lowpass signals are then quantized and encoded using special techniques like, Pulse Code Modulation (PCM). After the encoding stage, the signals are multiplexed and sent out along the communication network. When the signal reaches the receiver, the inverse operations are applied to the signal to get it back to its original state.[1] A speech scrambling system was developed at Bell Laboratories in the 1970s by Subhash Kak and Nikil Jayant.[2] In this system permutation matrices were used to scramble coded representations (such as Pulse Code Modulation and variants) of the speech data. Motorola developed a voice encryption system called Digital Voice Protection (DVP) as part of their first generation of voice encryption techniques. DVP uses a self-synchronizing encryption technique known as cipher feedback (CFB). The basic DVP algorithm is capable of 2.36 x 10^{21} different "keys" based on a key length of 32 bits."[3] The extremely high number of possible keys associated with the early DVP algorithm, makes the algorithm very robust and gives a high level of security. As with other symmetric keyed encryption systems, the encryption key is required to decrypt the signal with a special decryption algorithm.

43.2 Digital

A digital secure voice usually includes two components, a digitizer to convert between speech and digital signals and an encryption system to provide confidentiality. It is difficult in practice to send the encrypted signal over the same voiceband communication circuits used to transmit unencrypted voice, e.g. analog telephone lines or mobile radios, due to bandwidth expansion.

This has led to the use of Voice Coders (vocoders) to achieve tight bandwidth compression of the speech signals. NSA's STU-III, KY-57 and SCIP are examples of systems that operate over existing voice circuits. The STE system, by contrast, requires wide bandwidth ISDN lines for its normal mode of operation. For encrypting GSM and VoIP, which are digital anyway, the standard protocol ZRTP could be used as an end-to-end encryption technology.

Secure voice's robustness greatly benefits from having the voice data compressed into very low bit-rates by special component called speech coding, voice compression or voice coder (also known as vocoder). The old secure voice compression standards include (CVSD, CELP, LPC-10e and MELP, where the latest standard is the state of the art MELPe algorithm.

43.3 Digital methods using voice compression: MELP or MELPe

The MELPe or enhanced-MELP (Mixed Excitation Linear Prediction) is a United States Department of Defense speech coding standard used mainly in military applications and satellite communications, secure voice, and secure radio devices. Its development was led and supported by NSA, and NATO. The US government's MELPe secure voice standard is also known as MIL-STD-3005, and the NATO's MELPe secure voice standard is also known as STANAG−4591.

The 2400 bit/s MELP was created by Texas Instruments, and first standardized in 1997 and was known as MIL-STD-3005. Between 1998 and 2001, a new MELP-based vocoder was created at half the rate (i.e. 1200 bit/s) and substantial enhancements were added to the MIL-STD-3005 by SignalCom (later acquired by Microsoft) and AT&T Corporation, which included (a) additional new vocoder at half the rate (i.e. 1200 bit/s), (b) substantially improved encoding (analysis), (c) substantially improved decoding (synthesis), (d) Noise-Preprocessing for removing background noise, (e) transcoding between the 2400 bit/s and 1200 bit/s bitstreams. This fairly significant development was aimed to create a new coder at half the rate and have it interoperable with the old MELP standard.

This enhanced-MELP (also known as MELPe) was adopted as the new MIL-STD-3005 in 2001 in form of annexes and supplements made to the original MIL-STD-3005. The significant breakthrough of the 1200 bit/s MELPe enables the same quality as the old 2400 bit/s MELP's at half the rate!

One of the greatest advantages of the new 2400 bit/s MELPe is that it shares the same bit format as MELP, and hence can interoperate with legacy MELP systems, but would deliver better quality at both ends. MELPe provides much better quality than all older military standards, especially in noisy environments such as battlefield and vehicles and aircraft.

In 2002, the US DoD MELPe was adopted also as NATO standard, known as STANAG−4591. As part of NATO testing for new NATO standard, MELPe was tested against other candidates such as France's HSX (Harmonic Stochastic eXcitation) and Turkey's SB-LPC (Split-Band Linear Predictive Coding), as well as the old secure voice standards such as FS1015 LPC-10e (2.4 kbit/s), FS1016 CELP (4.8 kbit/s) and CVSD (16 kbit/s). Subsequently, the MELPe won also the NATO competition, surpassing the quality of all other candidates as well as the quality of all old secure voice standards (CVSD, CELP and LPC-10e).

The NATO competition concluded that MELPe substantially improved performance (in terms of speech quality, intelligibility, and noise immunity), while reducing throughput requirements. The NATO testing also included interoperability tests, used over 200 hours of speech data, and was conducted by 3 test laboratories world wide.

In 2005, a new 600 bit/s rate MELPe vocoder was added to the NATO standard STANAG-4591 by Thales (France), and there are more advanced efforts to lower the bitrates to 300 bit/s and even 150 bit/s.[4]

43.4   Other

In Aleksandr Solzhenitsyn's novel First Circle the character Volodin's recorded phone call is traced to him since it is not properly encrypted. Its decipherment makes use of spectral analysis.

43.5   See also

- Scrambler

- MELPe

- MELP

- Cryptography

- Crypto phone

- Pseudorandom noise

- SIGSALY

- SCIP

- Secure telephone

- Secure Terminal Equipment

- VINSON

- VoIP VPN

- NSA encryption systems

- ZRTP

- Fishbowl (secure phone)

43.6   References

[1] Owens, F. J. (1993). *Signal Processing of Speech*. Houndmills: MacMillan Press. ISBN 0-333-51922-1.

[2] Kak, S. and Jayant, N.S., Speech encryption using waveform scrambling. Bell System Technical Journal, vol. 56, pp. 781-808, May-June 1977.

[3] http://www.akardam.net/moto/docs/mirrored/encryption.pdf

[4] Nichols, Randall K. & Lekkas, Panos C. (2002). "Speech cryptology". *Wireless Security: Models, Threats, and Solutions*. New York: McGraw-Hill. ISBN 0-07-138038-8.

Chapter 44

Security Protocols Open Repository

SPORE, the **Security Protocols Open Repository**, is an online library of security protocols with comments and links to papers. Each protocol is downloadable in a variety of formats, including rules for use with automatic protocol verification tools. All protocols are described using BAN logic or the style used by Clark and Jacob, and their goals. The database includes details on formal proofs or known attacks, with references to comments, analysis & papers. A large number of protocols are listed, including many which have been shown to be insecure.

It is a continuation of the seminal work by John Clark and Jeremy Jacob.[1]

They seek contributions for new protocols, links and comments.

44.1 See also

- Cryptography

- Symmetric-key algorithm

- Public-key cryptography

- Cryptographic nonce

- List of cryptography topics.

- Short and long lists of cryptographers.

- Important books, papers, and open problems in cryptography.

44.2 External links

- SPORE - Security Protocols Open Repository

44.3 References

[1] A Survey of Authentication Protocol Literature: Version 1.0, the original 1997 paper by John Clark and Jeremy Jacob

Chapter 45

Server Name Indication

Server Name Indication (**SNI**) is an extension to the TLS computer networking protocol[1] by which a client indicates which hostname it is attempting to connect to at the start of the handshaking process. This allows a server to present multiple certificates on the same IP address and TCP port number and hence allows multiple secure (HTTPS) websites (or any other Service over TLS) to be served off the same IP address without requiring all those sites to use the same certificate. It is the conceptual equivalent to HTTP/1.1 name-based virtual hosting, but for HTTPS. The desired hostname is not encrypted,[2] so an eavesdropper can see which site is being requested.

To make use of SNI practical, the vast majority of users must use web browsers that implement it. Users whose browsers do not implement SNI are presented with a default certificate and hence are likely to receive certificate warnings, unless the server is equipped with a wildcard certificate that matches the name of the website.

45.1 Background of the problem

When making a TLS connection the client requests a digital certificate from the web server; once the server sends the certificate, the client examines it and compares the name it was trying to connect to with the name(s) included in the certificate. If a match occurs the connection proceeds as normal. If a match is not found the user may be warned of the discrepancy and the connection may abort as the mismatch may indicate an attempted man-in-the-middle attack. However, some applications allow the user to bypass the warning to proceed with the connection, with the user taking on the responsibility of trusting the certificate and, by extension, the connection.

It is possible for one certificate to cover multiple hostnames. The X.509 v3 specification introduced the *subjectAltName* field which allows one certificate to specify more than one domain and the usage of wildcards in both the common name and *subjectAltName* fields. However it may be impractical—or even impossible, due to lack of a full list

of all names in advance—to obtain a single certificate that covers all names a server will be responsible for. As such a server that is responsible for multiple hostnames is likely to need to present a different certificate for each name (or small group of names). Since 2005, CAcert has run experiments on different methods of using TLS on virtual servers.[3] Most of the experiments are unsatisfactory and impractical. For example, it is possible to use *subjectAltName* to contain multiple domains controlled by one person[4] in a single certificate. Such "unified communications certificates" must be reissued every time the list of domains changes.

Name-based virtual hosting allows multiple DNS hostnames to be hosted by a single server (usually a web server) on the same IP address. To achieve this the server uses a hostname presented by the client as part of the protocol (for HTTP the name is presented in the host header). However, when using HTTPS the TLS handshake happens before the server sees any HTTP headers. Therefore, it is not possible for the server to use the information in the HTTP host header to decide which certificate to present and as such only names covered by the same certificate can be served from the same IP address.

In practice, this means that an HTTPS server can only serve one domain (or small group of domains) per IP address for secured browsing. Assigning a separate IP address for each site increases the cost of hosting, since requests for IP addresses must be justified to the regional internet registry and IPv4 addresses are now in short supply. The result is that many websites are effectively constrained from using secure communications over IPv4. IPv6 address space is not in short supply so websites served using IPv6 are unaffected by this issue.

45.2 How SNI fixes the problem

SNI addresses this issue by having the client send the name of the virtual domain as part of the TLS negotiation.[2] This enables the server to select the correct virtual domain early

108

and present the browser with the certificate containing the correct name. Therefore, with clients and servers that implement SNI, a server with a single IP address can serve a group of domain names for which it is impractical to get a common certificate.

SNI was added to the IETF's Internet RFCs in June 2003 through RFC 3546, *Transport Layer Security (TLS) Extensions*. The latest version of the standard is RFC 6066.

45.3 Implementation

In 2004, a patch for adding TLS/SNI into OpenSSL was created by the EdelKey project.[5] In 2006, this patch was then ported to the development branch of OpenSSL, and in 2007 it was back-ported to OpenSSL 0.9.8.

For an application program to implement SNI, the TLS library it uses must implement it and the application must pass the hostname to the TLS library. Further complicating matters, the TLS library may either be included in the application program or be a component of the underlying operating system. Because of this, some browsers implement SNI when running on any operating system, while others implement it only when running on certain operating systems.

45.3.1 Web browsers[6]

- Internet Explorer 7 or later, on Windows Vista or higher. Not in any Internet Explorer version on Windows XP or Windows Server 2003 because SNI in Internet Explorer depends upon the SChannel system component shipped with Windows Vista.[7]

- Mozilla Firefox 2.0 or later

- Opera 8.0 (2005) or later (the TLS 1.1 protocol must be enabled)[8]

- Opera Mobile at least version 10.1 beta on Android

- Google Chrome (Vista or higher. XP on Chrome 6 or newer[9] OS X 10.5.7 or higher on Chrome 5.0.342.1 or newer)

- Safari 3.0 or later (Mac OS X 10.5.6 or higher and Windows Vista or higher)

- Konqueror/KDE 4.7 or later [10]

- MobileSafari in Apple iOS 4.0 or later[11]

- Android default browser on Honeycomb (v3.x) or newer[12]

- BlackBerry 10 and BlackBerry Tablet OS default browser

- Windows Phone 7 or later[13]

- MicroB on Maemo

- Odyssey on MorphOS

45.3.2 Servers

- Apache 2.2.12 or later using mod_ssl[14][15][16] or alternatively with mod_gnutls[17]

- Microsoft Internet Information Server IIS 8[18] or later.

- Nginx with an accompanying OpenSSL built with SNI capability

- Apache Traffic Server 3.2.0 or later.

- Radware Alteon ADC

- A10 Networks Thunder ADC 2.7.2 or later

- Cherokee if compiled with TLS capability

- Versions of lighttpd 1.4.x and 1.5.x with patch,[19] or 1.4.24+ without patch[20]

- F5 Networks Local Traffic Manager running version 11.1 or later [21]

- Ping Access 3.0 or later

- Hiawatha (web server) 8.6 or later

- LiteSpeed 4.1 or later

- Pound 2.6 or later[22]

- PageKite[23] tunneling reverse proxy[24]

- Saetta Web Server[25] via OpenSSL

- LBL®LoadBalancer 9.1 or later[26]

- Citrix NetScaler 9.2 or later (9.3 Enhanced)[27]

- Radware Alteon ADC running AlteonOS 28.1 or later [28]

- HAProxy 1.5 or later [29]

- EVO Mail Server [30]

- Fortinet FortiWeb WAF 5.3 or later [31]

- Avuna 1.3.1+[32]

45.3.3 Libraries

- Mozilla NSS 3.11.1[33] client-side only

- OpenSSL

    - 0.9.8f (released 11 October 2007) - not compiled in by default, can be compiled in with config option '--enable-tlsext'.

    - 0.9.8j (released 7 January 2009) through 1.0.0 (released 29 March 2010) - compiled in by default

- GnuTLS[34]

- wolfSSL (previously CyaSSL) - not compiled in by default, can be compiled in with config option '--enable-sni' or '--enable-tlsx'.[35]

- mbed TLS (previously PolarSSL) since 1.2.0 - compiled in by default

- libcurl / cURL since 7.18.1 (released 30 March 2008) when compiled against an SSL/TLS toolkit with SNI

- Python 2.7.9, 3.2 and above (ssl, urllib[2] and httplib modules) [36]

- Qt 4.8[37]

- Oracle Java 7 JSSE [38]

- Apache HttpComponents 4.3.2 [39]

- wget 1.14[40]

- Android 2.3 (Gingerbread) has partial support if application uses HttpsURLConnection class.[41]

- Android 4.2 (Jellybean MR1) exposes SNI support on raw sockets via its SSLCertificateSocketFactory class.[42]

- IO::Socket::SSL (Perl/CPAN module, client support since version 1.56,[43] server support since 1.83[43])

- Pike 7.9.5 (SSL module) [44]

- MatrixSSL (client and server)[45]

- stunnel (client and server)[46]

- Go (client and server)[47]

45.4 No support

The following combinations do not implement SNI:

45.4.1 Client side

- WebClient service (for WebDAV) included in any Windows version

- Internet Explorer 6 or earlier and any IE version on Windows XP or earlier

- Safari on Windows XP or earlier

- BlackBerry OS 7.1 or earlier

- Windows Mobile up to 6.5[48]

- Android default browser on Android 2.x[49] (Fixed in Honeycomb for tablets and Ice Cream Sandwich for phones)

- wget before 1.14

- Nokia Browser for Symbian

- Opera Mobile for Symbian at least on Series60

- SAP (according to some consultants)

45.4.2 Server side

- IBM HTTP Server [50][51]

- Apache Tomcat 8 or earlier[52]

45.4.3 Libraries

- Qt client side up to 4.7[37]

- Mozilla NSS server side [53][54]

- Java before 1.7

- Python 2.x (except 2.7.9), 3.0, 3.1 (ssl, urllib[2] and httplib modules) [36]

45.5 References

[1] "Server Name Indication". *Transport Layer Security (TLS) Extensions*. IETF. p. 8. sec. 3.1. RFC 3546. https://tools.ietf.org/html/rfc3546#section-3.1.

[2] "TLS Server Name Indication". *Paul's Journal*.

[3] "CAcert VHostTaskForce". *CAcert Wik*.

[4] "What is a Multiple Domain (UCC) SSL Certificate?".

[5] "EdelKey Project".

[6] Brand, Kaspar (2009-03-29). "TLS SNI Test Site".

[7] http://blogs.msdn.com/b/ieinternals/archive/2009/12/07/certificate-name-mismatch-warnings-and-server-name-indication.aspx

[8] Opera (2005-02-25). "Changelog for Opera [8] Beta 2 for Windows". *Opera Changelogs for Windows*. Opera.com. Archived from the original (via Archive.org) on 2005-11-23. Readded experimental support for TLS Extensions and TLS 1.1. Setting is disabled by default.

[9] "Google Chrome, Issue 43142, Use SSLClientSocketNSS on Windows by default". 2010-10-29.

[10] "Bug 122433: Server Name Identification support". Bugs.kde.org. Retrieved 2011-07-18.

[11] Kehrer, Paul. "SNI in iOS 4.0". Retrieved 22 November 2010.

[12] "Issue 12908 - android Https - websites that support Server Name Indication (SNI) don't work". code.google.com. Retrieved 23 August 2011.

[13] Server Name Indication (SNI) with IIS 8 (Windows Server 2012) - Unleashed - Site Home - MSDN Blogs

[14] "Bug 34607: Support for Server Name Indication". Apache Software Foundation.

[15] "Revision 776281: adding support for Server Name Indication to the Apache 2.2.x branch". Apache Software Foundation.

[16] "CHANGES: Server Name Indication support is listed under the changes for Apache 2.2.12". Apache Software Foundation.

[17] Notaras, George (2007-08-10). "SSL-enabled Name-based Apache Virtual Hosts with mod_gnutls".

[18] "What's New in IIS 8". weblogs.asp.net. Retrieved 2012-04-01.

[19] "#386 (TLS servername extension (SNI) for namebased TLS-vhosts)". Trac.lighttpd.net. Retrieved 2011-03-08.

[20] "1.4.24 - now with TLS SNI and money back guarantee". Blog.lighttpd.net. Retrieved 2011-03-08.

[21] "sol13452: Configuring a virtual server to serve multiple HTTPS sites using TLS Server Name Indication (SNI) feature". ask.f5.com. Retrieved 2012-03-26.

[22] "Apsis Gmbh". apsis.ch. Retrieved 2011-08-18.

[23] "Open Source PageKite". PageKite.net. Retrieved 2012-04-18.

[24] "SSL/TLS back-ends, endpoints and SNI". PageKite.net. Retrieved 2012-04-18.

[25] Features of Saetta web server

[26] "Unified Reverse Proxy". TCOGROUP.

[27] "Configuring an SSL Virtual Server for Secure Hosting of Multiple Sites". Citrix. Retrieved 5 July 2012.

[28] "Introducing AlteonOS 28.1! - Knowledge Base - Radware". kb.radware.com. Retrieved 2012-08-20.

[29] "HAProxy - The Reliable, High Performance TCP/HTTP Load Balancer".

[30] "EVO Mail Server with TLS SNI support".

[31] "FortiWeb 5.3.0 CLI Reference - server certificate sni". http://docs.fortinet.com/d/fortiweb-5-3-0-cli-html. Retrieved 2015-01-02.

[32] "Avuna HTTP SNI Built in Support".

[33] "NSS 3.11.1 Release Notes". Mozilla.org. Retrieved 2011-03-08.

[34] "TLS Extensions". Gnu.org. 2010-08-01. Retrieved 2011-03-08.

[35] "Using Server Name Indication (SNI) with CyaSSL". wolfSSL.com. 2013-05-24. Retrieved 2013-11-25.

[36] "Support TLS SNI extension in ssl module". Bugs.python.org. Retrieved 2011-03-08.

[37] "QTBUG-1352. It would be useful if QSslSocket supports TLS extensions such as Server Name Indication as per RFC 3546.". Qt Bug Tracker. Retrieved 2011-04-15.

[38] "Java SE 7 Release Security Enhancements". download.oracle.com. Retrieved 2012-12-12.

[39] "HttpComponents HttpClient 4.3.2 Released". Retrieved 2014-01-21.

[40] "News: GNU wget 1.14 released". *GNU Wget*. 2012-10-19.

[41] "Android's HTTP Clients". Google. Retrieved 3 August 2012.

[42] "SSLCertificateSocketFactory | Android Developers". Google. Retrieved 9 January 2015.

[43] "IO::Socket::SSL introduces client-side SNI".

[44] "Pike's SSL module". Retrieved 30 August 2013.

[45] "MatrixSSL news page".

[46] "stunnel man page".

[47] "Go TLS package".

[48] "Understanding Certificate Name Mismatches". Blogs.msdn.com. Retrieved 2011-03-08.

[49] "Android issue 1290 - Https websites that support Server Name Indication (SNI) don't work". Code.google.com. 2010-12-01. Retrieved 2011-12-13.

[50] "IBM HTTP Server SSL Questions and Answers". Publib.boulder.ibm.com. Retrieved 2011-03-08.

[51] "IHS 8 powered by Apache 2.2.x ?". Pub-lib.boulder.ibm.com. Retrieved 2011-03-08.

[52] "Stack Overflow post by a Tomcat Comitter". Retrieved 28 October 2014.

[53] "NSS Roadmap (as of 11 September 2009)". Wiki.mozilla.org. Retrieved 2011-03-08.

[54] "Implement TLS Server Name Indication for servers". Bugzilla@Mozilla. 2006-11-11. Retrieved 2012-10-30.

45.6 External links

- RFC 6066 (obsoletes RFC 4366 (which obsoleted RFC 3546))

- https://alice.sni.velox.ch/ Test client-side TLS SNI capability

Chapter 46

Signal (software)

Signal is a free and open-source encrypted voice calling and instant messaging application for iOS. It uses advanced end-to-end encryption protocols to secure all communications to other Signal users. Signal can be used to send and receive encrypted instant messages, group messages, attachments and media messages. Users can independently verify the identity of their messaging correspondents by comparing key fingerprints out-of-band. During calls, users can check the integrity of the data channel by checking if two words match on both ends of the call.

Signal is compatible with RedPhone and TextSecure on Android. All three apps are developed by Open Whisper Systems and are published under the GPLv3 license.

46.1 History

46.1.1 Whisper Systems and Twitter (2010–2011)

Signal is the iOS counterpart of RedPhone and TextSecure. The beta versions of RedPhone and TextSecure were first launched in May 2010 by Whisper Systems,[4] a startup company co-founded by security researcher Moxie Marlinspike and roboticist Stuart Anderson.[5][6] Whisper Systems also produced a firewall and tools for encrypting other forms of data.[5][7] All of these were proprietary enterprise mobile security software and were only available for Android.

In November 2011, Whisper Systems announced that it had been acquired by Twitter. The financial terms of the deal were not disclosed by either company.[8] The acquisition was done "primarily so that Mr. Marlinspike could help the then-startup improve its security".[9] Shortly after the acquisition, Whisper Systems' RedPhone service was made unavailable.[10] Some criticized the removal, arguing that the software was "specifically targeted [to help] people under repressive regimes" and that it left people like the Egyptians in "a dangerous position" during the events of the 2011 Egyptian revolution.[11]

Twitter released TextSecure as free and open-source software under the GPLv3 license in December 2011.[5][12][13][14] RedPhone was also released under the same license in July 2012.[15] Marlinspike later left Twitter and founded Open Whisper Systems[2] as a collaborative Open Source project for the continued development of TextSecure and RedPhone.[16]

46.1.2 Open Whisper Systems (2013–present)

Open Whisper Systems' website was launched in January 2013.[16]

Toward the end of July 2014, Open Whisper Systems announced plans to unify its RedPhone and TextSecure applications as Signal.[17] This announcement coincided with the initial release of Signal as a RedPhone counterpart for iOS. The developers said that their next steps would be to provide TextSecure instant messaging capabilities for iOS, unify the RedPhone and TextSecure applications on Android, and launch a web client.[17] It was the first iOS app to enable easy, strongly encrypted voice calls for free.[2][18]

In October 2014, researchers published a protocol analysis of TextSecure.[19] Among other findings, they presented an unknown key-share attack on the protocol, but in general, they found that the encrypted chat client is secure.[20]

Open Whisper Systems released Signal 2.0 with support for TextSecure private messaging in March 2015.[21][22]

46.2 Reception

In October 2014, the Electronic Frontier Foundation (EFF) included Signal in their updated surveillance self-defense guide.[23] In November 2014, "Signal / RedPhone" received a perfect score on the EFF's secure messaging scorecard;[24] they received points for having communications encrypted in transit, having communications encrypted with keys the providers don't have access to (end-to-end encryption), making it possible for users to independently verify their correspondent's identities, having past communications secure if the keys are stolen (forward secrecy), having their code open to independent review (open source), having their security designs well-documented, and having recent independent security audits.[24] As of 10 July 2015, "ChatSecure + Orbot", Cryptocat, TextSecure, Pidgin, Silent Phone, Silent Text, and Telegram's secret chats also have seven out of seven points on the scorecard.[24]

On December 28, 2014, *Der Spiegel* published slides from an internal NSA presentation dating to June 2012 in which the NSA deemed RedPhone on its own as a "major threat" to its mission, and when used in conjunction with other privacy tools such as Cspace, Tor, Tails, and TrueCrypt was ranked as "catastrophic," leading to a "near-total loss/lack of insight to target communications, presence..."[25][26]

Former NSA contractor Edward Snowden has endorsed Open Whisper Systems' applications on multiple occasions. In his keynote speech at SXSW in March 2014, he praised TextSecure and RedPhone for their ease-of-use.[27] During an interview with The New Yorker in October 2014, he recommended using "anything from Moxie Marlinspike and Open Whisper Systems".[28] During a remote appearance at an event hosted by Ryerson University and Canadian Journalists for Free Expression in March 2015, Snowden said that Signal is "very good" and that he knew the security model.[29] Asked about encrypted messaging apps during a Reddit AMA in May 2015, he recommended "Signal for iOS, Redphone/TextSecure for Android".[30][31]

In September 2015, the American Civil Liberties Union called on officials at the U.S. Capitol to ensure that lawmakers and staff members have secure communications technology.[32] One of the applications that the ACLU recommended in their letter to the Senate Sergeant at Arms and to the House Sergeant at Arms was Signal, writing:

> One of the most widely respected encrypted communication apps, Signal, from Open Whisper Systems, has received significant financial support from the U.S. government, has been audited by independent security experts, and is now widely used by computer security professionals, many of the top national security journalists, and public interest advocates. Indeed, members of the ACLU's own legal department regularly use Signal to make encrypted telephone calls.[33]

46.3   Features

Signal allows users to call other Signal users on iOS and RedPhone users on Android. All calls are made over a Wi-Fi or data connection and are free of charge, including long distance and international.[18] Signal also allows users to send group, text, picture, and video messages over a Wi-Fi or data connection to other Signal users on iOS and to TextSecure users on Android.

All communications to other Signal, RedPhone and TextSecure users are automatically end-to-end encrypted. The keys that are used to encrypt the user's communications are generated and stored at the endpoints (i.e. by users, not by servers). All three apps implement forward secrecy.[24][34]

Signal, RedPhone and TextSecure have built-in mechanisms for verifying that no man-in-the-middle attack has occurred. For calls, Signal and RedPhone display two words on the screen. If the words match on both ends of the call, the call is secure.[18][35] For messages, Signal and TextSecure users can compare key fingerprints (or scan QR codes) out-of-band.

46.4   Architecture

46.4.1   Encryption protocols

Further information: TextSecure § Encryption protocol

Signal instant messages are encrypted with the TextSecure encryption protocol developed by Open Whisper Systems.[21] They took the Off-the-Record Messaging (OTR) protocol and made some improvements to the deniability and forward secrecy aspects, and added a mechanism to allow the ephemeral key negotiation to work asynchronously.[36][37][38]

Signal voice calls are encrypted with the RedPhone encryption protocol, which is based on ZRTP (the VoIP encryption protocol developed by Phil Zimmermann) and SRTP.[2][39]

46.4.2   Servers

Further information: TextSecure § Servers

Signal calls are routed through Open Whisper Systems' servers. Open Whisper Systems has set up dozens of servers to handle the encrypted calls in more than 10 countries around the world to minimize latency.[2] According to the developers, Signal does not leave metadata about who called who and when because the servers do not keep call logs.[34]

All client-server communications are protected by TLS.[19][39] Messages are handled by a REST API and push messaging (both GCM and APN).[40]

46.4.3 Licensing

The complete source code of Signal is available on GitHub under a free software license. This enables interested parties to examine the code and help the developers verify that everything is behaving as expected. It also allows advanced users to compile their own copy of the application and compare it with the version that is distributed by Open Whisper Systems.[3]

46.5 Developers

Main article: Open Whisper Systems

Signal is developed by Open Whisper Systems, a nonprofit software group[41] that develops collaborative Open Source projects with a mission to "make private communication simple".[42] The group consists of a large community of volunteer Open Source contributors, as well as a small team of dedicated grant-funded developers.[42] Open Whisper Systems is funded by a combination of donations and grants, and all of its products are published as free and open-source software under the terms of the GNU General Public License (GPL) version 3.

The project has received financial support from, among others, the Freedom of the Press Foundation,[43] the Knight Foundation,[44] the Shuttleworth Foundation,[45] and the Open Technology Fund,[46] a U.S. government program that has also funded other privacy projects like the anonymity software Tor and the encrypted instant messaging website Cryptocat.

46.6 See also

- Comparison of instant messaging clients
- Comparison of VoIP software
- Internet privacy
- Secure communication

46.7 References

[1] Marlinspike, Moxie (2 Mar 2015). "Signal 2.0: Private messaging comes to the iPhone". Open Whisper Systems. Retrieved 2015-05-09.

[2] Greenberg, Andy (29 July 2014). "Your iPhone Can Finally Make Free, Encrypted Calls". Wired. Retrieved 18 January 2015.

[3] Open Whisper Systems. "Signal-iOS". *GitHub*. Retrieved 14 January 2015.

[4] "Announcing the public beta". Whisper Systems. 25 May 2010. Archived from the original on 30 May 2010. Retrieved 22 January 2015.

[5] Garling, Caleb (20 December 2011). "Twitter Open Sources Its Android Moxie | Wired Enterprise". Wired. Retrieved 21 December 2011.

[6] "Company Overview of Whisper Systems Inc.". Bloomberg Businessweek. Retrieved 2014-03-04.

[7] Greenberg, Andy (2010-05-25). "Android App Aims to Allow Wiretap-Proof Cell Phone Calls". Forbes. Retrieved 2014-02-28.

[8] Cheredar, Tom (28 November 2011). "Twitter acquires Android security startup Whisper Systems". VentureBeat. Retrieved 2011-12-21.

[9] Yadron, Danny (9 July 2015). "Moxie Marlinspike: The Coder Who Encrypted Your Texts". *The Wall Street Journal*. Retrieved 10 July 2015.

[10] Greenberg, Andy (2011-11-28). "Twitter Acquires Moxie Marlinspike's Encryption Startup Whisper Systems". Forbes. Retrieved 2011-12-21.

[11] Garling, Caleb (28 November 2011). "Twitter Buys Some Middle East Moxie | Wired Enterprise". Wired. Retrieved 21 December 2011.

[12] Aniszczyk, Chris (20 December 2011). "The Whispers Are True". *The Twitter Developer Blog*. Twitter. Archived from the original on 24 October 2014. Retrieved 22 January 2015.

[13] "TextSecure is now Open Source!". Whisper Systems. 20 December 2011. Archived from the original on 6 January 2012. Retrieved 22 January 2015.

[14] Pachal, Pete (2011-12-20). "Twitter Takes TextSecure, Texting App for Dissidents, Open Source". Mashable. Retrieved 2014-03-01.

[15] "RedPhone is now Open Source!". Whisper Systems. 18 July 2012. Archived from the original on 31 July 2012. Retrieved 22 January 2015.

[16] "A New Home". Open Whisper Systems. 2013-01-21. Retrieved 2014-03-01.

[17] Mimoso, Michael (29 July 2014). "New Signal App Brings Encrypted Calling to iPhone". Threatpost.

[18] Evans, Jon (29 July 2014). "Talk Private To Me: Free, Worldwide, Encrypted Voice Calls With Signal For iPhone". *TechCrunch*. AOL.

[19] Frosch, Tilman; Mainka, Christian; Bader, Christoph; Bergsma, Florian; Schwenk, Jörg; Holz, Thorsten. "How Secure is TextSecure?" (PDF). Horst Görtz Institute for IT Security, Ruhr University Bochum. Retrieved 4 November 2014.

[20] Pauli, Darren. "Auditors find encrypted chat client TextSecure is secure". The Register. Retrieved 4 November 2014.

[21] Lee, Micah (2015-03-02). "You Should Really Consider Installing Signal, an Encrypted Messaging App for iPhone". The Intercept. Retrieved 2015-03-03.

[22] Geuss, Megan (2015-03-03). "Now you can easily send (free!) encrypted messages between Android, iOS". Ars Technica. Retrieved 2015-03-03.

[23] "Surveillance Self-Defense. Communicating with Others". Electronic Frontier Foundation. 2014-10-23.

[24] "Secure Messaging Scorecard. Which apps and tools actually keep your messages safe?". Electronic Frontier Foundation. 4 November 2014.

[25] SPIEGEL Staff (28 December 2014). "Prying Eyes: Inside the NSA's War on Internet Security". *Der Spiegel*. Retrieved 23 January 2015.

[26] "Presentation from the SIGDEV Conference 2012 explaining which encryption protocols and techniques can be attacked and which not" (PDF). *Der Spiegel*. 28 December 2014. Retrieved 23 January 2015.

[27] Eddy, Max (11 March 2014). "Snowden to SXSW: Here's How To Keep The NSA Out Of Your Stuff". PC Magazine: SecurityWatch. Retrieved 2014-03-16.

[28] "The Virtual Interview: Edward Snowden - The New Yorker Festival". *YouTube*. The New Yorker. 11 October 2014. Retrieved 24 May 2015.

[29] Cameron, Dell (6 March 2015). "Edward Snowden tells you what encrypted messaging apps you should use". The Daily Dot. Retrieved 24 May 2015.

[30] Yuhas, Alan (21 May 2015). "NSA surveillance powers on the brink as pressure mounts on Senate bill – as it happened". The Guardian. Retrieved 24 May 2015.

[31] Beauchamp, Zack (21 May 2015). "The 9 best moments from Edward Snowden's Reddit Q&A". Vox Media. Retrieved 24 May 2015.

[32] Nakashima, Ellen (22 September 2015). "ACLU calls for encryption on Capitol Hill". *The Washington Post* (Nash Holdings LLC). Retrieved 22 September 2015.

[33] Macleod-Ball, Michael W.; Rottman, Gabe; Soghoian, Christopher (22 September 2015). "The Civil Liberties Implications of Insecure Congressional Communications and the Need for Encryption" (PDF). Washington, DC: American Civil Liberties Union. pp. 5–6. Retrieved 22 September 2015.

[34] Brandom, Russell (29 July 2014). "Signal brings painless encrypted calling to iOS". The Verge. Retrieved 26 January 2015.

[35] "Exactly how does Zfone and ZRTP protect against a man-in-the-middle (MiTM) attack?". The Zfone Project. Retrieved 25 January 2015.

[36] Pangburn, DJ (3 March 2014). "TextSecure Is the Easiest Encryption App To Use (So Far)". Motherboard. Retrieved 14 March 2014.

[37] Marlinspike, Moxie (22 August 2013). "Forward Secrecy for Asynchronous Messages". Open Whisper Systems. Retrieved 2014-03-01.

[38] Open Whisper Systems. "ProtocolV2". *GitHub*. Retrieved 21 January 2015.

[39] Marlinspike, Moxie (17 July 2012). "Encryption Protocols". *GitHub*. Retrieved 22 September 2015.

[40] Open Whisper Systems. "TextSecure-Server". *GitHub*. Retrieved 2 March 2014.

[41] Franceschi-Bicchierai, Lorenzo (18 November 2014). "WhatsApp messages now have Snowden-approved encryption on Android". Mashable. Retrieved 23 January 2015.

[42] Open Whisper Systems. "About us". Retrieved 18 January 2015.

[43] "Open Whisper Systems". Freedom of the Press Foundation. Retrieved 18 January 2015.

[44] "TextSecure". Knight Foundation. Retrieved 5 January 2015.

[45] "Moxie Marlinspike". Shuttleworth Foundation. Retrieved 14 January 2015.

[46] "Projects". Open Technology Fund. Retrieved 14 January 2015.

46.8   External links

- *Signal* on iTunes Preview

- *Signal-iOS* on GitHub

- Open Whisper Systems. The developers' homepage.

Chapter 47

SMTPS

SMTPS (Simple Mail Transfer Protocol Secure) refers to a method for securing SMTP with transport layer security. It is intended to provide authentication of the communication partners, as well as data integrity and confidentiality.

SMTPS is not a proprietary protocol and not an extension of SMTP. It is just a way to secure SMTP at the transport layer.

This means that the client and server speak normal SMTP at the application layer, but the connection is secured by SSL or TLS. This happens when the connection is established before any mail data has been exchanged. Since whether or not to use SSL or TLS is not negotiated by the peers, SMTPS services are usually reachable on a dedicated port of their own.

Originally, in early 1997, the Internet Assigned Numbers Authority registered 465 for SMTPS. [1] By the end of 1998, this was revoked when STARTTLS had been specified.[2] With STARTTLS, the same port can be used with or without TLS. SMTP was seen as particularly important, because clients of this protocol are often other mail servers, which can not know whether a server they wish to communicate with will have a separate port for TLS.[3] The port 465 is now registered for Source-Specific Multicast audio and video.[4][5]

In 2014, many services continue to offer the deprecated SMTPS interface on port 465 in addition to (or instead of) the RFC-compliant message submission interface on the port 587 defined by RFC 6409.[6] Service providers that maintain port 465 do so because[7] older Microsoft applications (including Entourage v10.0 and its successor, Outlook for Mac 2011) do not support STARTTLS,[8] and thus not the SMTP submission standard (ESMTPS on port 587). The only way for service providers to offer those clients an encrypted connection is to maintain port 465.

47.1 References

[1] "NEW DRAFT: Regularizing Port Numbers for SSL". w3. 1997-02-07. Retrieved 2013-07-27.

[2] Paul Hoffman (1998-11-12). "Revoking the smtps TCP port". Internet Mail Consortium. Retrieved 2009-09-16.

[3] Paul Hoffman (1997-06-01). "Do we need IMAP / TLS or POP / TLS?". Internet Mail Consortium. Retrieved 2009-09-16.

[4] "Port Numbers". Internet Assigned Numbers Authority. 2009-09-14. Retrieved 2009-09-16.

[5] "SSM". Cisco Systems. Retrieved 2009-09-16.

[6] "Re-mishap in Gmail". Heise Online. 2009-09-24. Retrieved 2009-09-25.

[7] "SMTP mail settings". The Art Farm. Retrieved 28 April 2013.

[8] "Postfix TLS support". Retrieved 28 April 2013.

Chapter 48

Sqlnet.ora

In database computing, **sqlnet.ora** is a plain-text configuration file that contains the information (like tracing options, encryption, route of connections, external naming parameters etc.) on how both Oracle server and Oracle client have to use Oracle Net (formerly Net8 or SQL*Net) capabilities for networked database access.

48.1 Location

The sqlnet.ora file typically resides in $ORACLE_HOME/network/admin on UNIX platforms and %ORACLE_HOME%\NETWORK\ADMIN on Windows operating systems.[1]

48.2 Sample file

NAMES.DIRECTORY_PATH= (LDAP, TNSNAMES, HOSTNAME)   NAMES.DEFAULT_DOMAIN = ORACLE.COM   TRACE_LEVEL_CLIENT = OFF SQLNET.EXPIRE_TIME = 30 SQL-NET.IDENTIX_FINGERPRINT_DATABASE = FINGRDB AUTOMATIC_IPC = ON SQLNET.EXPIRE_TIME = 0   SQL-NET.AUTHENTICATION_SERVICES = (ALL) SQL-NET.CRYPTO_CHECKSUM_CLIENT = ACCEPTED TNSPING.TRACE_DIRECTORY = /oracle/traces

48.3 Profile parameters

This section lists and describes some **sqlnet.ora** file parameters.

48.4 References

[1] http://www.ecst.csuchico.edu/~{}melody/courses/ Fall2001CSCI379/DOC/network.815/a67440/appb.htm

[1]

48.5 External links

[2]

[1] file:///C:/oracle/Docs/11gR2/E11882_01/network.112/ e10835/sqlnet.htm#BIIFHBFJ

[2] http://docs.oracle.com/cd/E11882_01/network.112/ e10835/sqlnet.htm#NETRF183

Chapter 49

STU-I

This article is about a secure telephone. For Stu as a common name, see Stuart (disambiguation).
For other uses, see STU (disambiguation).

The **STU-I**, like its successors sometimes known as a

STU-I secure telephone desk set. Electronics were housed in a separate cabinet.

"stew phone", was a secure telephone developed by the U.S. National Security Agency for use by senior U.S. government officials in the 1970s.

49.1   See also

- KY-3

- Navajo I

- STU-II

- STU-III

- SCIP

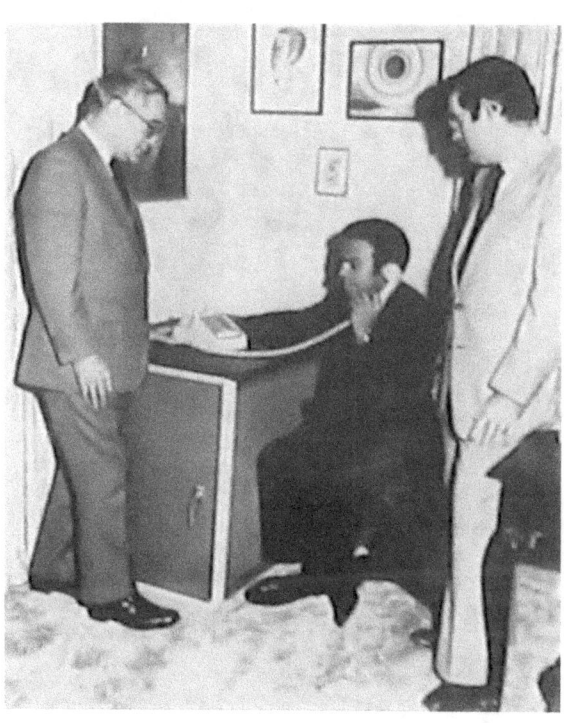

STU-I cabinet with desk set on top. The person talking is U.N. Ambassador Andrew Young, calling from New York City during the Israel-Egypt peace talks in the Carter administration.

49.2   External links

- Delusion.org - National Cryptologic Museum pictures

Chapter 50

STU-II

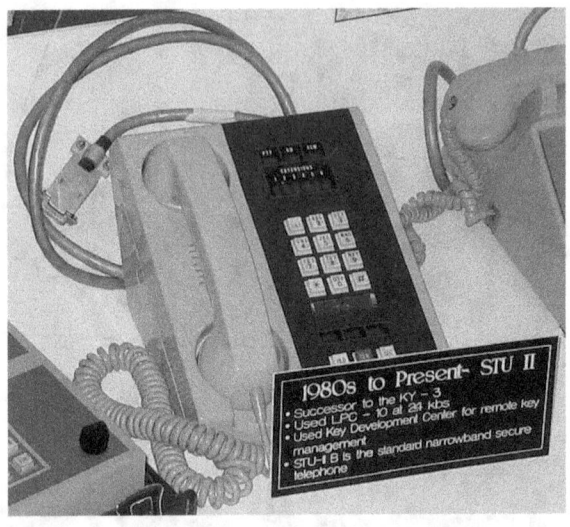

STU-II secure telephone desk set. Electronics were housed in a separate cabinet.

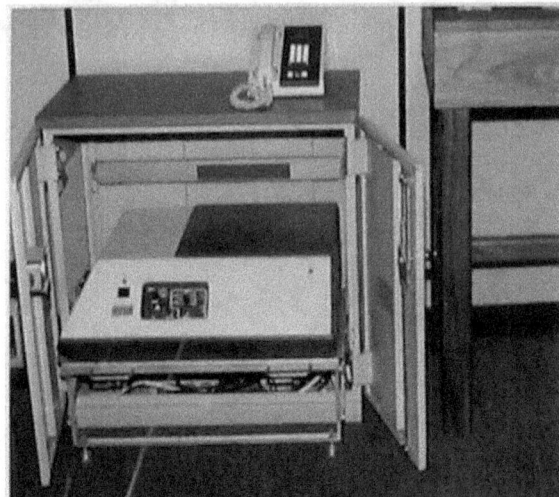

STU-II cabinet with desk set on top.

50.1   External links

- Delusion.org - National Cryptologic Museum pictures
- Pictures of president Reagan using a STU-II phone

50.2   See also

- STU-III
- SCIP

The **STU-II** is a secure telephone developed by the U.S. National Security Agency. It permitted up to six users to have secure communications, on a time-shared (e.g.: rotating) basis. It was made by ITT Defense Communications, Nutley, New Jersey. An OEM partner was Northern Telecom.

According to information on display in 2005 at the NSA's National Cryptologic Museum, the STU-II was in use from the 1980s to the present. It uses the linear predictive coding algorithm LPC-10 at 2.4 kilobits/second to digitize voice, and the "Key Distribution Center" (KDC) for key management. The display also stated that the STU-II B is the standard narrow band secure telephone.

STU-II replaced the STU-I, KY-3 and the Navajo I. The last was a secure telephone in a briefcase, of which 110 were built in the 1980s for use by senior government officials when traveling. The Navaho I also used LPC-10.

Some 10 000 STU-II units were produced.

Chapter 51

Temporal Key Integrity Protocol

Temporal Key Integrity Protocol or **TKIP** /tiˈkɪp/ was a stopgap security protocol used in the IEEE 802.11 wireless networking standard. TKIP was designed by the IEEE 802.11i task group and the Wi-Fi Alliance as an interim solution to replace WEP without requiring the replacement of legacy hardware. This was necessary because the breaking of WEP had left WiFi networks without viable link-layer security, and a solution was required for already deployed hardware. TKIP is no longer considered secure and was deprecated in the 2012 revision of the 802.11 standard.[1]

51.1 Background

On October 31, 2002, the Wi-Fi Alliance endorsed TKIP under the name Wi-Fi Protected Access (WPA).[2] The IEEE endorsed the final version of TKIP, along with more robust solutions such as 802.1X and the AES based CCMP, when they published IEEE 802.11i-2004 on 23 July 2004.[3] The Wi-Fi Alliance soon afterwards adopted the full specification under the marketing name WPA2.[4]

TKIP was deprecated by the IEEE in January 2009.[1]

51.2 Technical details

TKIP and the related WPA standard implement three new security features to address security problems encountered in WEP protected networks. First, TKIP implements a key mixing function that combines the secret root key with the initialization vector before passing it to the RC4 initialization. WEP, in comparison, merely concatenated the initialization vector to the root key, and passed this value to the RC4 routine. This permitted the vast majority of the RC4 based WEP related key attacks.[5] Second, WPA implements a sequence counter to protect against replay attacks. Packets received out of order will be rejected by the access point. Finally, TKIP implements a 64-bit Message Integrity Check (MIC).[6]

To be able to run on legacy WEP hardware with minor upgrades, TKIP uses RC4 as its cipher. TKIP also provides a rekeying mechanism. TKIP ensures that every data packet is sent with a unique encryption key.

Key mixing increases the complexity of decoding the keys by giving an attacker substantially less data that has been encrypted using any one key. WPA2 also implements a new message integrity code, MIC. The message integrity check prevents forged packets from being accepted. Under WEP it was possible to alter a packet whose content was known even if it had not been decrypted.

51.3 Security

TKIP uses the same underlying mechanism as WEP, and consequently is vulnerable to a number of similar attacks. The message integrity check, per-packet key hashing, broadcast key rotation, and a sequence counter discourage many attacks. The key mixing function also eliminates the WEP key recovery attacks.

Notwithstanding these changes, the weakness of some of these additions have allowed for new, although narrower, attacks.

51.3.1 Packet Spoofing and Decryption

TKIP is vulnerable to a MIC key recovery attack that, if successfully executed, permits an attacker to transmit and decrypt arbitrary packets on the network being attacked.[7] The current publicly available TKIP-specific attacks do not reveal the Pairwise Master Key or the Pairwise Temporal Keys. On November 8, 2008, Martin Beck and Erik Tews released a paper detailing how to recover the MIC key and transmit a few packets.[8] This attack was improved by Mathy Vanhoef and Frank Piessens in 2013, where they increase the amount of packets an attacker can transmit, and show how an attacker can also decrypt arbitrary packets.[7]

The basis of the attack is an extension of the WEP chop-chop attack. Because WEP uses a cryptographically insecure checksum mechanism (CRC32), an attacker can guess individual bytes of a packet, and the wireless access point will confirm or deny whether or not the guess is correct. If the guess is correct, the attacker will be able to detect the guess is correct and continue to guess other bytes of the packet. However, unlike the chop-chop attack against a WEP network, the attacker must wait for at least 60 seconds after an incorrect guess (a successful circumvention of the CRC32 mechanism) before continuing the attack. This is because although TKIP continues to use the CRC32 checksum mechanism, it implements an additional MIC code named Michael. If two incorrect Michael MIC codes are received within 60 seconds, the access point will implement countermeasures, meaning it will rekey the TKIP session key, thus changing future keystreams. Accordingly, attacks on TKIP will wait an appropriate amount of time to avoid these countermeasures. Because ARP packets are easily identified by their size, and the vast majority of the contents of this packet would be known to an attacker, the number of bytes an attacker must guess using the above method is rather small (approximately 14 bytes). Beck and Tews estimate recovery of 12 bytes is possible in about 12 minutes on a typical network, which would allow an attacker to transmit 3–7 packets of at most 28 bytes.[8] Vanhoef and Piessens improved this technique by relying on fragmentation, allowing an attacker to transmit arbitrary many packets, each at most 112 bytes in size.[7] The Vanhoef-Piessens attacks also can be used to decrypt arbitrary packets of the attack's choice.

An attacker already has access to the entire ciphertext packet. Upon retrieving the entire plaintext of the same packet, the attacker has access to the keystream of the packet, as well as the MIC code of the session. Using this information the attacker can construct a new packet and transmit it on the network. To circumvent the WPA implemented replay protection, the attacks use QoS channels to transmit these newly constructed packets. An attacker able to transmit these packets may be able to implement any number of attacks, including ARP poisoning attacks, denial of service, and other similar attacks, with no need of being associated with the network.

51.3.2   Royal Holloway attack

A group of security researchers at the Information Security Group at Royal Holloway, University of London reported a theoretical attack on TKIP which exploits the underlying RC4 encryption mechanism. TKIP uses a similar key structure to WEP with the low 16-bit value of a sequence counter (used to prevent replay attacks) being expanded into the 24-bit "IV", and this sequence counter always increment on every new packet. An attacker can use this key structure to improve existing attacks on RC4. In particular, if the same data is encrypted multiple times, an attacker can learn this information from only 2^{24} connections.[9][10][11] While they claim that this attack is on the verge of practicality, only simulations were performed, and the attack has not been demonstrated in practice.

51.4   Legacy

ZDNet reported on June 18, 2010 that WEP & TKIP would soon be disallowed on Wi-Fi devices by the Wi-Fi alliance.[12] However, a recent survey in 2013 showed that it is still in widespread use.[13]

51.5   See also

- Wireless network interface controller

- CCMP

- Wi-Fi Protected Access

- IEEE 802.11i-2004

51.6   References

[1] "802.11mb Issues List v12" (excel). 20 Jan 2009. p. CID 98. The use of TKIP is deprecated. The TKIP algorithm is unsuitable for the purposes of this standard

[2] "Wi-Fi Alliance Announces Standards-Based Security Solution to Replace WEP". *Wi-Fi Alliance*. 2002-10-31. Retrieved 2007-12-21.

[3] "IEEE 802.11i-2004: Amendment 6: Medium Access Control (MAC) Security Enhancements" (pdf). IEEE Standards. 2004-07-23. Retrieved 2007-12-21.

[4] "Wi-Fi Alliance Introduces Next Generation of Wi-Fi Security". *Wi-Fi Alliance*. 2004-09-01. Retrieved 2007-12-21.

[5] Edney, Jon; Arbaugh, William A. (2003-07-15). *Real 802.11 Security: Wi-Fi Protected Access and 802.11i*. Addison Wesley Professional. ISBN 0-321-13620-9.

[6] IEEE-SA Standards Board. Wireless LAN Medium Access Control (MAC) and Physical Layer (PHY) Specifications. *Communications Magazine*, IEEE, 2007.

[7] Vanhoef, Mathy; Piessens, Frank (May 2013). "Practical Verification of WPA-TKIP Vulnerabilities" (PDF). *Proceedings of the 8th ACM SIGSAC symposium on Information, computer and communications security*. ASIA CCS '13: 427–436. doi:10.1145/2484313.2484368.

[8] Martin Beck & Erik Tews, "Practical attacks against WEP and WPA", available at .

[9] AlFardan; et al. (2013-07-08). "On the Security of RC4 in TLS and WPA" (PDF). Information Security Group, Royal Holloway, University of London.

[10] Paterson; et al. (2014-03-01). "Plaintext Recovery Attacks Against WPA/TKIP" (PDF). Information Security Group, Royal Holloway, University of London.

[11] Paterson; et al. (2014-03-01). "Big Bias Hunting in Amazonia: Large-Scale Computation and Exploitation of RC4 Biases (Invited Paper)". Information Security Group, Royal Holloway, University of London.

[12] Wi-Fi Alliance to dump WEP and TKIP ... not soon enough

[13] Vanhoef, Mathy; Piessens, Frank (May 2013). "Practical Verification of WPA-TKIP Vulnerabilities" (PDF). *Proceedings of the 8th ACM SIGSAC symposium on Information, computer and communications security*. ASIA CCS '13: 427–436. doi:10.1145/2484313.2484368.

Chapter 52

TLS termination proxy

An **TLS termination proxy** is a proxy server that is used by an institution to handle incoming TLS connections, decrypting the TLS and passing on the unencrypted request to the institution's other servers (it is assumed that the institution's own network is secure so the user's session data does not need to be encrypted on that part of the link). TLS termination proxies are used to reduce the load on the main servers by offloading the cryptographic processing to another machine, and to support servers that do not support TLS, like Varnish.

52.1 Servers capabable of acting as an TLS termination proxy

- Apache HTTP Server

- HAProxy

- Nginx

- Pound (networking)

- Squid (software)

Wikipedia uses Nginx as its TLS termination proxy.[1]

52.2 References

[1] "Wikitech: HTTPS". Wikitech.wikimedia.org. 3 October 2011. Retrieved 3 December 2011.

Chapter 53

TLS-PSK

Transport Layer Security pre-shared key ciphersuites (**TLS-PSK**) is a set of cryptographic protocols that provide secure communication based on pre-shared keys (PSKs). These pre-shared keys are symmetric keys shared in advance among the communicating parties.

There are several ciphersuites: The first set of ciphersuites uses only symmetric key operations for authentication. The second set uses a Diffie-Hellman key exchange authenticated with a pre-shared key. The third set combines public key authentication of the server with pre-shared key authentication of the client.

Usually, Transport Layer Security (TLS) uses public key certificates or Kerberos for authentication. TLS-PSK uses symmetric keys, shared in advance among the communicating parties, to establish a TLS connection. There are several reasons to use PSKs:

- Using pre-shared keys can, depending on the ciphersuite, avoid the need for public key operations. This is useful if TLS is used in performance-constrained environments with limited CPU power.

- Pre-shared keys may be more convenient from a key management point of view. For instance, in closed environments where the connections are mostly configured manually in advance, it may be easier to configure a PSK than to use certificates. Another case is when the parties already have a mechanism for setting up a shared secret key, and that mechanism could be used to "bootstrap" a key for authenticating a TLS connection.

53.1 Standards

- RFC 4279: "Pre-Shared Key Ciphersuites for Transport Layer Security (TLS)".

- RFC 4785: "Pre-Shared Key (PSK) Ciphersuites with NULL Encryption for Transport Layer Security (TLS)".

- RFC 5487: "Pre-Shared Key Cipher Suites for TLS with SHA-256/384 and AES Galois Counter Mode".

- RFC 5489: "ECDHE_PSK Cipher Suites for Transport Layer Security (TLS)".

53.2 See also

- AES Galois Counter Mode (GCM)

- Elliptic curve Diffie–Hellman (ECDHE)

- Null encryption

- SHA-256

53.3 References

Chapter 54

TLS-SRP

Transport layer security Secure Remote Password (TLS-SRP) ciphersuites are a set of cryptographic protocols that provide secure communication based on passwords, using an SRP password-authenticated key exchange.

There are two classes of TLS-SRP ciphersuites: The first class of cipher suites uses only SRP authentication. The second class uses SRP authentication and public key certificates together for added security.

Usually, TLS uses only public key certificates for authentication. TLS-SRP uses a value derived from a password (the SRP verifier) and a salt, shared in advance among the communicating parties, to establish a TLS connection. There are several reasons to use TLS-SRP:

- Using password-based authentication does not require reliance on certificate authorities.

- The end user does not need to check the URL being certified. If the server does not know the password equivalent data then the connection simply cannot be made. This prevents Phishing.

- Password authentication is less prone than certificate authentication to certain types of configuration mistakes, such as expired certificates or mismatched common name fields.

- TLS-SRP provides mutual authentication (the client and server both authenticate each other), while TLS with server certificates only authenticates the server to the client. Client certificates can authenticate the client to the server, but it may be easier for a user to remember a password than to install a certificate.

54.1 Implementations

TLS-SRP is implemented in GnuTLS,[1] OpenSSL as of release 1.0.1, Apache mod_gnutls and mod_ssl, cURL, TLS Lite and SecureBlackbox.

54.2 Standards

- RFC 2945: "The SRP Authentication and Key Exchange System".

- RFC 5054: "Using the Secure Remote Password (SRP) Protocol for TLS Authentication".

54.3 See also

- Transport Layer Security

54.4 References

[1] GnuTLS Manual, Authentication using SRP

Chapter 55

Tor (anonymity network)

This article is about the software and network. For the organization, see The Tor Project, Inc

Tor is free software for enabling anonymous communication. The name is an acronym derived from the original software project name *The Onion Router*.[6] Tor directs Internet traffic through a free, worldwide, volunteer network consisting of more than six thousand relays[7] to conceal a user's location and usage from anyone conducting network surveillance or traffic analysis. Using Tor makes it more difficult for Internet activity to be traced back to the user: this includes "visits to Web sites, online posts, instant messages, and other communication forms".[8] Tor's use is intended to protect the personal privacy of users, as well as their freedom and ability to conduct confidential communication by keeping their Internet activities from being monitored. An extract of a Top Secret appraisal by the National Security Agency (NSA) characterized Tor as "the King of high-secure, low-latency Internet anonymity" with "no contenders for the throne in waiting",[9] and the Parliamentary Office of Science and Technology deemed it, with approximately 2.5 million users daily "by far the most popular anonymous internet communication system." [10] Furthermore, a July 2015 NATO analysis opines that "the use of anonymisation technologies such as Tor will continue to thrive. Despite the attention that Tor has received worldwide, the technical and legal questions surrounding it remain relatively unexplored." [11]

Onion routing is implemented by encryption in the application layer of a communication protocol stack, nested like the layers of an onion. Tor encrypts the data, including the destination IP address, multiple times and sends it through a virtual circuit comprising successive, randomly selected Tor relays. Each relay decrypts a layer of encryption to reveal only the next relay in the circuit in order to pass the remaining encrypted data on to it. The final relay decrypts the innermost layer of encryption and sends the original data to its destination without revealing, or even knowing, the source IP address. Because the routing of the communication is partly concealed at every hop in the Tor circuit, this method eliminates any single point at which the communicating peers can be determined through network surveillance that relies upon knowing its source and destination.

An adversary unable to defeat the strong anonymity that Tor provides may try to de-anonymize the communication by other means. One way this may be achieved is by exploiting vulnerable software on the user's computer.[12] The NSA has a technique that targets outdated Firefox browsers codenamed EgotisticalGiraffe,[13] and targets Tor users in general for close monitoring under its XKeyscore program.[14][15] Attacks against Tor are an active area of academic research,[16][17] which is welcomed by the Tor Project itself.[18]

55.1 History

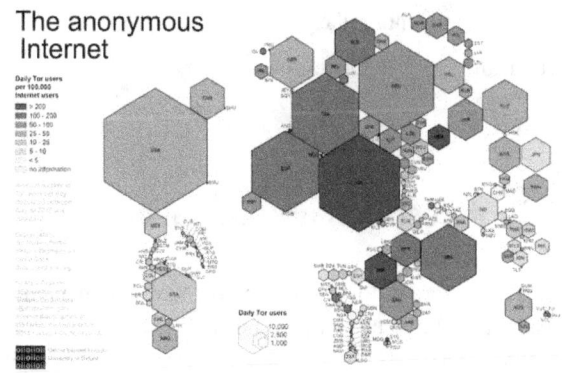

A cartogram illustrating Tor usage

The core principle of Tor, "onion routing", was developed in the mid-1990s by United States Naval Research Laboratory employees, mathematician Paul Syverson and computer scientists Michael G. Reed and David Goldschlag, with the purpose of protecting U.S. intelligence communications online. Onion routing was further developed by DARPA in 1997.[19][20][21]

The alpha version of Tor, developed by Syverson and computer scientists Roger Dingledine and Nick Mathewson[22] and then called The Onion Routing project, or TOR project, launched on 20 September 2002.[1][23] On 13 August 2004, Syverson, Dingledine and Mathewson presented "Tor: The Second-Generation Onion Router" at the 13th USENIX Security Symposium.[24] In 2004, the Naval Research Laboratory released the code for Tor under a free licence, and the Electronic Frontier Foundation (EFF) began funding Dingledine and Mathewson to continue its development.[22]

In December 2006, Dingledine, Mathewson and five others founded The Tor Project, a Massachusetts-based 501(c)(3) research-education nonprofit organization responsible for maintaining Tor.[25] The EFF acted as The Tor Project's fiscal sponsor in its early years, and early financial supporters of The Tor Project included the U.S. International Broadcasting Bureau, Internews, Human Rights Watch, the University of Cambridge, Google, and Netherlands-based Stichting NLnet.[26][27][28][29][30]

In November 2014 there was speculation in the aftermath of Operation Onymous that a Tor weakness has been exploited. A representative of Europol was secretive about the method used, saying: "*This is something we want to keep for ourselves. The way we do this, we can't share with the whole world, because we want to do it again and again and again*"[31] A BBC source cited a 'technical breakthrough'[32] that allowed the tracking of the physical location of servers, and the number of sites that police initially claimed to have infiltrated led to speculation that a weakness in the Tor network had been exploited. This possibility was downplayed by Andrew Lewman, a representative of the not-for-profit Tor project, suggesting that execution of more traditional police work was more likely.[33][34]

Further information: Operation Onymous

55.2   Reception and impact

Tor has been praised for providing privacy and anonymity to vulnerable Internet users such as political activists fearing surveillance and arrest, ordinary web users seeking to circumvent censorship, and people who have been threatened with violence or abuse by stalkers.[35][36] The U.S. National Security Agency (NSA) has called Tor "the king of high-secure, low-latency Internet anonymity",[12] and BusinessWeek magazine has described it as "perhaps the most effective means of defeating the online surveillance efforts of intelligence agencies around the world".[37] Other media have described Tor as "a sophisticated privacy tool",[38] "easy to use"[39] and "so secure that even the world's most sophisticated electronic spies haven't figured

out how to crack it".[40]

In March 2011, The Tor Project received the Free Software Foundation's 2010 Award for Projects of Social Benefit. The citation read, "Using free software, Tor has enabled roughly 36 million people around the world to experience freedom of access and expression on the Internet while keeping them in control of their privacy and anonymity. Its network has proved pivotal in dissident movements in both Iran and more recently Egypt."[41]

In 2012, *Foreign Policy* magazine named Dingledine, Mathewson, and Syverson among its Top 100 Global Thinkers "for making the web safe for whistleblowers".[42]

In 2013, Jacob Appelbaum described Tor as a "part of an ecosystem of software that helps people regain and reclaim their autonomy. It helps to enable people to have agency of all kinds; it helps others to help each other and it helps you to help yourself. It runs, it is open and it is supported by a large community spread across all walks of life."[43]

In June 2013, whistleblower Edward Snowden used Tor to send information about PRISM to *The Washington Post* and *The Guardian*.[44]

In 2014, the Russian government offered a $111,000 contract to "study the possibility of obtaining technical information about users and users' equipment on the Tor anonymous network".[45][46]

Advocates for Tor say it supports freedom of expression, including in countries where the Internet is censored, by protecting the privacy and anonymity of users. The mathematical underpinnings of Tor lead it to be characterized as acting "like a piece of infrastructure, and governments naturally fall into paying for infrastructure they want to use".[47]

The project was originally developed on behalf of the U.S. intelligence community and continues to receive U.S. government funding, and has been criticized as "more resembl[ing] a spook project than a tool designed by a culture that values accountability or transparency".[22] As of 2012, 80% of The Tor Project's $2M annual budget came from the United States government, with the U.S. State Department, the Broadcasting Board of Governors, and the National Science Foundation as major contributors,[48] "to aid democracy advocates in authoritarian states".[14] The Swedish government and other organizations provided the other 20%, including NGOs and thousands of individual sponsors.[29][49] Dingledine said that the United States Department of Defense funds are more similar to a research grant than a procurement contract. Tor executive director Andrew Lewman said that even though it accepts funds from the U.S. federal government, the Tor service did not collaborate with the NSA to reveal identities of users.[50]

Critics say Tor is not as secure as it claims,[51] pointing to U.S. law enforcement's investigations and shutdowns of Tor-using sites such as web-hosting company Freedom Hosting and online marketplace Silk Road.[22] In October 2013, after analyzing documents leaked by Edward Snowden, the Guardian reported that the NSA had repeatedly tried to crack Tor and had failed to break its core security, although it had had some success attacking the computers of individual Tor users.[12] The Guardian also published a 2012 NSA classified slide deck, entitled "Tor Stinks", which said: "We will never be able to de-anonymize all Tor users all the time", but "with manual analysis we can de-anonymize a very small fraction of Tor users".[52] When Tor users are arrested, it is typically due to human error, not to the core technology being hacked or cracked.[53] On 7 November 2014, for example, a joint operation by the FBI, ICE Homeland Security investigations and European Law enforcement agencies led to 17 arrests and the seizure of 27 sites containing 400 pages.[54] A late 2014 report by *Der Spiegel* using a new cache of Edward Snowden leaks revealed, however, that as of 2012 the NSA deemed Tor on its own as a "major threat" to its mission, and when used in conjunction with other privacy tools such as OTR, Cspace, ZRTP, RedPhone, Tails, and TrueCrypt was ranked as "catastrophic," leading to a "near-total loss/lack of insight to target communications, presence..."[55][56]

In October 2014 The Tor Project hired the public relations firm Thomson Communications in order to improve its public image (particularly regarding the terms "Dark Net" and "hidden services," which are widely viewed as being problematic) and to educate journalists about the technical aspects of Tor.[57]

On June 2015 the special rapporteur from the United Nation's Office of the High Commissioner for Human Rights specifically mentioned Tor in the context of the debate in the U.S. of allowing so-called backdoors in encryption programs for law enforcement purposes [58] in an interview for *The Washington Post.*

On July 2015 the Tor Project announced an alliance with the Library Freedom Project to establish exit nodes in public libraries.[59][60] The pilot program, which established a middle relay running on the excess bandwidth afforded by the Kilton Library in Lebanon, New Hampshire, making it the first library in the U.S. to host a Tor node, was briefly put on hold due to pressure from the Department of Homeland Security, but was re-established on September 15, 2015.[61]

On August 2015 an IBM security research group, called "X-Team", put out a quarterly report that advised companies to block Tor on security grounds, citing a "steady increase" in attacks from Tor exit nodes as well as botnet traffic.[62][63]

55.3 Usage

Further information: Dark Web

Tor enables users to surf the Internet, chat and send instant messages anonymously, and is used by a wide variety of people for both licit and illicit purposes.[65] Tor has for example been used by criminal enterprises, hacktivism groups, and law enforcement agencies at cross purposes, sometimes simultaneously;[66][67] likewise, agencies within the U.S. government variously fund Tor (the U.S. State Department), the National Science Foundation, and (through the Broadcasting Board of Governors, which itself partially funded Tor until October 2012), Radio Free Asia, and seek to subvert it.[12][68]

Tor is not meant to completely solve the issue of anonymity on the web. Instead, it simply focuses on protecting the transportation of data so that certain sites cannot trace back the data to a given location. It is still possible for sites to backtrack to a location. Tor is not designed to erase a user's tracks but to simply make it less likely for sites to trace back to them.[69]

Tor is also used for illegal activities, e.g., to gain access to censored information, to organize political activities,[70] or to circumvent laws against criticism of heads of state.

Tor has been described by *The Economist*, in relation to Bitcoin and the Silk Road, as being "a dark corner of the web".[71] It has been targeted by both the American NSA and the British GCHQ signals intelligence agencies, albeit with marginal success,[12] and more successfully by the British National Crime Agency in its Operation Notarise.[72] At the same time, GCHQ has been using a tool named SHADOWCAT for "end-to-end encrypted access to VPS over SSH using the TOR network".[73][74] Tor can be used for anonymous defamation, unauthorized news leaks of sensitive information and copyright infringement, distribution of illegal sexual content,[75][76][77] selling controlled substances,[78] weapons, and stolen credit card numbers,[79] money laundering,[80] bank fraud,[81] credit card fraud, identity theft and the exchange of counterfeit currency;[82] the black market utilizes the Tor infrastructure, at least in part, in conjunction with Bitcoin.[66]

In its complaint against Ross William Ulbricht of the Silk Road the FBI acknowledged that Tor has "known legitimate uses".[83][84] According to CNET, Tor's anonymity function is "endorsed by the Electronic Frontier Foundation and other civil liberties groups as a method for whistleblowers and human rights workers to communicate with journalists".[85] EFF's Surveillance Self-Defense guide includes a description of where Tor fits in a larger strategy for protecting privacy and anonymity.[86]

In 2014, the EFF's Eva Galperin told BusinessWeek magazine that "Tor's biggest problem is press. No one hears about that time someone wasn't stalked by their abuser. They hear how somebody got away with downloading child porn."[40]

The Tor Project states that Tor users include "normal people" who wish to keep their Internet activities private from websites and advertisers, people concerned about cyberspying, users who are evading censorship such as activists and journalists, and military professionals. As of November 2013, Tor had about four million users.[87] According to the Wall Street Journal, in 2012 about 14% of Tor's traffic connected from the United States, with people in "Internet-censoring countries" as its second-largest user base.[88] Tor is increasingly used by victims of domestic violence and the social workers and agencies that assist them. It has also been used to prevent digital stalking, which has increased due to the prevalence of digital media in contemporary online life.[89] Along with SecureDrop, Tor is used by news organizations such as the Guardian, the *The New Yorker*, ProPublica and *The Intercept* to protect the privacy of whistleblowers.[90]

In March 2015 the Parliamentary Office of Science and Technology released a briefing which stated that "There is widespread agreement that banning online anonymity systems altogether is not seen as an acceptable policy option in the U.K." and that ""Even if it were, there would be technical challenges." The report further noted that Tor "plays only a minor role in the online viewing and distribution of indecent images of children" (due in part to its inherent latency); its usage by the Internet Watch Foundation, the utility of its hidden services for whistleblowers, and its circumvention of the Great Firewall of China were touted.[10]

Tor's executive director, Andrew Lewman, also said in August 2014 that agents of the NSA and the GCHQ have anonymously provided Tor with bug reports.[91]

The Tor Project's FAQ offers supporting reasons for EFF's endorsement:

> Criminals can already do bad things. Since they're willing to break laws, they already have lots of options available that provide better privacy than Tor provides....
>
> Tor aims to provide protection for ordinary people who want to follow the law. Only criminals have privacy right now, and we need to fix that....
>
> So yes, criminals could in theory use Tor, but they already have better options, and it seems unlikely that taking Tor away from the world will stop them from doing their bad things. At the same time, Tor and other privacy measures can

fight identity theft, physical crimes like stalking, and so on.[92]

55.4 Operation

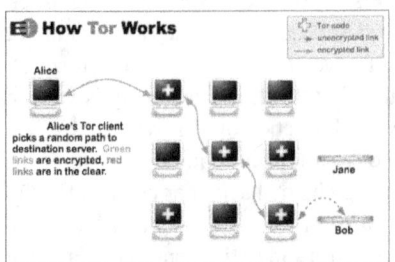

Infographic about how Tor works, by EFF

Tor aims to conceal its users' identities and their online activity from surveillance and traffic analysis by separating identification and routing. It is an implementation of onion routing, which encrypts and then randomly bounces communications through a network of relays run by volunteers around the globe. These onion routers employ encryption in a multi-layered manner (hence the onion metaphor) to ensure perfect forward secrecy between relays, thereby providing users with anonymity in network location. That anonymity extends to the hosting of censorship-resistant content by Tor's anonymous hidden service feature.[24] Furthermore, by keeping some of the entry relays (bridge relays) secret, users can evade Internet censorship that relies upon blocking public Tor relays.[93]

Because the IP address of the sender and the recipient are not *both* in cleartext at any hop along the way, anyone eavesdropping at any point along the communication channel cannot directly identify both ends. Furthermore, to the recipient it appears that the last Tor node (called the exit node), rather than the sender, is the originator of the communication.

55.4.1 Originating traffic

A Tor user's SOCKS-aware applications can be configured to direct their network traffic through a Tor instance's SOCKS interface. Tor periodically creates virtual circuits through the Tor network through which it can multiplex and onion-route that traffic to its destination. Once inside a Tor network, the traffic is sent from router to router along the circuit, ultimately reaching an exit node at which point the cleartext packet is available and is forwarded on to its original destination. Viewed from the destination, the traffic appears to originate at the Tor exit node.

Tor's application independence sets it apart from most other

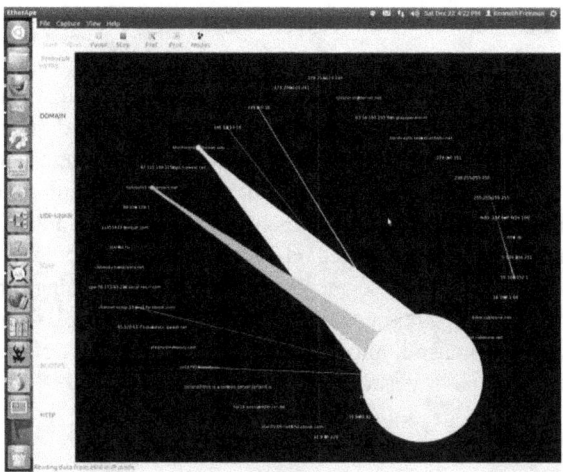

A visual depiction of the traffic between some Tor relay nodes from the open-source packet sniffing program EtherApe

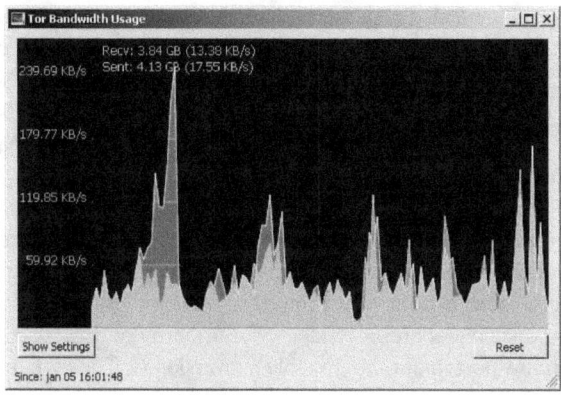

A Tor non-exit relay with a maximum output of 239.69 KB/s

anonymity networks: it works at the Transmission Control Protocol (TCP) stream level. Applications whose traffic is commonly anonymised using Tor include Internet Relay Chat (IRC), instant messaging, and World Wide Web browsing.

55.4.2 Hidden services

See also: List of Tor hidden services

Tor can also provide anonymity to websites and other servers. Servers configured to receive inbound connections only through Tor are called hidden services. Rather than revealing a server's IP address (and thus its network location), a hidden service is accessed through its onion address. The Tor network understands these addresses and can route data to and from hidden services, even those hosted behind firewalls or network address translators (NAT), while preserving the anonymity of both parties. Tor is necessary to

access hidden services.[94]

Hidden services have been deployed on the Tor network since 2004.[95] Other than the database that stores the hidden-service descriptors,[96] Tor is decentralized by design; there is no direct readable list of all hidden services, although a number of hidden services catalog publicly known onion addresses.

Because hidden services do not use exit nodes, connection to a hidden service is encrypted end-to-end and not subject to eavesdropping. There are, however, security issues involving Tor hidden services. For example, services that are reachable through Tor hidden services *and* the public Internet are susceptible to correlation attacks and thus not perfectly hidden. Other pitfalls include misconfigured services (e.g. identifying information included by default in web server error responses), uptime and downtime statistics, intersection attacks, and user error.[96][97]

Hidden services could be also accessed from a standard web browser without client-side connection to the Tor network, using services like Tor2web.[98]

Popular sources of dark web .onion links include Pastebin, Twitter, Reddit and other Internet forums.[99]

Further information: Dark Web

55.4.3 Arm status monitor

Arm logo

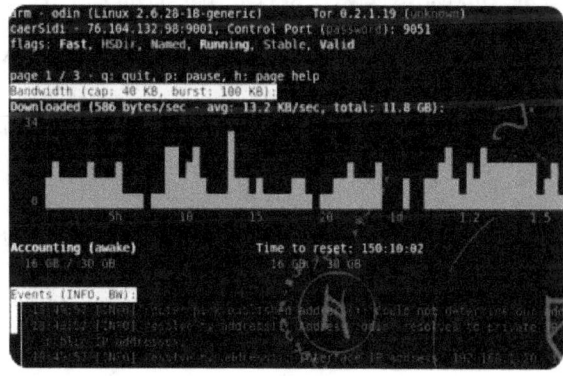

Arm's header panel and bandwidth graph.

The **anonymizing relay monitor** (**arm**) is a command-line status monitor written in Python for Tor.[100][101][102] This functions much like top does for system usage, providing real time statistics for:

- resource usage (bandwidth, cpu, and memory usage)

- general relaying information (nickname, fingerprint, flags, or/dir/controlports)

- event log with optional regex filtering and deduplication

- connections correlated against tor's consensus data (ip, connection types, relay details, etc.)

- torrc configuration file with syntax highlighting and validation

Most of arm's attributes are configurable through an optional armrc configuration file. It runs on any platform supported by curses including Linux, Mac OS X, and other Unix-like variants.

The project began in the summer of 2009,[103][104] and since July 18, 2010 it has been an official part of the tor project. It is free software, available under the GNU General Public License.

55.5 Weaknesses

Like all current low-latency anonymity networks, Tor cannot and does not attempt to protect against monitoring of traffic at the boundaries of the Tor network (i.e., the traffic entering and exiting the network). While Tor does provide protection against traffic analysis, it cannot prevent traffic confirmation (also called *end-to-end correlation*).[105][106]

In spite of known weaknesses and attacks listed here, Tor and the alternative network system JonDonym (Java Anon Proxy, JAP) are considered more resilient than alternatives such as VPNs. Were a local observer on an ISP or WLAN to attempt to analyze the size and timing of the encrypted data stream going through the VPN, Tor, or JonDo system, the latter two would be harder to analyze, as demonstrated by a 2009 study.[107]

Researchers from the University of Michigan developed a network scanner allowing identification of 86% of live Tor "bridges" with a single scan.[108]

55.5.1 Eavesdropping

Autonomous System (AS) eavesdropping

If an autonomous system (AS) exists on both path segments from a client to entry relay and from exit relay to destination, such an AS can statistically correlate traffic on the entry and exit segments of the path and potentially infer the destination with which the client communicated. In 2012, LASTor proposed a method to predict a set of potential ASes on these two segments and then avoid choosing this path during path selection algorithm on client side. In this paper, they also improve latency by choosing shorter geographical paths between client and destination.[109]

Exit node eavesdropping

In September 2007, Dan Egerstad, a Swedish security consultant, revealed that he had intercepted usernames and passwords for e-mail accounts by operating and monitoring Tor exit nodes.[110] As Tor cannot encrypt the traffic between an exit node and the target server, any exit node is in a position to capture traffic passing through it that does not use end-to-end encryption such as Secure Sockets Layer (SSL) or Transport Layer Security (TLS). While this may not inherently breach the anonymity of the source, traffic intercepted in this way by self-selected third parties can expose information about the source in either or both of payload and protocol data.[111] Furthermore, Egerstad is circumspect about the possible subversion of Tor by intelligence agencies:[112]

> If you actually look in to where these Tor nodes are hosted and how big they are, some of these nodes cost thousands of dollars each month just to host because they're using lots of bandwidth, they're heavy-duty servers and so on. Who would pay for this and be anonymous?

In October 2011, a research team from ESIEA claimed to have discovered a way to compromise the Tor network by decrypting communication passing over it.[113][114] The technique they describe requires creating a map of Tor network nodes, controlling one third of them, and then acquiring their encryption keys and algorithm seeds. Then, using these known keys and seeds, they claim the ability to decrypt two encryption layers out of three. They claim to break the third key by a statistical-based attack. In order to redirect Tor traffic to the nodes they controlled, they used a denial-of-service attack. A response to this claim has been published on the official Tor Blog stating that these rumours of Tor's compromise are greatly exaggerated.[115]

55.5.2 Traffic-analysis attack

Steven J. Murdoch and George Danezis from University of Cambridge presented an article at the 2005 IEEE Symposium on security and privacy on traffic-analysis techniques that allow adversaries with only a partial view of the network to infer which nodes are being used to relay the anonymous streams.[116] These techniques greatly reduce the anonymity provided by Tor. Murdoch and Danezis have also shown that otherwise unrelated streams can be linked back to the same initiator. This attack, however, fails to reveal the identity of the original user.[116] Murdoch has been working with and has been funded by Tor since 2006.

55.5.3 Tor exit node block

Operators of Internet sites have the ability to prevent traffic from Tor exit nodes or to offer reduced functionality to Tor users. For example, it is not generally possible to edit Wikipedia when using Tor or when using an IP address that also is used by a Tor exit node, due to the use of the TorBlock MediaWiki extension, unless an exemption is obtained. The BBC blocks the IP addresses of all known Tor relays from its iPlayer service—including guards, relays, and exit nodes—regardless of geographic location. Bridge relays are not affected.

55.5.4 Bad apple attack

In March 2011, researchers with the Rocquencourt French Institute for Research in Computer Science and Automation (Institut national de recherche en informatique et en automatique, INRIA), documented an attack that is capable of revealing the IP addresses of BitTorrent users on the Tor network. The "bad apple attack" exploits Tor's design and takes advantage of insecure application use to associate the simultaneous use of a secure application with the IP address of the Tor user in question. One method of attack depends on control of an exit node or hijacking tracker responses, while a secondary attack method is based in part on the statistical exploitation of distributed hash table tracking.[117] According to the study:[117]

> This attack against Tor consists of two parts: (a) exploiting an insecure application to reveal the source IP address of, or trace, a Tor user and (b) exploiting Tor to associate the use of a secure application with the IP address of a user (revealed by the insecure application). As it is not a goal of Tor to protect against application-level attacks, Tor cannot be held responsible for the first part of this attack. However, because Tor's design makes it possible to associate streams originating from secure application with traced users, the second part of this attack is indeed an attack against Tor. We call the second part of this attack the bad apple attack. (The name of this attack refers to the saying "one bad apple spoils the bunch". We use this wording to illustrate that one insecure application on Tor may allow to trace other applications.)

The results presented in the bad apple attack research paper are based on an attack in the wild launched against the Tor network by the authors of the study. The attack targeted six exit nodes, lasted for 23 days, and revealed a total of 10,000 IP addresses of active Tor users. This study is particularly significant because it is the first documented attack designed to target P2P file-sharing applications on Tor.[117] BitTorrent may generate as much as 40% of all traffic on Tor.[118] Furthermore, the bad apple attack is effective against insecure use of any application over Tor, not just BitTorrent.[117]

55.5.5 Some protocols expose IP addresses

Researchers from the French Institute for Research in Computer Science and Automation (INRIA) showed that the Tor dissimulation technique in BitTorrent can be bypassed by attackers controlling a Tor exit node. The study was conducted by monitoring six exit nodes for a period of 23 days. Researches used three attack vectors:[119]

Inspection of BitTorrent control messages Tracker announces and extension protocol handshakes may optionally contain client IP address. Analysis of collected data revealed that 35% and 33% of messages, respectively, contained addresses of clients.[119]:3

Hijacking trackers' responses Due to lack of encryption or authentication in communication between tracker and peer, typical man-in-the-middle attacks allow attackers to determine peer IP addresses and even verify the distribution of content. Such attacks work when Tor is used only for tracker communication.[119]:4

Exploiting distributed hash tables (DHT) This attack exploits the fact that distributed hash table (DHT) connections through Tor are impossible, so an attacker is able to reveal a target's IP address by looking it up in the DHT even if the target uses Tor to connect to other peers.[119]:4–5

With this technique, researchers were able to identify other streams initiated by users, whose IP addresses were revealed.[119]

55.5.6   Sniper attack

Jensen *et al.*, describe a DDoS attack targeted at the TOR node software, as well as defenses against that attack and its variants. The attack works using a colluding client and server, and filling the queues of the exit node until the node runs out of memory, and hence can serve no other (genuine) clients. By attacking a significant proportion of the exit nodes this way, an attacker can degrade the network and increase the chance of targets using nodes controlled by the attacker.[120]

55.5.7   Heartbleed bug

The Heartbleed OpenSSL bug disrupted the Tor network for several days in April 2014 while private keys were renewed. The Tor Project recommended that Tor relay operators and hidden service operators revoke and generate fresh keys after patching OpenSSL, but noted that Tor relays use two sets of keys and that Tor's multi-hop design minimizes the impact of exploiting a single relay.[121] 586 relays later found to be susceptible to the Heartbleed bug were taken off-line as a precautionary measure.[122][123][124][125]

55.6   Implementations

The main implementation of Tor is written primarily in the C programming language and consists of approximately 340,000 lines of source code.[5]

55.6.1   Tor Browser

Tor Browser, previously known as **Tor Browser Bundle** (TBB), is the flagship product of the Tor Project. It consists of a modified Mozilla Firefox ESR web browser, the TorButton, TorLauncher, NoScript and HTTPS Everywhere Firefox extensions and the Tor proxy.[128][129] It can be run from removable media and is available for Windows, Mac OS X, and Linux.[130]

The Tor Browser automatically starts Tor background processes and routes traffic through the Tor network. Upon termination of a session the browser deletes privacy-sensitive data such as HTTP cookie and the browsing history.[129]

Following a series of global surveillance disclosures, Stuart Dredge (The Guardian) recommended using Tor Browser to avoid eavesdropping and retain privacy on the Internet.[131]

Firefox / JavaScript anonymity attack

In August 2013, it was discovered that the Firefox browsers in many older versions of the Tor Browser Bundle were vulnerable to a JavaScript attack, as NoScript was not enabled by default.[132] This attack was being exploited to send users' MAC and IP addresses and Windows computer names to the attackers.[133][134][135] News reports linked this to a United States Federal Bureau of Investigation (FBI) operation targeting Freedom Hosting's owner, Eric Eoin Marques, who was arrested on a provisional extradition warrant issued by a United States court on 29 July. The FBI is seeking to extradite Marques out of Ireland to Maryland on four charges — distributing, conspiring to distribute, and advertising child pornography — as well as aiding and abetting advertising of child pornography. The warrant alleges that Marques is "the largest facilitator of child porn on the planet".[136][137] The FBI acknowledged the attack in a 12 September 2013 court filing in Dublin;[138] further technical details from a training presentation leaked by Edward Snowden showed that the codename for the exploit was *EgotisticalGiraffe*.[139]

The FBI, in Operation Torpedo, has been targeting Tor hidden servers since 2012, such as in the case of Aaron McGrath, who was sentenced to 20 years for running three hidden Tor servers containing child pornography.[140]

55.6.2   Third-party applications

Vuze (formerly Azureus) BitTorrent client,[141] Bitmessage anonymous messaging system,[142] and TorChat instant messenger include Tor support.

The Guardian Project is actively developing a free and open-source suite of application programs and firmware for the Android operating system to improve the security of mobile communications.[143] The applications include ChatSecure instant messaging client,[144] Orbot Tor implementation,[145] Orweb (discontinued) privacy-enhanced mobile browser,[146][147] Orfox the mobile counterpart of the Tor Browser, ProxyMob Firefox add-on[148] and ObscuraCam.[149]

55.6.3   Security-focused operating systems

Several security-focused operating systems like GNU/Linux distributions including Hardened Linux From Scratch, Incognito, Liberté Linux, Qubes OS, Tails, Tor-ramdisk and Whonix, make extensive use of Tor.[150]

55.7 See also

- .onion
- Anonymous P2P
- Anonymous web browsing
- Crypto-anarchism
- Darknet
- Dark Web
- Deep Web (search indexing)
- Freedom of information
- I2P
- Internet censorship
- Internet privacy
- Privoxy
- Proxy server
- Tor2web

55.8 References

[1] Dingledine, Roger (20 September 2002). "pre-alpha: run an onion proxy now!". *or-dev* (Mailing list). Retrieved 17 July 2008.

[2] Mathewson, Nick (12 July 2015). "Tor 0.2.6.10 is released". *Tor Project Blog*. Retrieved 13 July 2015.

[3] Dingledine, Roger (7 April 2015). "Tor 0.2.5.12 and 0.2.6.7 are released". *tor-announce* (Mailing list). Retrieved 7 April 2015.

[4] Mathewson, Nick (27 July 2015). "Tor 0.2.7.2-alpha is released". *nickm's blog*. Tor Project. Retrieved 1 August 2015.

[5] "Tor". *Open HUB*. Retrieved 20 September 2014.

[6] Li, Bingdong; Erdin, Esra; Güneş, Mehmet Hadi; Bebis, George; Shipley, Todd (14 June 2011). "An Analysis of Anonymity Usage". In Domingo-Pascual, Jordi; Shavitt, Yuval; Uhlig, Steve. *Traffic Monitoring and Analysis: Third International Workshop, TMA 2011, Vienna, Austria, April 27, 2011, Proceedings*. Berlin: Springer-Verlag. pp. 113–116. ISBN 978-3-642-20304-6. Retrieved 6 August 2012.

[7] "Tor Network Status". Retrieved 14 February 2015.

[8] Glater, Jonathan D. (25 January 2006). "Privacy for People Who Don't Show Their Navels". *The New York Times*. Retrieved 13 May 2011.

[9] "Tor: 'The king of high-secure, low-latency anonymity'". *The Guardian*. 4 October 2013. Retrieved 5 October 2013.

[10] "U.K. Parliament says banning Tor is unacceptable and impossible". *The Daily Dot*. Retrieved 19 April 2015.

[11] https://ccdcoe.org/sites/default/files/multimedia/pdf/TOR_Anonymity_Network.pdf

[12] Ball, James; Schneier, Bruce; Greenwald, Glenn (4 October 2013). "NSA and GCHQ target Tor network that protects anonymity of web users". *The Guardian*. Retrieved 5 October 2013.

[13] "'Peeling back the layers of Tor with EgotisticalGiraffe'". *The Guardian*. 4 October 2013. Retrieved 5 October 2013.

[14] J. Appelbaum, A. Gibson, J. Goetz, V. Kabisch, L. Kampf, L. Ryge (3 July 2014). "NSA targets the privacy-conscious". *Panorama* (Norddeutscher Rundfunk). Retrieved 4 July 2014.

[15] http://daserste.ndr.de/panorama/xkeyscorerules100.txt

[16] "Tor developers vow to fix bug that can uncloak users". *Ars Technica*.

[17] "Free Haven's Selected Papers in Anonymity".

[18] "Tor Research Home".

[19] Fagoyinbo, Joseph Babatunde (2013-05-24). *The Armed Forces: Instrument of Peace, Strength, Development and Prosperity*. AuthorHouse. ISBN 9781477226476. Retrieved 29 August 2014.

[20] Leigh, David; Harding, Luke (2011-02-08). *WikiLeaks: Inside Julian Assange's War on Secrecy*. PublicAffairs. ISBN 1610390628. Retrieved 29 August 2014.

[21] Ligh, Michael; Adair, Steven; Hartstein, Blake; Richard, Matthew (2010-09-29). *Malware Analyst's Cookbook and DVD: Tools and Techniques for Fighting Malicious Code*. John Wiley & Sons. ISBN 9781118003367. Retrieved 29 August 2014.

[22] Levine, Yasha (16 July 2014). "Almost everyone involved in developing Tor was (or is) funded by the US government". *Pando Daily*. Retrieved 30 August 2014.

[23] "Tor FAQ: Why is it called Tor?". *Tor Project*. Retrieved 1 July 2011.

[24] Dingledine, Roger; Mathewson, Nick; Syverson, Paul (13 August 2004). "Tor: The Second-Generation Onion Router". *Proc. 13th USENIX Security Symposium*. San Diego, California. Retrieved 17 November 2008.

[25] "Tor Project: Core People". *Tor Project*. Retrieved 17 July 2008.

[26] "Tor Project Form 990 2008" (PDF). *Tor Project*. Tor Project. 2009. Retrieved 30 August 2014.

[27] "Tor Project Form 990 2007" (PDF). *Tor Project*. Tor Project. 2008. Retrieved 30 August 2014.

[28] "Tor Project Form 990 2009" (PDF). *Tor Project*. Tor Project. 2010. Retrieved 30 August 2014.

[29] "Tor: Sponsors". *Tor Project*. Retrieved 11 December 2010.

[30] Krebs, Brian (8 August 2007). "Attacks Prompt Update for 'Tor' Anonymity Network". *Washington Post*. Retrieved 27 October 2007.

[31] Greenberg, Andy (7 November 2014). "Global Web Crackdown Arrests 17, Seizes Hundreds Of Dark Net Domains". Retrieved 9 August 2015.

[32] Wakefield, Jane (7 November 2014). "Huge raid to shut down 400-plus dark net sites". Retrieved 9 August 2015.

[33] Patrick Howell O'Neill (7 November 2014). "The truth behind Tor's confidence crisis". The Daily Dot. Retrieved 10 November 2014.

[34] Shawn Knight (7 November 2014). "Operation Onymous seizes hundreds of darknet sites, 17 arrested globally". Techspot. Retrieved 8 November 2014.

[35] Brandom, Russell (9 May 2014). "Domestic violence survivors turn to Tor to escape abusers". *The Verge*. Retrieved 30 August 2014.

[36] Gurnow, Michael (1 July 2014). "Seated Between Pablo Escobar and Mahatma Gandhi: The Sticky Ethics of Anonymity Networks". *Dissident Voice*. Retrieved 17 July 2014.

[37] Lawrence, Dune (23 January 2014). "The Inside Story of Tor, the Best Internet Anonymity Tool the Government Ever Built". *Businessweek magazine*. Retrieved 30 August 2014.

[38] Zetter, Kim (1 June 2010). "WikiLeaks Was Launched With Documents Intercepted From Tor". *Wired*. Retrieved 30 August 2014.

[39] Lee, Timothy B. (10 June 2013). "Five ways to stop the NSA from spying on you". *Washington Post*. Retrieved 30 August 2014.

[40] "Not Even the NSA Can Crack the State Department's Favorite Anonymous Service". *Foreign Policy*. 24 October 2014. Retrieved 30 August 2014.

[41] "2010 Free Software Awards announced". *Free Software Foundation*. Retrieved 23 March 2011.

[42] Wittmeyer, Alicia P.Q. (26 November 2012). "The FP Top 100 Global Thinkers". *Foreign Policy*. Archived from the original on 28 November 2012. Retrieved 28 November 2012.

[43] Sirius, R. U. (11 March 2013). "Interview uncut: Jacob Appelbaum". *theverge.com*.

[44] Gaertner, Joachim (1 July 2013). "Darknet – Netz ohne Kontrolle". *Das Erste* (in German). Retrieved 28 August 2013.

[45] Gallagher, Sean (25 July 2014). "Russia publicly joins war on Tor privacy with $111,000 bounty". *Ars Technica*. Retrieved 26 July 2014.

[46] Lucian, Constantin (25 July 2014). "Russian government offers huge reward for help unmasking anonymous Tor users". *PC World*. Retrieved 26 July 2014.

[47] "Clearing the air around Tor". *PandoDaily*.

[48] McKim, Jenifer B. (8 March 2012). "Privacy software, criminal use". *The Boston Globe*. Archived from the original on 12 March 2012.

[49] Fowler, Geoffrey A. (17 December 2012). "Tor: an anonymous, and controversial, way to web-surf". *Wall Street Journal*. Retrieved 19 May 2013.

[50] Fung, Brian (6 September 2013). "The feds pay for 60 percent of Tor's development. Can users trust it?". *The Switch* (Washington Post). Retrieved 6 February 2014.

[51] "Tor is Not as Safe as You May Think". *Infosecurity magazine*. 2 September 2013. Retrieved 30 August 2014.

[52] "'Tor Stinks' presentation – read the full document". *The Guardian*. 4 October 2014. Retrieved 30 August 2014.

[53] "/silk-road-tor-arrests/". *The Daily Dot*.

[54] "Dark net experts trade theories on 'de-cloaking' after raids". 7 November 2014. Retrieved 12 November 2014.

[55] SPIEGEL Staff (28 December 2014). "Prying Eyes: Inside the NSA's War on Internet Security". *Der Spiegel*. Retrieved 23 January 2015.

[56] "Presentation from the SIGDEV Conference 2012 explaining which encryption protocols and techniques can be attacked and which not" (PDF). *Der Spiegel*. 28 December 2014. Retrieved 23 January 2015.

[57] "Can Tor solve its PR problem?". *The Daily Dot*. Retrieved 19 April 2015.

[58] Andrea Peterson (28 May 2015). "U.N. report: Encryption is important to human rights — and backdoors undermine it". *Washington Post*.

[59] "Tor Exit Nodes in Libraries - Pilot (phase one)". *torproject.org*. Retrieved 15 September 2015.

[60] "Library Freedom Project". *libraryfreedomproject.org*. Retrieved 15 September 2015.

[61] http://www.vnews.com/photos/inthenews/18620952-95/despite-law-enforcement-concerns-lebanon-board-will-reactivate-privac

[62] "IBM Tells Companies To Block Tor Anonymisation Network". *TechWeekEurope UK*. Retrieved 15 September 2015.

[63] https://public.dhe.ibm.com/common/ssi/ecm/wg/en/wgl03086usen/WGL03086USEN.PDF

[64] Owen, Gareth. "Dr Gareth Owen: Tor: Hidden Services and Deanonymisation". Retrieved 20 June 2015.

[65] Zetter, Kim (17 May 2005). "Tor Torches Online Tracking". *Wired*. Retrieved 30 August 2014.

[66] Gregg, Brandon (30 April 2012). "How online black markets work". *CSO Online*. Retrieved 6 August 2012.

[67] Morisy, Michael (8 June 2012). "Hunting for child porn, FBI stymied by Tor undernet". *Muckrock*. Retrieved 6 August 2012.

[68] Lawrence, Dune (23 January 2014). "The Inside Story of Tor, the Best Internet Anonymity Tool the Government Ever Built". *Bloomberg Businessweek*. Retrieved 28 April 2014.

[69] The Tor Project, Inc. "Tor". *torproject.org*.

[70] Cochrane, Nate (2 February 2011). "Egyptians turn to Tor to organise dissent online". *SC Magazine*. Retrieved 10 December 2011.

[71] "Bitcoin: Monetarists Anonymous". *The Economist*. 29 September 2012. Retrieved 19 May 2013.

[72] Boiten, Eerke; Hernandez-Castro, Julio (28 July 2014). "Can you really be identified on Tor or is that just what the cops want you to believe?". Phys.org.

[73] "JTRIG Tools and Techniques". The Intercept. 14 Jul 2014.

[74] "document from an internal GCHQ wiki lists tools and techniques developed by the Joint Threat Research Intelligence Group". documentcoud.org. 5 July 2012. Retrieved 30 July 2014.

[75] Bode, Karl (12 March 2007). "Cleaning up Tor". *Broadband.com*. Retrieved 28 April 2014.

[76] Jones, Robert (2005). *Internet forensics*. O'Reilly. p. 133. ISBN 0-596-10006-X.

[77] Chen, Adrian (11 June 2012). "'Dark Net' Kiddie Porn Website Stymies FBI Investigation". *Gawker*. Retrieved 6 August 2012.

[78] Chen, Adrian (1 June 2011). "The Underground Website Where You Can Buy Any Drug Imaginable". *Gawker*. Retrieved 20 April 2012.

[79] Steinberg, Joseph (8 January 2015). "How Your Teenage Son or Daughter May Be Buying Heroin Online". *Forbes*. Retrieved 6 February 2015.

[80] Goodin, Dan (16 April 2012). "Feds shutter online narcotics store that used TOR to hide its tracks". *Ars Technica*. Retrieved 20 April 2012.

[81] "Treasury Dept: Tor a Big Source of Bank Fraud — Krebs on Security". *krebsonsecurity.com*.

[82] "How a $3.85 latte paid for with a fake $100 bill led to counterfeit kingpin's downfall". *Ars Technica*. Retrieved 19 April 2015.

[83] Turner, Serrin (27 September 2013). "Sealed compaint" (PDF). *United States of America v. Ross William Ulbricht*. Archived from the original (PDF) on 2 October 2013.

[84] Higgins, Parker (2013-10-03). "In the Silk Road Case, Don't Blame the Technology". *Electronic Frontier Foundation*. Retrieved 2013-12-22.

[85] Soghoian, Chris (16 September 2007). "Tor anonymity server admin arrested". *CNET News*. Retrieved 17 January 2011.

[86] "Surveillance Self-Defense: Tor". *Electronic Frontier Foundation*. Retrieved 28 April 2014.

[87] Dredge, Stuart (5 November 2013). "What is Tor? A beginner's guide to the privacy tool". *Guardian*. Retrieved 30 August 2014.

[88] Fowler, Geoffrey A. (17 December 2012). "Tor: An Anonymous, And Controversial, Way to Web-Surf". *Wall Street Journal*. Retrieved 30 August 2014.

[89] LeVines, George (7 May 2014). "As domestic abuse goes digital, shelters turn to counter-surveillance with Tor". *Boston Globe*. Retrieved 8 May 2014.

[90] "The Guardian introduces SecureDrop for document leaks". *Nieman Journalism Lab*. 5 June 2014. Retrieved 30 August 2014.

[91] http://www.bbc.com/news/technology-28886462 NSA & GCHQ "leak Tor bugs" alleges developer.

[92] "Doesn't Tor enable criminals to do bad things?". *Tor Project*. Retrieved 28 August 2013.

[93] "Tor: Bridges". *Tor Project*. Retrieved 9 January 2011.

[94] "Configuring Hidden Services for Tor". *Tor Project*. Retrieved 9 January 2011.

[95] Øverlier, Lasse; Syverson, Paul (21 June 2006). "Locating Hidden Servers" (PDF). *Proceedings of the 2006 IEEE Symposium on Security and Privacy*. IEEE Symposium on Security and Privacy. Oakland, CA: IEEE CS Press. p. 1. doi:10.1109/SP.2006.24. ISBN 0-7695-2574-1. Retrieved 9 November 2013.

[96] "Tor: Hidden Service Protocol, Hidden services". *Tor Project*. Retrieved 9 January 2011.

[97] Goodin, Dan (10 September 2007). "Tor at heart of embassy passwords leak". *TheRegister*. Retrieved 20 September 2007.

[98] Zetter, Kim (12 December 2008). "New Service Makes Tor Anonymized Content Available to All". wired.com. Retrieved 22 February 2014.

[99] Koebler, Jason (23 February 2015). "The Closest Thing to a Map of the Dark Net: Pastebin". Retrieved 14 July 2015.

[100] "Official Website".

[101] "Tor Project: Arm". *torproject.org.*

[102] "Ubuntu Manpage: arm - Terminal Tor status monitor". *Manpages.ubuntu.com.*

[103] "Summer Conclusion (ARM Project)". *torproject.org.* Retrieved 19 April 2015.

[104] interview by Brenno Winter

[105] Dingledine, Roger (18 February 2009). "One cell is enough to break Tor's anonymity". *Tor Project.* Retrieved 9 January 2011.

[106] "TheOnionRouter/TorFAQ".   Retrieved 18 September 2007. Tor (like all current practical low-latency anonymity designs) fails when the attacker can see both ends of the communications channel

[107] Herrmann, Dominik; Wendolsky, Rolf; Federrath, Hannes (13 November 2009). "Website Fingerprinting: Attacking Popular Privacy Enhancing Technologies with the Multinomial Naïve-Bayes Classifier" (PDF). *Proceedings of the 2009 ACM Cloud Computing Security Workshop (CCSW).* Cloud Computing Security Workshop. New York, USA: Association for Computing Machinery. Retrieved 2 September 2010.

[108] Judge, Peter (20 August 2013). "Zmap's Fast Internet Scan Tool Could Spread Zero Days In Minutes". *TechWeek Europe.* Retrieved 28 April 2014.

[109] Akhoondi, Masoud; Yu, Curtis; Madhyastha, Harsha V. (May 2012). *LASTor: A Low-Latency AS-Aware Tor Client* (PDF). IEEE Symposium on Security and Privacy. Oakland, USA. Retrieved 28 April 2014.

[110] Zetter, Kim (10 September 2007). "Rogue Nodes Turn Tor Anonymizer Into Eavesdropper's Paradise". *Wired.* Retrieved 16 September 2007.

[111] Lemos, Robert (8 March 2007). "Tor hack proposed to catch criminals". *SecurityFocus.*

[112] Gray, Patrick (13 November 2007). "The hack of the year". *Sydney Morning Herald.* Retrieved 28 April 2014.

[113] "Tor anonymizing network Compromised by French researchers". *The Hacker News.* Thehackernews.com. 24 October 2011. Retrieved 10 December 2011.

[114] "Announcement on 01net.com" (in French). Retrieved 17 October 2011.

[115] phobos (24 October 2011). "Rumors of Tor's compromise are greatly exaggerated". *Tor Project.* Retrieved 20 April 2012.

[116] Murdoch, Steven J.; Danezis, George (19 January 2006). "Low-Cost Traffic Analysis of Tor" (PDF). Retrieved 21 May 2007.

[117] Le Blond, Stevens; Manils, Pere; Chaabane, Abdelberi; Ali Kaafar, Mohamed; Castelluccia, Claude; Legout, Arnaud; Dabbous, Walid (March 2011). *One Bad Apple Spoils the Bunch: Exploiting P2P Applications to Trace and Profile Tor Users* (PDF). 4th USENIX Workshop on Large-Scale Exploits and Emergent Threats (LEET '11). National Institute for Research in Computer Science and Control. Retrieved 13 April 2011.

[118] McCoy, Damon; Bauer, Kevin; Grunwald, Dirk; Kohno, Tadayoshi; Sicker, Douglas (2008). "Shining Light in Dark Places: Understanding the Tor Network" (PDF). *Proceedings of the 8th International Symposium on Privacy Enhancing Technologies.* 8th International Symposium on Privacy Enhancing Technologies. Berlin, Germany: Springer-Verlag. pp. 63–76. doi:10.1007/978-3-540-70630-4_5. ISBN 978-3-540-70629-8.

[119] Manils, Pere; Abdelberri, Chaabane; Le Blond, Stevens; Kaafar, Mohamed Ali; Castelluccia, Claude; Legout, Arnaud; Dabbous, Walid (April 2010). *Compromising Tor Anonymity Exploiting P2P Information Leakage* (PDF). 7th USENIX Symposium on Network Design and Implementation. arXiv:1004.1461. Bibcode:2010arXiv1004.1461M.

[120] Jansen, Rob; Tschorsch, Florian; Johnson, Aaron; Scheuermann, Björn (2014). *The Sniper Attack: Anonymously Deanonymizing and Disabling the Tor Network* (PDF). 21st Annual Network & Distributed System Security Symposium. Retrieved 28 April 2014.

[121] Dingledine, Roger (7 April 2014). "OpenSSL bug CVE-2014-0160". *Tor Project.* Retrieved 28 April 2014.

[122] Dingledine, Roger (16 April 2014). "Rejecting 380 vulnerable guard/exit keys". *tor-relays* (Mailing list). Retrieved 28 April 2014.

[123] Lunar (16 April 2014). "Tor Weekly News — April 16th, 2014". *Tor Project.* Retrieved 28 April 2014.

[124] Gallagher, Sean (18 April 2014). "Tor network's ranks of relay servers cut because of Heartbleed bug". *Ars Technica.* Retrieved 28 April 2014.

[125] Mimoso, Michael (17 April 2014). "Tor begins blacklisting exit nodes vulnerable to Heartbleed". *Threat Post.* Retrieved 28 April 2014.

[126] Koppen, Georg (27 August 2015). "Tor Browser 5.0.2 is released". *Tor Project Blog.* Retrieved 29 August 2015.

[127] Koppen, Georg (28 August 2015). "Tor Browser 5.5a2 is released". *Tor Project Blog.* Retrieved 29 August 2015.

[128] Perry, Mike; Clark, Erinn; Murdoch, Steven (15 March 2013). "The Design and Implementation of the Tor Browser [DRAFT]". *Tor Project.* Retrieved 28 April 2014.

[129] Alin, Andrei (2 December 2013). "Tor Browser Bundle Ubuntu PPA". *Web Upd8*. Retrieved 28 April 2014.

[130] Knight, John (1 September 2011). "Tor Browser Bundle-Tor Goes Portable". *Linux Journal*. Retrieved 28 April 2014.

[131] Dredge, Stuart (5 November 2013). "What is Tor? A beginner's guide to the privacy tool". *The Guardian*. Retrieved 28 April 2014.

[132] "'Peeling back the layers of Tor with EgotisticalGiraffe' – read the document". *Guardian*. 4 October 2013.

[133] Samson, Ted (5 August 2013). "Tor Browser Bundle for Windows users susceptible to info-stealing attack". *InfoWorld*. Retrieved 28 April 2014.

[134] Poulsen, Kevin (8 May 2013). "Feds Are Suspects in New Malware That Attacks Tor Anonymity". *Wired*. Retrieved 29 April 2014.

[135] Owen, Gareth. "FBI Malware Analysis". Retrieved 6 May 2014.

[136] Best, Jessica (21 January 2014). "Man branded 'largest facilitator of child porn on the planet' remanded in custody again". *Daily Mirror*. Retrieved 29 April 2014.

[137] Dingledine, Roger (5 August 2013). "Tor security advisory: Old Tor Browser Bundles vulnerable". *Tor Project*. Retrieved 28 April 2014.

[138] Poulsen, Kevin (13 September 2013). "FBI Admits It Controlled Tor Servers Behind Mass Malware Attack". *Wired*. Retrieved 22 December 2013.

[139] Schneier, Bruce (4 October 2013). "Attacking Tor: how the NSA targets users' online anonymity". *The Guardian*. Retrieved 22 December 2013.

[140] "Visit the Wrong Website, and the FBI Could End Up in Your Computer". *WIRED*.

[141] "Tor". Vuze. Retrieved 3 March 2010.

[142] "Bitmessage FAQ". *Bitmessage*. Retrieved 17 July 2013.

[143] "About". *The Guardian Project*. Retrieved 10 May 2011.

[144] "ChatSecure: Private Messaging". *The Guardian Project*. Retrieved 20 September 2014.

[145] "Orbot: Mobile Anonymity + Circumvention". *The Guardian Project*. Retrieved 10 May 2011.

[146] "Orweb: Privacy Browser". *The Guardian Project*. Retrieved 10 May 2011.

[147] n8fr8 (2015-06-30). "Orfox: Aspiring to bring Tor Browser to Android". *guardianproject.info*. Retrieved 2015-08-17. Our plan is to actively encourage users to move from Orweb to Orfox, and stop active development of Orweb, even removing to from the Google Play Store.

[148] "ProxyMob: Firefox Mobile Add-on". *The Guardian Project*. Retrieved 10 May 2011.

[149] "Obscura: Secure Smart Camera". *The Guardian Project*. Retrieved 19 September 2014.

[150] Жуков, Антон (15 December 2009). "Включаем Tor на всю катушку" [Make Tor go the whole hog]. *Xakep*. Retrieved 28 April 2014.

55.9 Footnotes

- Anonymity Bibliography Retrieved: 21 May 2007

- Schneier, Bruce. *Applied Cryptography*. ISBN 0-471-11709-9.

- Schneier, Bruce. *Email Security*. ISBN 0-471-05318-X.

- Bacard, Andre. *Computer Privacy Handbook*. ISBN 1-56609-171-3.

55.10 External links

- Official website

  - Official blog

- Tor: Hidden Services and Deanonymisation [31c3] on YouTube

Chapter 56

Tox (protocol)

Tox is a free and open-source, peer-to-peer, encrypted instant messaging and video calling protocol. The stated goal of the project is to provide secure yet easily accessible communication for everyone.[1]

56.1 History

The initial commit to GitHub was pushed on June 23, 2013, by a user named irungentoo.[2] Pre-alpha testing binaries were made available for users from February 3, 2014, onwards, and nightly builds of Tox are published by the Jenkins Automatron.[3] On July 12, 2014, Tox entered alpha and a redesigned download page was created for the occasion.[4] The Tox Foundation took part in Google Summer of Code 2014 and 2015.

56.2 Features

Users are assigned a public and private key, and they connect to each other directly in a fully distributed, peer-to-peer network. Users have the ability to message friends, join chat rooms with friends or strangers, and send each other files. Everything is encrypted using the NaCl library.

The mainstream clients aim to provide support for messaging, group messaging, voice and video calling, voice and video conferencing, typing indicators, message read-receipts, file sharing, profile encryption, and desktop streaming. Additional features can be implemented by any client as long as they are supported by the core protocol. Features that are not related to the core networking system are left up to the client.

56.3 Architecture

56.3.1 Core

The Tox core is a library establishing the protocol and API. User front-ends, or clients, are built on the top of the core. Anyone can create a client utilizing the core. A technical report describing the design of the Core written by the core developer irungentoo and updated occasionally is available publicly as of August 2014 on the Tox Foundation's Jenkins.[5]

56.3.2 Protocol

The core of Tox is an implementation of the Tox protocol, an example of the application layer of the OSI model and arguably the presentation layer. Implementations of the Tox protocol not done by the project exist, an example of one being Xot.[6]

Tox uses the Opus lossy audio coding format for audio streaming and VP8 video compression format for video streaming.

56.3.3 Clients

A client is a program that uses the Tox core library to communicate with other users of the Tox protocol. Various clients are available for a wide range of systems; the following list is incomplete.[7]

There is also a Tox protocol plugins for Pidgin[15] and Miranda NG.[16]

56.4 Reception

Tox received some significant publicity in its early conceptual stage, catching the attention of global online tech news sites.[17][18][19][20] On August 15, 2013, Tox was number five on GitHub's top trending list.[21] Concerns about metadata leaks were raised, and developers responded by implementing Onion routing for the friend-finding process.[22] Tox was accepted into the Google Summer of Code as a Mentoring Organization in 2014 and 2015.[23][24]

56.5 See also

- Anonymous P2P

- RetroShare - RetroShare is free software for encrypted filesharing, serverless email, instant messaging, chatrooms, and BBS, based on a friend-to-friend network built on GPG (GNU Privacy Guard).

56.6 External links

- Official website

56.7 References

[1] "Secure Messaging for Everyone". *Tox*. Retrieved 6 August 2015.

[2] "Initial commit". GitHub. Retrieved 18 February 2014.

[3] "Dashboard [Jenkins]". Retrieved 11 March 2014.

[4] "The day we've all been waiting for". Tox Foundation. Retrieved 31 August 2014.

[5] "Technical Report" (PDF). Tox Foundation. Retrieved 31 August 2014.

[6] "Xot". GitHub. Retrieved 6 May 2014.

[7] "Client". *Tox*. Retrieved 6 August 2015.

[8] "Antidote". Retrieved 6 August 2015.

[9] "Antox". Tox-Wiki. Retrieved 6 August 2015.

[10] "Cyanide". OpenRepos.net. Retrieved 26 May 2015.

[11] "qTox". Tox-Wiki. Retrieved 6 August 2015.

[12] Papastamos, Dimitris. "ratox: FIFO based tox client". *2f30.org*. Retrieved 1 October 2014.

[13] "Toxic". Tox-Wiki. Retrieved 6 August 2015.

[14] "Tox". Tox-Wiki. Retrieved 6 August 2015.

[15] "tox-prpl – Tox Protocol Plugin For Pidgin". Retrieved 17 September 2015.

[16] "Tox protocol". *Miranda NG Official Community Forum*. watcher. Retrieved 17 September 2015.

[17] Kar, Saroj (5 August 2013). "Tox: A Replacement For Skype And Your Privacy?". *Silicon Angle*. Retrieved 19 February 2014.

[18] Grüner, Sebastian (30 July 2013). "Skype-Alternative Freier und sicherer Videochat mit Tox" [More free and secure video chat with Tox]. *Golem.de* (in German). Retrieved 19 February 2014.

[19] "Проект Tox развивает свободную альтернативу Skype" [Tox project develops free Skype replacement]. *opennet.ru* (in Russian). 30 July 2013. Retrieved 19 February 2014.

[20] Nitschke, Manuel (2 August 2013). "Skype-Alternative Tox zum Ausprobieren" [Tox Skype replacement tested]. *heise.de* (in German). Retrieved 19 February 2014.

[21] Asay, Matt (15 August 2013). "GitHub's new 'Trending' Feature Lets You See The Future". *ReadWrite.com*. Retrieved 19 February 2014.

[22] "Prevent_Tracking.txt". *GitHub*. Retrieved 20 February 2014.

[23] "Project Tox". *GSoC 2014*. Retrieved 7 March 2015.

[24] "Project Tox". *GSoC 2015*. Retrieved 7 March 2015.

Chapter 57

Transport Layer Security

Transport Layer Security (**TLS**) and its predecessor, **Secure Sockets Layer** (**SSL**), both of which are frequently referred to as 'SSL', are cryptographic protocols designed to provide communications security over a computer network.[1] Several versions of the protocols are in widespread use in applications such as web browsing, email, Internet faxing, instant messaging, and voice-over-IP (VoIP). Major web sites (including Google, Youtube, Facebook and many others) use TLS to secure all communications between their servers and web browsers.

The primary goal of the TLS protocol is to provide privacy and data integrity between two communicating computer applications. [1]:3 When secured by TLS, connections between a client (e.g. a web browser) and a server (e.g. wikipedia.org) will have one or more of the following properties:

- The connection is private because symmetric cryptography is used to encrypt the data transmitted. The keys for this symmetric encryption are generated uniquely for each connection and are based on a secret negotiated at the start of the session (see Handshake Protocol). The server and client negotiate the details of which encryption algorithm and cryptographic keys to use before the first byte of data is transmitted (see Algorithm). The negotiation of a shared secret is both secure (the negotiated secret is unavailable to eavesdroppers and cannot be obtained, even by an attacker who places himself in the middle of the connection) and reliable (no attacker can modify the communications during the negotiation without being detected).

- The identity of the communicating parties can be authenticated using public key cryptography. This authentication can be made optional, but is generally required for at least one of the parties (typically the server).

- The connection is reliable because each message transmitted includes a message integrity check using a message authentication code to prevent undetected loss or alteration of the data during transmission.[1]:3

In addition to the properties above, careful configuration of TLS can provide additional privacy-related properties such as forward secrecy, ensuring that any future disclosure of encryption keys cannot be used to decrypt any TLS communications recorded in the past.[2]

TLS supports many different methods for exchanging keys, encrypting data, and authenticating message integrity (see Algorithm). As a result, secure configuration of TLS involves many configurable parameters, and not all choices provide all of the privacy-related properties described in the list above (see authentication and key exchange table, cipher security table, and data integrity table).

Attempts have been made to subvert aspects of the communications security that TLS seeks to provide and the protocol has been revised several times to address these security threats (see Security). Web browsers have also been revised by their developers to defend against potential security weaknesses after these were discovered (see TLS/SSL support history of web browsers.)

The TLS protocol is composed of two layers: the TLS Record Protocol and the TLS Handshake Protocol.

TLS is a proposed Internet Engineering Task Force (IETF) standard, first defined in 1999 and updated in RFC 5246 (August 2008) and RFC 6176 (March 2011). It is based on the earlier SSL specifications (1994, 1995, 1996) developed by Netscape Communications[3] for adding the HTTPS protocol to their Navigator web browser.

Since early 2015, Wikipedia has used TLS to secure by default all access by readers and editors to its web servers.

57.1 Description

The TLS protocol allows client-server applications to communicate across a network in a way designed to prevent eavesdropping and tampering.

Since protocols can operate either with or without TLS (or SSL), it is necessary for the client to indicate to the server the setup of a TLS connection. There are two main ways of achieving this. One option is to use a different port number for TLS connections (for example, port 443 for HTTPS). The other is for the client to use a protocol-specific mechanism (for example, STARTTLS for mail and news protocols) to request that the server switch the connection to TLS.

Once the client and server have agreed to use TLS, they negotiate a stateful connection by using a handshaking procedure.[4] During this handshake, the client and server agree on various parameters used to establish the connection's security:

- The handshake begins when a client connects to a TLS-enabled server requesting a secure connection and presents a list of supported cipher suites (ciphers and hash functions).

- From this list, the server picks a cipher and hash function that it also supports and notifies the client of the decision.

- The server usually then sends back its identification in the form of a digital certificate. The certificate usually contains the server name, the trusted certificate authority (CA) and the server's public encryption key.

- The client may contact the server that issued the certificate (the trusted CA as above) and confirm the validity of the certificate before proceeding.

- In order to generate the session keys used for the secure connection, the client encrypts a random number with the server's public key and sends the result to the server. Only the server should be able to decrypt it, with its private key.

- From the random number, both parties generate a 'master secret' and then negotiate a session key for encryption and decryption.

This concludes the handshake and begins the secured connection, which is encrypted and decrypted with the key material until the connection closes. If any one of the above steps fail, the TLS handshake fails, and the connection is not created.

The Internet Protocol Suite places TLS and SSL as tools into the application layer, while the OSI model characterizes them as being initialized in Layer 5 (session layer) and operating in Layer 6 (presentation layer). The session layer employs a handshake using an asymmetric cipher in order to establish cipher settings and a shared key for a session; the presentation layer encrypts the rest of the communication

using a symmetric cipher and the session key. TLS and SSL may be characterized to work on behalf of the underlying transport layer protocol, which carries encrypted data.

57.2 History and development

57.2.1 Secure Network Programming

Early research efforts towards transport layer security included the Secure Network Programming (SNP) application programming interface (API), which in 1993 explored the approach of having a secure transport layer API closely resembling Berkeley sockets, to facilitate retrofitting preexisting network applications with security measures.[5]

57.2.2 SSL 1.0, 2.0 and 3.0

Netscape developed the original SSL protocols.[6] Version 1.0 was never publicly released because of serious security flaws in the protocol; version 2.0, released in February 1995, "contained a number of security flaws which ultimately led to the design of SSL version 3.0".[7] SSL version 3.0, released in 1996, represented a complete redesign of the protocol, produced by Paul Kocher working with Netscape engineers Phil Karlton and Alan Freier, with a reference implementation by Christopher Allen and Tim Dierks of Consensus Development. Newer versions of SSL/TLS are based on SSL 3.0. The 1996 draft of SSL 3.0 was published by IETF as a historical document in RFC 6101.

Dr. Taher Elgamal, chief scientist at Netscape Communications from 1995 to 1998, is recognized as the "father of SSL".[8][9]

As of 2014 the 3.0 version of SSL is considered insecure as it is vulnerable to the POODLE attack that affects all block ciphers in SSL; and RC4, the only non-block cipher supported by SSL 3.0, is also feasibly broken as used in SSL 3.0.[10]

SSL 2.0 is deprecated in RFC 6176.

SSL 3.0 is deprecated in RFC 7568.

57.2.3 TLS 1.0

TLS 1.0 was first defined in RFC 2246 in January 1999 as an upgrade of SSL Version 3.0, and written by Christopher Allen and Tim Dierks of Consensus Development. As stated in the RFC, "the differences between this protocol and SSL 3.0 are not dramatic, but they are significant enough to preclude interoperability between TLS 1.0 and SSL 3.0". TLS 1.0 does include a means by which a TLS implementation can downgrade the connection to SSL 3.0, thus weakening security.[11]:1–2

57.2.4   TLS 1.1

TLS 1.1 was defined in RFC 4346 in April 2006.[12] It is an update from TLS version 1.0. Significant differences in this version include:

- Added protection against cipher-block chaining (CBC) attacks.

  - The implicit initialization vector (IV) was replaced with an explicit IV.
  - Change in handling of padding errors.

- Support for IANA registration of parameters.[11]:2

57.2.5   TLS 1.2

TLS 1.2 was defined in RFC 5246 in August 2008. It is based on the earlier TLS 1.1 specification. Major differences include:

- The MD5-SHA-1 combination in the pseudorandom function (PRF) was replaced with SHA-256, with an option to use cipher suite specified PRFs.

- The MD5-SHA-1 combination in the finished message hash was replaced with SHA-256, with an option to use cipher suite specific hash algorithms. However the size of the hash in the finished message is still truncated to 96 bits.

- The MD5-SHA-1 combination in the digitally signed element was replaced with a single hash negotiated during handshake, which defaults to SHA-1.

- Enhancement in the client's and server's ability to specify which hash and signature algorithms they will accept.

- Expansion of support for authenticated encryption ciphers, used mainly for Galois/Counter Mode (GCM) and CCM mode of Advanced Encryption Standard encryption.

- TLS Extensions definition and Advanced Encryption Standard cipher suites were added.[11]:2

All TLS versions were further refined in RFC 6176 in March 2011 removing their backward compatibility with SSL such that TLS sessions will never negotiate the use of Secure Sockets Layer (SSL) version 2.0.

57.2.6   TLS 1.3 (draft)

As of September 2015, TLS 1.3 is a working draft, and details are provisional and incomplete.[13][14] It is based on the earlier TLS 1.2 specification. Major differences from TLS 1.2 include:

- Removed support for weak and lesser used named curves

- Removed support for MD5 and SHA-224 hashes with signatures

- Required digital signatures even when a previous configuration is used

- Integration of HKDF and the semi-ephemeral DH proposal

- Replacement of resumption with PSK and tickets

- Relegation of ClientKeyShare to an appendix

- Support for 1-RTT handshakes and initial support for 0-RTT

- Dropped support for many insecure or obsolete features including, compression, renegotiation, non-AEAD ciphers, static RSA and DH key exchange, custom DHE groups, point format negotiation, Change Cipher Spec protocol, Hello message UNIX time, and the length field AD input to AEAD ciphers

- Prohibition of SSL or RC4 negotiation for backwards compatibility

- Integrated usage of session hash

- The record layer version number has been frozen and deprecated for improved backwards compatibility

- Movement of some security related algorithm details from an appendix to the specification

57.3   Digital Certificates

A digital certificate certifies the ownership of a public key by the named subject of the certificate. This allows others (relying parties) to rely upon signatures or on assertions made by the private key that corresponds to the certified public key.

57.3.1 Certificate Authorities

In this model of trust relationships, a CA is a trusted third party - trusted both by the subject (owner) of the certificate and by the party relying upon the certificate.

According to Netcraft, who monitor active TLS certificates, the market-leading CA has been Symantec since the beginning of their survey (or VeriSign before the authentication services business unit was purchased by Symantec). Symantec currently accounts for just under a third of all certificates and 44% of the valid certificates used by the 1 million busiest websites, as counted by Netcraft.[15]

As a consequence of choosing X.509 certificates, certificate authorities and a public key infrastructure are necessary to verify the relation between a certificate and its owner, as well as to generate, sign, and administer the validity of certificates. While this can be more beneficial than verifying the identities via a web of trust, the 2013 mass surveillance disclosures made it more widely known that certificate authorities are a weak point from a security standpoint, allowing man-in-the-middle attacks (MITM).[16][17]

57.4 Algorithm

See also: Cipher suite

57.4.1 Key exchange or key agreement

Before a client and server can begin to exchange information protected by TLS, they must securely exchange or agree upon an encryption key and a cipher to use when encrypting data (see Cipher). Among the methods used for key exchange/agreement are: public and private keys generated with RSA (denoted TLS_RSA in the TLS handshake protocol), Diffie-Hellman (TLS_DH), ephemeral Diffie-Hellman (TLS_DHE), Elliptic Curve Diffie-Hellman (TLS_ECDH), ephemeral Elliptic Curve Diffie-Hellman (TLS_ECDHE), anonymous Diffie-Hellman (TLS_DH_anon),[1] pre-shared key (TLS_PSK)[18] and Secure Remote Password (TLS_SRP).[19]

The TLS_DH_anon and TLS_ECDH_anon key agreement methods do not authenticate the server or the user and hence are rarely used because those are vulnerable to Man-in-the-middle attack. Only TLS_DHE and TLS_ECDHE provide forward secrecy.

Public key certificates used during exchange/agreement also vary in the size of the public/private encryption keys used during the exchange and hence the robustness of the security provided. In July 2013, Google announced that it would no longer use 1024 bit public keys and would switch instead to 2048 bit keys to increase the security of the TLS encryption it provides to its users.[20]

57.4.2 Cipher

See also: Cipher suite, Block cipher and Cipher security summary

Notes

[1] RFC 5746 must be implemented in order to fix a renegotiation flaw that would otherwise break this protocol.

[2] If libraries implement fixes listed in RFC 5746, this will violate the SSL 3.0 specification, which the IETF cannot change unlike TLS. Fortunately, most current libraries implement the fix and disregard the violation that this causes.

[3] the BEAST attack breaks all block ciphers (CBC ciphers) used in SSL 3.0 and TLS 1.0 unless mitigated by the client and/or the server. See #Web browsers.

[4] The POODLE attack breaks all block ciphers (CBC ciphers) used in SSL 3.0 unless mitigated by the client and/or the server. See #Web browsers.

[5] AEAD ciphers (such as GCM and CCM) can be used in only TLS 1.2.

[6] CBC ciphers can be attacked with the Lucky Thirteen attack if the library is not written carefully to eliminate timing side channels.

[7] Although the key length of 3DES is 168 bits, effective security strength of 3DES is only 112 bits,[29] which is below the recommended minimum of 128 bits.[30]

[8] IDEA and DES have been removed from TLS 1.2.[31]

[9] 40 bits strength of cipher suites were designed to operate at reduced key lengths in order to comply with US regulations about the export of cryptographic software containing certain strong encryption algorithms (see Export of cryptography from the United States). These weak suites are forbidden in TLS 1.1 and later.

[10] Use of RC4 in all versions of TLS is prohibited by RFC 7465. (Due RC4 attacks weaken or break RC4 used in SSL/TLS)

[11] authentication only, no encryption

57.4.3 Data integrity

Message authentication code (MAC) is used for data integrity. HMAC is used for CBC mode of block ciphers and stream ciphers. AEAD is used for Authenticated encryption such as GCM mode and CCM mode.

57.5 Applications and adoption

In applications design, TLS is usually implemented on top of any of the Transport Layer protocols, encapsulating the application-specific protocols such as HTTP, FTP, SMTP, NNTP and XMPP. Historically it has been used primarily with reliable transport protocols such as the Transmission Control Protocol (TCP). However, it has also been implemented with datagram-oriented transport protocols, such as the User Datagram Protocol (UDP) and the Datagram Congestion Control Protocol (DCCP), usage which has been standardized independently using the term Datagram Transport Layer Security (DTLS).

57.5.1 Websites

A prominent use of TLS is for securing World Wide Web traffic between the website and the browser carried by HTTP to form HTTPS.[33] Notable applications are electronic commerce and asset management.

Notes

[1] see #Cipher table below

[2] see #Web browsers and #Attacks against TLS/SSL sections

57.5.2 Web browsers

Further information: Comparison of web browsers

As of September 2015, the latest versions of all major web browsers support TLS 1.0, 1.1, and 1.2, and have them enabled by default. However, not all supported Microsoft operating systems support the latest version of IE. Additionally many operating systems currently support multiple versions of IE, but this will change according to Microsoft's Internet Explorer Support Lifecycle Policy FAQ, "beginning January 12, 2016, only the most current version of Internet Explorer available for a supported operating system will receive technical support and security updates." The page then goes on to list the latest supported version of IE at that date for each operating system. The next critical date would be when an operating system reaches the end of life stage, which is in Microsoft's Windows lifecycle fact sheet.

There are still problems on several browser versions:

- TLS 1.1 and 1.2 supported, but disabled by default: Internet Explorer (8–10 for Windows 7 / Server 2008 R2, 10 for Windows 8 / Server 2012, IE Mobile 10 for Windows Phone 8)

- TLS 1.1 and 1.2 not supported: Internet Explorer (7–9 for Windows Vista / Server 2008)

Mitigations against known attacks are not enough yet:

- Mitigations against POODLE attack: Some browsers already prevent fallback to SSL 3.0; however, this mitigation needs to be supported by not only clients, but also servers. Disabling SSL 3.0 itself, implementation of "anti-POODLE record splitting", or denying CBC ciphers in SSL 3.0 is required.

  - Google Chrome: Complete (TLS_FALLBACK_SCSV is implemented since version 33, fallback to SSL 3.0 is disabled since version 39, SSL 3.0 itself is disabled by default since version 40. Support of SSL 3.0 itself was dropped since version 44.)

  - Mozilla Firefox: Complete (SSL 3.0 itself is disabled by default and fallback to SSL 3.0 are disabled since version 34, TLS_FALLBACK_SCSV is implemented since version 35. In ESR, SSL 3.0 itself is disabled by default and TLS_FALLBACK_SCSV is implemented since ESR 31.3. Support of SSL 3.0 itself is dropped since version 39.)

  - Internet Explorer: Partial (Only in version 11, SSL 3.0 is disabled by default since April 2015. Version 10 and older are still vulnerable against POODLE.)

  - Opera: Complete (TLS_FALLBACK_SCSV is implemented since version 20, "anti-POODLE record splitting", which is effective only with client-side implementation, is implemented since version 25, SSL 3.0 itself is disabled by default since version 27. Support of SSL 3.0 itself will be dropped since version 31.)

  - Safari: Complete (Only on OS X 10.8 and later and iOS 8, CBC ciphers during fallback to SSL 3.0 is denied, but this means it will use RC4, which is not recommended as well. Support of SSL 3.0 itself is dropped on OS X 10.11 and later and iOS 9.)

- Mitigation against RC4 attacks:

  - Internet Explorer for Windows 7 / Server 2008 R2 and for Windows 8 / Server 2012 have set the priority of RC4 to lowest.

  - Google Chrome disabled RC4 except as a fallback since version 43. Chrome is expected to disable RC4 completely in early 2016.

  - Opera disabled RC4 except as a fallback since version 30.

- Firefox disabled RC4 except as a fallback since version 36. Firefox is expected to allow offering RC4 cipher-suites only for hosts which only support RC4 cipher suites and are registered on the whitelist by Mozilla since version 43, and to disable RC4 completely since version 44.

- Internet Explorer 11 for Windows 8.1 / Server 2012 R2 and Mobile 11 for Windows Phone 8.1, and Edge for Windows 10 disable RC4 except as a fallback if no other enabled algorithm works (Internet Explorer for Windows 7 / Server 2008 R2 and for Windows 8 / Server 2012 can also disable RC4 except as a fallback through registry settings). Edge and IE 11 are expected to disable RC4 completely in early 2016.

- Mitigation against FREAK attack:

  - The Android Browser of Android 4 and older are still vulnerable to the FREAK attack.

  - Currently supported versions of Internet Explorer Mobile are still vulnerable to the FREAK attack.

  - Google Chrome, Internet Explorer (desktop), Safari (desktop & mobile), and Opera (mobile) have FREAK mitigations in place.

  - Mozilla Firefox on all platforms and Google Chrome on Windows were not affected by FREAK.

- view

- talk

- edit

Notes

[1] Does the browser have mitigations or is not vulnerable for the known attacks. Note actual security depends on other factors such as negotiated cipher, encryption strength etc (see #Cipher table).

[2] Whether a user or administrator can choose the protocols to be used or not. If yes, several attacks such as BEAST (vulnerable in SSL 3.0 and TLS 1.0) or POODLE (vulnerable in SSL 3.0) can be avoided.

[3] Whether EV SSL and DV SSL (normal SSL) can be distinguished by indicators (green lock icon, green address bar, etc.) or not.

[4] e.g. 1/n-1 record splitting.

[5] e.g. Disabling header compression in HTTPS/SPDY.

[6]
- Complete mitigations; disabling SSL 3.0 itself, "anti-POODLE record splitting". "Anti-POODLE record splitting" is effective only with client-side implementation and valid according to the SSL 3.0 specification, however, it may also cause compatibility issues due to problems in server-side implementations.

- Partial mitigations; disabling fallback to SSL 3.0, TLS_FALLBACK_SCSV, disabling cipher suites with CBC mode of operation. If the server also supports TLS_FALLBACK_SCSV, the POODLE attack will fail against this combination of server and browser, but connections where the server does not support TLS_FALLBACK_SCSV and does support SSL 3.0 will still be vulnerable. If disabling cipher suites with CBC mode of operation in SSL 3.0, only cipher suites with RC4 are available, RC4 attacks become easier.

- When disabling SSL 3.0 manually, POODLE attack will fail.

[7]
- Complete mitigation; disabling cipher suites with RC4.

- Partial mitigations to keeping compatibility with old systems; setting the priority of RC4 to lower.

[8] Google Chrome (and Chromium) supports TLS 1.0, and TLS 1.1 from version 22 (it was added, then dropped from version 21). TLS 1.2 support has been added, then dropped from Chrome 29.[42][43][44]

[9] Uses the TLS implementation provided by BoringSSL for Android, OS X, and Windows[45] or by NSS for Linux. Google is switching the TLS library used in Chrome to BoringSSL from NSS completely.

[10] configure enabling/disabling of each protocols via setting/option (menu name is dependent on browsers)

[11] configure the maximum and the minimum version of enabling protocols with command-line option

[12] TLS_FALLBACK_SCSV is implemented.[53] Fallback to SSL 3.0 is disabled since version 39.[54]

[13] In addition to TLS_FALLBACK_SCSV and disabling a fallback to SSL 3.0, SSL 3.0 itself is disabled by default.[54]</ref> in RFC 5746, this will violate the SSL 3.0 specification, which the IETF cannot change unlLowest priority tely, most current libraries implement the fix and disregard the violation that thisVulnerable (except Windows) [#BEAST attack|BEAST]] attack breaks all block ciphers (CBC ciphers) used in SSL 3.0Vulnerable 0 unless mitigated by the client and/or the server. See [[#Web browsers]].</ref><ref Yes[n 14]

[14] configure the minimum version of enabling protocols via chrome://flags[58] (the maximum version can be configured with command-line option)</ref> <b /><ref group="n" name="rfc5746"/><ref group 41, 42 e="BEAST"/> ! TLS 1.1
<ref group="n" name="rfc5746"/> ! TLS

1.2
<ref group=No name="rfc5746"/> ! TLS 1.3
<small>(Draft)</small> |- !   rowspan="14"|[[Block cipDisabled by default [Block cipher mode of operation|mode of operation]] ![[Advanced Encryption Standard|AYes   [[Galois/Counter   Model|GCM]]<ref name=aes-gcm>RFC   5288,   RFC   5289</ref><ref group="n" Yes ="aead">[[AEAD block cipher modes of operation|AEAD]] ciphers (such as [[Galois/CountYes ode|GCM]] and [[CCM mode|CCM]]) can be used in only TLS 1.2.</ref> | rowspan="3"| 256Yes

(only   desktop) || {{N/a}} || {{Good|Secure}} || {{Good|Secure}} || rowspan="9"| Defined for TLS 1.2 Yes FCs |- ![[Advanced Encryption Standard|AES]] [[CCM   mode|CCM]]<ref name=aes-ccm>RFC 6655, needs ECC compatible OS[39]   {{N/a}} || {{N/a}} || {{N/a}} || {{Good|Secure}} || {{Good|Secure}} |- ![[Advanced EnNot affected dard|AES]] [[Cipher block chaining|CBC]]<ref group="n" name="Lucky13"/> | {{N/a}} || Mitigated {{Depends|Depends on mitigations}} || {{Good|Secure}} || {{Good|Secure}} || {{N/a}} Mitigated llia (cipher)|Camellia]] [[Galois/Counter Mode|GCM]]<ref name=camellia-gcm>RFC   6367</ref><Lowest   priority me="aead"/> | rowspan="2"| 256, 128 | {{N/a}} || {{N/a}} || {{N/a}} || {{N/a}} || {{GMitigated }} || {{Good|Secure}} |- ![[Camellia (cipher)|Camellia]] [[Cipher block chaining|CBCVulnerable =camellia-cbc>RFC 5932, RFC 6367</ref><ref group="n" name="Lucky13"/> | {{N/a}} || {{Yes[n 14] ns } || {{Good|Secure}} || {{Good|Secure}} || 43 /a}} |- ![[ARIA (cipher)|ARIA]] [[Galois/Counter   Mode|GCM]]<ref   name=aria/><ref groupNo " name="aead"/> | rowspan="2"| 256, 128 | {{N/a}} || {{N/a}} || {{N/a}} || {{N/a}} ||Disabled by default {{Good|Secure}} |- ![[ARIA (cipher)|ARIA]] [[Cipher   block   chaining|CBC]]<ref   name=ariYes C 6209</ref><ref group="n" name="Lucky13"/> | {{N/a}} || {{N/a}} || {{Depends|DependsYes mitigations}} || {{Good|Secure}} || {{Good|Secure}} || {{N/a}} |- ![[SEED (cipher)|SEYes [[Cipher block chaining|CBC]]<ref name="seed-cbc">RFC   4162</ref><ref group="n" name=Yes

(only desktop) || {{Depends|Depends on mitigations}} || {{Good|Secure}} || {{Good|Secure}} || {{N/aYes - ! [[Triple DES|3DES EDE]] [[Cipher block chaining|CBC]]<ref group="n" name="Lucky13">CBCneeds ECC compatible OS[39] k]] if the library is not written carefully to eliminate timing side channels.</ref> Not affected roup="n"|name="3des"|Although the key length of 3DES is 168 bits, effective security Mitigated f 3DES is only 112 bits,<ref name=NIST_SP_800-57>{{cite web|url=http://csrc.nist.gov/Mitigated   ns/nistpubs/800-57/sp800-57-Part1-revised2_Mar08-2007.pdf|title=NIST Special Publication 8Only as fallback
[n 15][59]   w.ssllabs.com/projects/best-practices/index. html|title=SSL/TLS   Deployment   Best   PractiMitigated =Qualys SSL Labs|accessdate=2 June 2015}}</ref>}} | {{Bad|Insecure}} || {{Bad|InsecVulnerable   Depends|Low strength, Depends on mitigations}} || {{Depends|Low   strength}} || {{DepenYes[n 14]   T 8147-89]] [[Block cipher mode 44, 45 tion#Counter

(CTR)|CNT]]<ref na 46 ostlink/> | 256 | {{N/a}} || {{N/a}} || {{Good|Secure}} || {{Good|Secure}} || {{Good|No ure}} || || Proposed in RFC drafts |- ![[International Data Encryption Algorithm|IDENo[60] ations}} || {{Good|Secure}} || {{N/a}} || {{N/a}} || rowspan="2"| Removed from TLS 1.Yes !rowspan="2"| [[Data Encryption Standard|DES]] [[Cipher block chaining|CBC]]<ref groYes n" name="Lucky13"/><ref group="n" name="removal_from_tls1.2"/> | {{0}}56 | {{Bad|InseYes }} || {{Bad|Insecure}} || {{Bad|Insecure}} || {{Bad|Insecure}} ||{{N/a}} || {{N/a}} Yes

(only desktop) T">40 bits strength of cipher suites were designed to operate at reduced key lengths Yes rder to comply with US regulations about the export of cryptographic software containing cneeds ECC compatible OS[39] ptography from the United States]]). These weak suites are forbidden in TLS 1.1 and lNot affected {{Bad|Insecure}} || {{Bad|Insecure}} || {{Bad|Insecure}} || {{N/a}} || {{N/a}} || {{Mitigated owspan="2"| Forbidden in TLS 1.1 and later |- ![[RC2]] [[Cipher block chaining|CBC]]<Not affected name="Lucky13"/> | {{0}}40<ref group="n" name="EXPORT"/> | {{Bad|Insecure}} || {{Bad|InseOnly as fallback
[n 15] }} |- ! rowspan="3"|[[Stream cipher]] ![[ChaCha20]]-[[Poly1305]]<ref   name="chacha20poMitigated /tools.ietf.org/html/draft-ietf-tls-chacha20-poly1305 draft-ietf-tls-chacha20-poly130Mitigated[61]   span="2"| [[RC4]]<ref group="n" name="RC4">Use of RC4 in all versions of TLS is prohibitedTemporary
[n 11] n r break RC4 use Google Android OS Browser
[62] Insecure}} || {{Bad|Insecure}} || {{Bad|In Android 1.0, 1.1, 1.5, 1.6, 2.0–2.1, 2.2–2.2.3 {{Bad|Insecure}} || {{Bad|Insecure}} || {{Bad|Insecure}} || {{Bad|Insecure}} || DefinNo for TLS 1.2 in RFCs |} ;Notes {{reflist|group="n"}}   ===Data integrity=== [[Message Enabled by default]] (MAC) is used for data integrity. [[HMAC]] is used for [[Cipher block chaining|CBCYes ode of block ciphers and stream ciphers. [[AEAD block cipher modes of operation|AEADNo is used for [[Authenticated encryption]] such as [[Galois/Counter Mode|GCM mode]] anNo [CCM mode]]. {{Anchor|integrity-table}} {| class="wikitable" style="text-align:center" |+ Data integrity ! Algorithm !! SSL 2.0 !! SSL 3.0 Unknown .0 !! TLS 1.1 !! TLS 1.2 !! TLS 1.3
<small>(Draft)</small> !! Status |- ! [[HMANo -[[MD5]] | {{Yes}} || {{Yes}} || {{Yes}} || {{Yes}} || {{Yes}} || || rowspan="4" | DNo ned for TLS 1.2 in RFCs |- ! [[HMAC]]-[[SHA-1|SHA1]] | {{No}} || {{Yes}} || {{Yes}} || {{Yes}} || {{Yes}} || |- !   [[HMAC]]-[[SHA-2|SHA256/3Unknown {No}} || {{No}} || {{No}} || {{No}} || {{Yes}} || |- ! [[AEAD block cipher modes of operation|AEAD]] | {{No}} || {{No}} || {{No}} || {{No}}Unknown s}} || |- ! [[GOST 28147-89|GOST 28147-89 IMIT]]<ref name=gostlink/> | {{No}} || {{NVulnerable s}} || {{Yes}} || {{Yes}} || || rowspan="2" | Proposed in RFC drafts |- ! [[GOST (haVulnerable)|GOST R 34.11-94]]<ref name=gostlink/> | {{No}} || {{No}} || {{Yes}} || {{Yes}} || Vulnerable |} ==Applications and adoption== In applications design, TLS is usually implementedVulnerable any of

the [[Transport Layer]] protocols, encapsulating the application-specific proNo ol such as [[Hypertext Transfer Protocol|HTT Android 2.3–2.3.7, 3.0–3.2.6, 4.0–4.0.4, 4.1–4.3.1 ence Protocol|XMPP]]. Historically it has been used primarily with reliable transportNo otocols such as the [[Transmission Control Protocol]] (TCP). However, it has also beEnabled by default datagram-oriented transport protocols, such as the [[User Datagram Protocol]] (UDP) Yes the [[Datagram Congestion Control Protocol]] (DCCP), usage which has been standardizNo independently using the term [[Datagram Transport Layer Security]] (DTLS). ===WebsiNo === A prominent use of TLS is for securing [[World Wide Web]] traffic between the [[website]] and the [[web browser|browser]] carried by HTUnknown rm [[https|HTTPS]].<ref>{{cite web |url=https://www.instantssl.com/ssl-certificate-prYes[38] e= HTTP VS HTTPS |accessdate=2015-02-12}}</ref> Notable applications are [[electronic commsince Android OS 3.0[63] ember 3, 2015. {{cite web|url=https://www.trustworthyinternet.org/ssl-pulse/|title=SSL Pulse: Survey of the SSL Implementation of the Most Unknown Web Sites|accessdate=2015-09-10}}</ref> !Security<ref name="trustworthy_ssl_pulse"/><ref name="community.qualys">{{cite web|url=https://comUnknown ualys.com/blogs/securitylabs/2013/03/19/rc4-in-tls-is-broken-now-what|accessdate=201Vulnerable lisher=Qualsys Security Labs|author=ivanr|title=RC4 in TLS is Broken: Now What?}}</rVulnerable 2.0 |10.8% (−0.4%) |{{Bad|Insecure}} |- !SSL 3.0 |33.8% (−1.2%) |{{Bad|Insecure<refVulnerable le_pdf" />}} |- !TLS 1.0 |99.2% (−0.1%) |{{Depends|Depends on cipher<ref group="n" nVulnerable s">see [[#Cipher]] table below</ref> and client mitigations<ref group="n" name="mitiNo io s">see [[#Web browsers]] and [[#Attacks a Android 4.4–4.4.4 TLS 1.1 |64.3% (+1.4%) |{{Depends|Depends on cipher<ref group="n" name="ciphers"/> anNo lient mitigations<ref group="n" name="mitigations"/>}} |- !TLS 1.2 |66.5% (+1.4%) |{En-abled by default cipher<ref group="n" name="ciphers"/> and client mitigations<ref group="n" name="mitYes ions"/>}} |- !TLS 1.3
<small>(Draft)</small> | {{N/a}} | |} ;Notes {{reflist|group="nDisabled by default === {{Further|Comparison of web browsers}} {{As of|2015|09}}, the latest versions of all mDisabled by default pport TLS 1.0, 1.1, and 1.2, and have them enabled by default. However, not all supported Microsoft operating systems support the latest veUnknown IE. Additionally many operating systems currently support multiple versions of IE, bYes his will change according to Microsoft's [https://support.microsoft.com/en-us/gp/micrYes[39] er Internet Explorer Support Lifecycle Policy FAQ], "beginning January 12, 2016, only the most current version of Internet Explorer availabUnknown supported operating system will receive technical support and security updates." The page then goes on to list the latest supported versiUnknown at that date for each operating system. The next critical date would be when an opVulnerable tem reaches the end of life stage, which is in Microsoft's [http://windows.microsoftVulnerable windows/lifecycle Windows lifecycle fact sheet]. There are still problems on severaVulnerable ersions: * TLS 1.1 and 1.2 supported, but disabled by default: Internet Explorer (8–Vulnerable ows 7 / Server 2008 R2, 10 for Windows 8 / Server 2012, IE Mobile 10 for Windows PhoNo 8) * TLS 1.1 and 1.2 not supported: Internet Android 5.0-5.0.2 r 2008) Mitigations against known attacks are not enough yet: * Mitigations against No POODLE attack|POODLE attack]]: Some browsers already prevent fallback to SSL 3.0; hoEnabled by default ion needs to be supported by not only clients, but also servers. Disabling SSL 3.0 iYes f, implementation of "anti-POODLE record splitting", or denying CBC ciphers in SSL 3.Yes[64] itself was dropped since version 44.) ** Mozilla Firefox: Complete (SSL 3.0 itself iYes[64] lback to SSL 3.0 are disabled since version 34, TLS_FALLBACK_SCSV is implemented since version 35. In ESR, SSL 3.0 itself is disabled by deUnknown d TLS_FALLBACK_SCSV is implemented since ESR 31.3. Support of SSL 3.0 itself is droppYes ince version 39.) ** Internet Explorer: Partial (Only in version 11, SSL 3.0 is disabYes by default since April 2015. Version 10 and older are still vulnerable against POODLE.) ** Opera: Complete (TLS_FALLBACK_SCSV is implementeUnknown version 20, "anti-POODLE record splitting", which is effective only with client-side implementation, is implemented since version 25, SSL 3Unknown f is disabled by default since version 27. Support of SSL 3.0 itself will be dropped-Vulnerable ion 31.) ** Safari: Complete (Only on OS X 10.8 and later and iOS 8, CBC ciphers durVulnerable k to SSL 3.0 is denied, but this means it will use RC4, which is not recommended as Vulnerable rt of SSL 3.0 itself is dropped on OS X 10.11 and later and iOS 9.) * Mitigation agaVulnerable attacks|RC4 attacks]]: ** Internet Explorer for Windows 7 / Server 2008 R2 and forNo nd ws 8 / Server 2012 have set the priority Android 5.1-5.1.1 led RC4 except as a fallback since version 43. Chrome is expected to disable RC4 compNo ely in early 2016. ** Opera disabled RC4 except as a fallback since version 30. ** FiNo ox disabled RC4 except as a fallback since version 36. Firefox is expected to allow oYes ing RC4 cipher-suites only for hosts which only support RC4 cipher suites and are regYes red on the whitelist by Mozilla since version 43, and to disable RC4 completely sinceYes sion 44. ** Internet Explorer 11 for Windows 8.1 / Server 2012 R2 and Mobile 11 for Windows Phone 8.1, and Edge for Windows 10 disable RC4 Unknown s a fallback if no other enabled algorithm works (Internet Explorer for Windows 7 / SYes r 2008 R2 and for Windows 8 / Server 2012 can also disable RC4 except as a fallback tYes gh registry settings). Edge and IE 11 are expected to disable RC4 completely in early 2016. * Mitigation against [[#FREAK|FREAK attack]]: *Unknown droid Browser of Android 4 and older are still vulnerable to the FREAK attack. ** Currently supported versions of Internet Explorer Mobile Unknown l vulnerable to the FREAK attack. ** Google Chrome, Internet Explorer (desktop), SafaNot affected mobile), and Opera (mobile) have FREAK mitigations in place. ** Mozilla Firefox on Vulnerable ms and Google Chrome on Windows were not affected by FREAK. {{clear}} {{TLS/SSL suppMitigated y of web browsers}} ===Libraries=== {{main|Comparison of TLS implementations}} MostVulnerable S programming libraries are [[free and open source software]]. * [[Botan (programminNo ib ary)|Botan]], a BSD-licensed cryptographic Android 6.0]: a portable open source cryptography library (includes TLS/SSL implementation) * [[No phi (programming language)|Delphi]] programmers may use a library called [[Internet DNo ctl|Indy]] which utilizes [[OpenSSL]]. * [[GnuTLS]]: a free implementation (LGPL licenYes * [[Java Secure Socket Extension]]: a [[Java (programming language)|Java]] implementYes n included in the [[Java Runtime Environment]] supports TLS 1.1 and 1.2 from Java 7, Yes ough is disabled by default for client, and enabled by default for server.<ref>{{cite web |author=Oracle |url=http://docs.oracle.com/javaseUnknown technotes/guides/security/SunProviders.html|title=Java Cryptography Architecture OracYes roviders Documentation |accessdate=2012-08-16}}</ref> Java 8 supports TLS 1.1 and 1.2Yes bled on both the client and server by default.<ref>{{cite web |author=Oracle |url=//docs.oracle.com/javase/8/docs/technotes/guides/securityUnknown ments-8.html|title=JDK 8 Security Enhancements |accessdate=2015-02-25}}</ref> * [[LibreSSL]]: a fork of OpenSSL by OpenBSD project. * [[MatUnknown : a dual licensed implementation * [[mbed TLS]] (previously PolarSSL): A tiny SSL libNot affected tation for embedded devices that is designed for ease of use * [[Network Security Services]]: [[FIPS 140]] validated open source library * Unknown L]]: a free implementation (BSD license with some extensions) * [[Security Support PrMitigated erface|SChannel]]: an implementation of SSL and TLS [[Microsoft Windows]] as part of its package. * [[Secure Transport]]: an implementatioUnknown and TLS used in [[OS X]] and [[iOS]] as part of their packages. * [[wolfSSL]] (previously CyaSSL): Embedded SSL/TLS Library with a strong Unknown s e Browser ze. {| class=" Version " Platforms SSL 2.0 (insecure) | SSL 3.0 (insecure) TLS 1.0 i TLS 1.1 ! TLS 1.2 (EV certificate 1 SHA-2 certificate 1 ECDSA certificate s BEAST r CRIME m POODLE (SSLv3) n RC4 g FREAK Logjam | Protocol selection by user |N }}<ref name="Bot Mozilla Firefox (Firefox for mobile)

[n 16] itle=[gnutls-devel] GnuTLS 3.4.0 released| 1.0 2015-04-08|acces Windows (XP SP2+)

OS X (10.6+)

Linux

Android (2.3+)

iOS (preview)

Firefox OS

Maemo

ESR only for:

Windows (XP SP2+)

OS X (10.6+)

Linux   e=2014-11-01|accessdate=2015-01-20}}</ref> | {{yes|No}}<ref name=libressl-2.3>{{citeEnabled by default

[65] {cref2|group=protocollibrary-table|a}} | {{yes|Disabled by default at compile time}}Enabled by default

[65]   xssl.org/news.html|title=MatrixSSL   -

News|accessdate=2014-11-09}}</ref> | {{yes}} | {Yes[65]] (previously PolarSSL) | {{Yes|No}} | {{No|Enabled by default}} | {{yes}} | {{yes}}No {{yes}} | |- ! [[Network Security Services]] | {{Yes|Disabled by default}}{{cref2|grNo =protocollibrary-table|a}} | {{Yes|Disabled by default}}<ref name=NSS-3.19>{{cite weNo rl=https://developer.mozilla.org/en-US/docs/Mozilla/Projects/NSS/NSS_3.19_release_notYes[38] er Network|title=NSS 3.19 release notes|publisher=Mozilla|accessdate=2015-05-06}}</rNo | {{yes}} | {{yes}}<ref name=NSS-3.14>{{cite web|url=https://developer.mozilla.org/eNot affected [66]   twork|title=NSS 3.15.1 release notes|publisher=Mozilla|accessdate=2013-08-10}}</ref> Not affected SSL]] | {{No|Enabled by default}} | {{No|Enabled by default}} | {{yes}} | {{yes}}<reVulnerable ssl-1.0.1-note>{{cite web| title = Major changes between OpenSSL 1.0.0h and OpenSSL Vulnerable ar 2012]| url = https://www.openssl.org/news/openssl-1.0.1-notes.html | date = 2012-0Not affected ate = 2015-01-20}}</ref> | {{yes}}<ref name=openssl-1.0.1-note /> | |- ! [[RSA BSAFEVulnerable =RSABSAFETECH>{{cite web| title = RSA BSAFE Technical Specification Comparison TablesYes[n 10] da a-sheet/11433-bsafe-tech-table.pdf}}</ref> 1.5 yes|No}} | {{No|Yes}} | {{yes}} | {{yes}} | {{yes}} | |- ! [[SChannel|SChannel XP / Enabled by default 003schannel>[https://msdn.microsoft.com/en-us/library/windows/desktop/aa380512%28v=vEnabled by default pher suites in Microsoft Windows XP and 2003]</ref> | {{Partial|Disabled by default bYes IE 7}} | {{No|Enabled by default}} | {{partial|Enabled by default by MSIE 7}} | {{noNo | {{no}} | |- ! [[SChannel|SChannel Vista / 2008]]<ref name=Vista2008schannel>[httpsNo msdn.microsoft.com/en-us/library/windows/desktop/ff468651%28v=vs.85%29.aspx SChannelNo pher Suites in Microsoft Windows Vista]</ref> | {{Yes|Disabled by default}} | {{No|EnYes d by default}} | {{yes}} | {{no}} | {{no}} | |- ! [[SChannel|SChannel 7 / 2008 R2]]<No name=Windows7schannel>[https://msdn.microsoft.com/en-us/library/windows/desktop/aa37Not affected 5%29.aspx TLS Cipher Suites in SChannel for Windows 7, 2008R2, 8, 2012]</ref> | {{YesNot affected default}} | {{partial|Disabled by default in MSIE 11}} | {{yes}} | {{partial|EnabledVulnerable by MSIE 11}} | {{partial|Enabled by default by MSIE 11}} | |- ! [[SChannel|SChannelVulnerable <ref name=Windows7schannel/> | {{Yes|Disabled by default}} | {{No|Enabled by default}Not affected {{partial|Disabled by default}} | {{partial|Disabled by default}} | |- ! [[SChannelVulnerable .1 / 2012 R2, 10]]<ref name=Windows7schannel/> | {{Yes|Disabled by default}} | {{partYes[n 10] { yes}} | {{yes}} | {{yes}} | |- ! Secure Tr 2 port OS X 10.2-10.7 / iOS 1-4 | {{No|Yes}} | {{No|Yes}} | {{yes}} | {{no}} | {{no}} |Disabled by default

[65][67] | |- ! Secure Transport OS X 10.11 / iOS 9 | {{Yes|No}} | {{Yes|No}}{{cref2|group=prEnabled by default |c}} | {{yes}} | {{yes}} | {{yes}} | |- ! SharkSSL | {{Yes|No}} | {{No|Enabled by defYes }} | {{yes}} | {{yes}} | {{yes}} | |- !

[[wolfSSL]] (previously CyaSSL) | {{Yes|No}}No {{Yes|Disabled by default}}<ref name=wolfSSL-3.6.6>{{cite web|url=http://wolfssl.comNo lf-SSL/Blog/Entries/2015/8/24_wolfSSL_3.6.6_is_Now_Available.htmaldries-Yesolfisla wolNo L 3.6.6 Released|date=2015-08-20|accessdate=2015-08-25}}</ref> | {{yes}} | {{yes}} | Yes s}} | |- |- class="sortbottom" ! Implementation ! SSL 2.0 (insecure) ! SSL 3.0 (insecYes[39] 1.1 ! TLS 1.2 ! TLS 1.3
<small>(Draft)</small> |} {{cnote2 begin | list-style=lowNot affected cnote2 | group=protocollibrary-table | a | SSL 2.0 client hello is supported even thoNot affected s not supported or is disabled because of the backward compatibilities.}} {{cnote2 |Vulnerable ocollibrary-table | b | SSL 3.0 support has been disabled by default as of Java 8 upVulnerable f>{{cite web|url=http://www.oracle. com/technetwork/java/javase/8u31-relnotes-2389094.Not affected va™ SE Development Kit 8, Update 31 Release Notes|accessdate=2015-01-22}}</ref>}} {{Vulnerable oup=protocollibrary-table | c | Secure Transport: SSL 2.0 was discontinued in OS X 10Yes[n 10] 0. 1 and iOS 9. TLS 1.1 and 1.2 are available 3–7 OS 5.0 and later, and OS X 10.8 and later.<ref>{{cite web|url=http://developer.apple.Disabled by default notes/tn2287/|work=iOS Developer Library|title=Technical Note TN2287: iOS 5 and TLS Enabled by default y Issues|publisher=Apple Inc.|accessdate=2012-05-03}}</ref>}}<ref>https://dev.ssllabsYes /ssltest/clients.html</ref> {{cnote2 end}} A paper presented at the 2012 [[AssociatNo for Computing Machinery|ACM]] [[Computer security conference|conference on computerNo d communications security]]<ref>{{cite book|author=Georgiev, Martin and Iyengar, SuboYes nd Jana, Suman and Anubhai, Rishita and Boneh, Dan and Shmatikov, Vitaly|title=The moYes angerous code in the world: validating SSL certificates in non-browser software. ProcYes ngs of the 2012 ACM conference on Computer and communications security|year=2012|isbnNot affected 651-4|url=//www.cs.utexas.edu/~{}shmat/shmat_ccs12. pdf|pages=38–49}}</ref> showed that Not affected ons used some of these SSL libraries incorrectly, leading to vulnerabilities. AccordVulnerable authors <blockquote>"the root cause of most of these vulnerabilities is the terribleVulnerable the APIs to the underlying SSL libraries. Instead of expressing high-level security pNot affected network tunnels such as confidentiality and authentication, these APIs expose low-leVulnerable of the SSL protocol to application developers. As a consequence, developers often usYes[n 10] g nd misunderstanding their manifold paramet 8–10 ESR 10 "</blockquote> ===Other uses=== The [[Simple Mail Transfer Protocol]] (SMTP) can alsNo[67] These applications use [[public key certificate]]s to verify the identity of endpoinEnabled by default e used to tunnel an entire network stack to create a [[virtual private network|VPN]],Yes is the case with [[OpenVPN]] and [[OpenConnect]]. Many vendors now marry TLS's encryNo on and authentication capabilities with authorization. There has also been substantiNo development since the late 1990s in creating client technology outside of the browserYes

enable support for client/server applications. When compared against traditional [[IPYes] VPN technologies, TLS has some inherent advantages in firewall and [[network dltes-Yes olfisl addtion|NAT]] traversal that make it easier to administer for large remote-access pNot affected TLS is also a standard method to protect [[Session Initiation Protocol]] (SIP) applicNot affected ng. TLS can be used to provide authentication and encryption of the SIP signaling asVulnerable th [[Voice over Internet Protocol|VoIP]] and other SIP-based applications.{{citationVulnerable e=December 2013}} ==Security== ===SSL 2.0=== SSL 2.0 is flawed in a variety of waysNot affected journal|url=http://www.sciencedirect.com/science/article/ pii/S0167404802003127|titleVulnerable urity of Today's Online Electronic Banking Systems|author=Joris Claessens, Valentin DYes[n 10] Va dewalle|journal=Computers & Security|volum 11–14 sue=3|year=2002|pages=253–265|doi=10.1016/S0167-4048(02)00312-7}}</ref> * Identical cNo tographic keys are used for [[message authentication]] and encryption. <!-- please eEnabled by default ects security --> (In SSL 3.0, MAC secrets may be larger than encryption keys, so mesYes s can remain tamper resistant even if encryption keys are broken.<ref name=RFC6101 /No * SSL 2.0 has a weak MAC construction that uses the MD5 hash function with a secretNo efix, making it vulnerable to [[length extension attack]]s. * SSL 2.0 does not have aYes rotection for the handshake, meaning a man-in-the-middle downgrade attack can go undeYes ed. * SSL 2.0 uses the TCP connection close to indicate the end of data. This means tYes truncation attacks are possible: the attacker simply forges a TCP FIN, leaving the reNot affected re of an illegitimate end of data message (SSL 3.0 fixes this problem by having an eVulnerable (SPDY)[48] domain certificate, which clashes with the standard feature of virtual hosting in WeVulnerable This means that most websites are practically impaired from using SSL. SSL 2.0 is dVulnerable default, beginning with [[Internet Explorer 7]],<ref>{{cite web|url=http://blogs.msdnNot affected ve/2005/10/22/483795.aspx|title=IEBlog: Upcoming HTTPS Improvements in Internet ExplVulnerable 2|accessdate=2007-11-25|last=Lawrence|first=Eric|date=2005-10-22|publisher=[[MicrosoYes[n 10] ef [[Mozilla Firefox]] 2,<ref>{{cite web|url 15–22 ESR 17.0–17.0.10 e=Bugzilla@Mozilla — Bug 236933 – Disable SSL2 and other weak ciphers|accessdateNo 07-11-25|publisher=[[Mozilla Corporation]]}}</ref> [[Opera (web browser)|Opera]] 9.5Enabled by default pera.com/docs/changelogs/windows/950/ "Opera 9.5 for Windows Changelog" at [[Opera.cYes : "Disabled SSL v2 and weak ciphers."</ref> and [[Safari (web browser)|Safari]]. AftNo it sends a TLS "ClientHello", if Mozilla Firefox finds that the server is unable to No plete the handshake, it will attempt to fall back to using SSL 3.0 with an SSL 3.0 "CYes tHello" in SSL 2.0 format to maximize the likelihood of successfully handshaking withYes er servers.<ref>{{cite web|url=https: //bugzilla.mozilla.org/show_bug.cgi?id=454759|tiYes Firefox still sends SSLv2 handshake even though the protocol is disabled|date=2008-09Not affected upport for SSL 2.0 (and weak [[40-bit encryption|40-bit]] and 56-bit ciphers) has beeMitigated completely from Opera as of version 10<!-—9.5=disabled, v10=removed, see changelogs-Vulnerable tp://www.opera.com/docs/changelogs/windows/1000/ "Opera 10 for Windows changelog"] aVulnerable om]]: "Removed support for SSL v2 and weak ciphers"</ref><ref>{{cite web|url=http://mNot affected ngve/blog/2007/04/30/10-years-of-ssl-in-opera |title=10 years of SSL in Opera —Vulnerable r's notes |accessdate=2007-11-25 |last=Pettersen |first=Yngve |date=2007-04-30 |publiYes[n 10] ar hiveurl=https://web.archive.org/2007101220 ESR 17.0.11 07/04/30/10-years-of-ssl-in-opera |archivedate=October 12, 2007 }}</ref> ===SSL 3.0=No SSL 3.0 improved upon SSL 2.0 by adding SHA-1–based ciphers and support for certificEnabled by default From a security standpoint, SSL 3.0 should be considered less desirable than TLS 1.Yes he SSL 3.0 cipher suites have a weaker key derivation process; half of the master keNo hat is established is fully dependent on the MD5 hash function, which is not resistaNo to collisions and is, therefore, not considered secure. Under TLS 1.0, the master keyYes t is established depends on both MD5 and SHA-1 so its derivation process is not curreYes considered weak. It is for this reason that SSL 3.0 implementations cannot be validaYes under FIPS 140-2.<ref>{{cite web |author=National Institute of Standards and TechnoloNot affected lementation Guidance for FIPS PUB 140-2 and the Cryptographic Module Validation ProgrMitigated ecember 2010 |url=http://csrc.nist.gov/groups/ STM/cmvp/documents/fips140-2/FIPS1402IVulnerable f> In October 2014, the vulnerability in the design of SSL 3.0 has been reported, which mLowest priority [68][69] ttack (see [[#POODLE attack]]). ===TLS=== TLS has a variety of security measures: * Not affected ainst a downgrade of the protocol to a previous (less secure) version or a weaker ciVulnerable * Numbering subsequent Application records with a sequence number and using this seqYes[n 10] ca ion code]]s (MACs). * Using a message dige 23 nhanced with a key (so only a key-holder can check the MAC). The [[HMAC]] constructioNo sed by most TLS cipher suites is specified in RFC 2104 (SSL 3.0 used a different hasEnabled by default message that ends the handshake ("Finished") sends a hash of all the exchanged handsYes messages seen by both parties. * The [[pseudorandomness|pseudorandom]] function splits thDisabled by default [70] these algorithms is found to be vulnerable. ===Attacks against TLS/SSL=== SignificaNo attacks against TLS/SSL are listed below: Note: On February 2015, IETF issued an infYes tional RFC<ref>{{cite web|title=RFC 7457 : Summarizing Known Attacks on Transport LayYes ecurity (TLS) and Datagram TLS (DTLS)|url=https://tools.ietf.org/html/rfc7457}}</ref>Yes marizing the various known attacks against TLS/SSL. ====Renegotiation attack==== A vNot affected of the renegotiation procedure was discovered in August 2009 that can lead to plainteMitigated on attacks against SSL 3.0 and all current versions of TLS. For example, it allows aVulnerable who

can hijack an https connection to splice their own requests into the beginning oVulnerable rsation the client has with the web server. The attacker can't actually decrypt the cNot affected communication, so it is different from a typical man-in-the-middle attack. A short-tVulnerable for web servers to stop allowing renegotiation, which typically will not require otheYes<ref name='aboutconfig' group='n'>configure the maximum and the minimum version of enabling protocols via about:config or add-on<ref name='FxSSLVersionControl'>SSL Version Control :: Add-ons for Firefox

[15] Only when no cipher suites with other than RC4 is available, cipher suites with RC4 will be used as a fallback.

[16] Uses the TLS implementation provided by NSS. As of Firefox 22, Firefox supports only TLS 1.0 despite the bundled NSS supporting TLS 1.1. Since Firefox 23, TLS 1.1 can be enabled, but was not enabled by default due to issues. Firefox 24 has TLS 1.2 support disabled by default. TLS 1.1 and TLS 1.2 have been enabled by default in Firefox 27 release.

[17]

[18] SSL 3.0 itself is disabled by default.[75] In addition, fallback to SSL 3.0 is disabled since version 34,[77] and TLS_FALLBACK_SCSV is implemented since 35.0 and ESR 31.3.[75][78]

[19] Offering RC4 cipher-suites is allowed only for hosts which only support RC4 cipher suites and are registered on the whitelist by Mozilla.[82][83]</ref> July 2013 |title=On the Security of RC4 in TLS and WPA |format=PDF |accessdate=2 SeptNot affected rl=http://www.isg.rhul. ac.uk/tls/RC4biases.pdf }}</ref><ref>{{cite conference |url=httMitigated senix.org/sites/default/files/conference/ protected-files/alfardan_sec13_slides.pdf |tYes[n 17] st Nadhem J. |last=AlFardan |first2=Daniel J. |last2=Bernstein |first3=Kenneth G. |last3=Paterson |first4=Bertram |last4=Poettering |44 st5=Jacob C. N. |last5=Schuldt |date=15 August 2013 |conference=22nd [[USENIX]] Security Symposium |format=PDF |accessdate=2 SepteESR 45 attacks against RC4 in TLS are feasible although not truly practical |page=51}}</refNo In July 2015, subsequent improvements in the attack make it increasingly practical toNo feat the security of RC4-encrypted TLS.<ref>{{cite web|last1=Goodin|first1=Dan|title=Yes -theoretical crypto attack against HTTPS now verges on practicality|url=http://arstecYes a.com/security/2015/07/ once-theoretical-crypto-attack-against-https-now-verges-on-praYes ality/|website=Ars Technical|publisher=Conde Nast|accessdate=16 July 2015}}</ref> AsYes y modern browsers have been designed to defeat BEAST attacks (except Safari for Mac OYes 10.7 or earlier, for iOS 6 or earlier, and for Windows; see [[#Web browsers]]), RC4 iYes longer a good choice for TLS 1.0. The CBC ciphers which were affected by the BEAST aNot affected past have become a more popular choice for protection.<ref name="best-practices"/> MMitigated Microsoft recommend disabling RC4 where

possible.<ref>{{cite weblurl=https://wiki.moNot affected urity/Server_Side_TLSltitle=Mozilla Security Server Side TLS Recommended ConfiguratioNot affected[n 20]

[20] All RC4 cipher-suites is disabled by default.[82]<ref name='RC4Fx-2'>"Intent to ship: RC4 disabled by default in Firefox 44". Retrieved 2015-10-18.

[21] IE uses the TLS implementation of the Microsoft Windows operating system provided by the SChannel security support provider. TLS 1.1 and 1.2 are disabled by default until IE11.[84][85]

[22] Windows NT 3.1 supports IE 1–2, Windows NT 3.5 supports IE 1–3, Windows NT 3.51 and Windows NT 4.0 supports IE 1–6

[23] Windows XP as well as Server 2003 and older only support weak ciphers like 3DES and RC4.[88] The weak ciphers of these SChannel version are not only used for IE. They are also used for other Microsoft products running on this OS, e.g like Office. Only Windows Server 2003 can get a manually update to support AES ciphers by KB948963[89]

[24] MS13-095 or MS14-049 for 2003 and XP-64 or SP3 for XP (32-bit)

[25] Internet Explorer Support Announcement[93]

[26] RC4 can be disabled except as a fallback (Only when no cipher suites with other than RC4 is available, cipher suites with RC4 will be used as a fallback.)[96]

[27] Fallback to SSL 3.0 is sites blocked by default in Internet Explorer 11 for Protected Mode.[98][99] SSL 3.0 is disabled by default in Internet Explorer 11 since April 2015.[100]

[28] Edge (formerly known as Project Spartan) is based on a fork of the Internet Explorer 11 rendering engine.

[29] Except Windows 10 LTSB 2015 (LongTermSupportBranch)[103]

[30] Could be disabled via registry editing but need 3rd Party tools to do this.[104]

[31] Opera 10 added support for TLS 1.2 as of Presto 2.2. Previous support was for TLS 1.0 and 1.1. TLS 1.1 and 1.2 are disabled by default (except for version 9[108] that enabled TLS 1.1 by default).

[32] SSL 3.0 is disabled by default remotely since October 15, 2014[117]

[33] TLS support of Opera 14 and above is same as that of Chrome, because Opera has migrated to Chromium backend (Opera 14 for Android is based on Chromium 26 with WebKit,[121] and Opera 15 and above are based on Chromium 28 and above with Blink[122]).

[34] TLS_FALLBACK_SCSV is implemented.[125]

[35] SSL 3.0 is enabled by default, with some mitigations against known vulnerabilities such as BEAST and POODLE implemented.[117]</ref> Yes Yes Yes Yes
(only desktop) Yes needs ECC compatible OS[39] Not affected Mitigated Mitigated
[n 36]

[36] In addition to TLS_FALLBACK_SCSV, "anti-POODLE record splitting" is implemented.[117]</ref> Lowest priority Vulnerable
(except Windows) Vulnerable Temporary
[n 11] 27 No Disabled by default
[58] Yes Yes Yes Yes
(only desktop) Yes needs ECC compatible OS[39] Not affected Mitigated Mitigated
[n 37] Qualys SSL report simulates Safari 5.1.9 connecting with TLS 1.0 not 1.1 or 1.2<ref>Qualys SSL Report: google.co.uk (simulation Safari 5.1.9 TLS 1.0)

[37] In addition to TLS_FALLBACK_SCSV and "anti-POODLE record splitting", SSL 3.0 itself is disabled by default.[58]</ref> Lowest priority Vulnerable
(except Windows) Vulnerable Yes[n 38] Safari 5 is the last version available for Windows. OS X 10.8 on have SecureTransport support for TLS 1.1 and 1.2<ref>Curl: Patch to add TLS 1.1 and 1.2 support & replace deprecated functions in SecureTransport

[38] configure the minimum version of enabling protocols via opera://flags[58] (the maximum version can be configured with command-line option)</ref>
(only desktop) 28, 29 No Disabled by default Yes Yes Yes Yes
(only desktop) Yes needs ECC compatible OS[39] Not affected Mitigated Mitigated Lowest priority Mitigated Vulnerable Yes[n 38]
(only desktop) 30 No Disabled by default Yes Yes Yes Yes
(only desktop) Yes needs ECC compatible OS[39] Not affected Mitigated Mitigated Only as fallback
[n 15][59] Mitigated Mitigated[118] Yes[n 38]
(only desktop) 31 32 No No[60] Yes Yes Yes Yes
(only desktop) Yes needs ECC compatible OS[39] Not affected Mitigated Not affected Only as fallback
[n 15][59] Mitigated Mitigated Temporary
[n 11] Browser Version Platforms SSL 2.0 (insecure) SSL 3.0 (insecure) TLS 1.0 TLS 1.1 TLS 1.2 EV certificate SHA-2 certificate ECDSA certificate BEAST CRIME POODLE (SSLv3) RC4 FREAK Logjam Protocol selection by user Apple Safari
[n 39] with unknown version,<ref>Apple (2009-06-10). "Features". Retrieved 2009-06-10.

[39] Safari uses the operating system implementation on Mac OS X, Windows (XP, Vista, 7)<ref>Adrian, Dimcev. "Common browsers/libraries/servers and the associated cipher suites implemented". *TLS Cipher Suites Project*.

[40] In September 2013, Apple implemented BEAST mitigation in OS X 10.8 (Mountain Lion), but it was not turned on by default resulting in Safari still being theoretically vulnerable to the BEAST attack on that platform.[127][128] BEAST

mitigation has been enabled by default from OS X 10.8.5 updated in February 2014.[129]

[41] Because Apple removed support for all CBC protocols in SSL 3.0 to mitigate POODLE[130][131], this leaves only RC4 which is also completely broken by the RC4 attacks in SSL 3.0.

[42] Mobile Safari and third-party software utilizing the system UIWebView library use the iOS operating system implementation, which supports TLS 1.2 as of iOS 5.0.[136][137][138]

57.5.3   Libraries

Main article: Comparison of TLS implementations

Most SSL and TLS programming libraries are free and open source software.

- Botan, a BSD-licensed cryptographic library written in C++.

- cryptlib: a portable open source cryptography library (includes TLS/SSL implementation)

- Delphi programmers may use a library called Indy which utilizes OpenSSL.

- GnuTLS: a free implementation (LGPL licensed)

- Java Secure Socket Extension: a Java implementation included in the Java Runtime Environment supports TLS 1.1 and 1.2 from Java 7, although is disabled by default for client, and enabled by default for server.[146] Java 8 supports TLS 1.1 and 1.2 enabled on both the client and server by default.[147]

- LibreSSL: a fork of OpenSSL by OpenBSD project.

- MatrixSSL: a dual licensed implementation

- mbed TLS (previously PolarSSL): A tiny SSL library implementation for embedded devices that is designed for ease of use

- Network Security Services: FIPS 140 validated open source library

- OpenSSL: a free implementation (BSD license with some extensions)

- SChannel: an implementation of SSL and TLS Microsoft Windows as part of its package.

- Secure Transport: an implementation of SSL and TLS used in OS X and iOS as part of their packages.

- wolfSSL (previously CyaSSL): Embedded SSL/TLS Library with a strong focus on speed and size.

1. ^ SSL 2.0 client hello is supported even though SSL 2.0 is not supported or is disabled because of the backward compatibilities.

2. ^ SSL 3.0 support has been disabled by default as of Java 8 update 31.[162]

3. ^ Secure Transport: SSL 2.0 was discontinued in OS X 10.8. SSL 3.0 was discontinued in OS X 10.11 and iOS 9. TLS 1.1 and 1.2 are available on iOS 5.0 and later, and OS X 10.8 and later.[163]

[164]

A paper presented at the 2012 ACM conference on computer and communications security[165] showed that few applications used some of these SSL libraries incorrectly, leading to vulnerabilities. According to the authors

> "the root cause of most of these vulnerabilities is the terrible design of the APIs to the underlying SSL libraries. Instead of expressing high-level security properties of network tunnels such as confidentiality and authentication, these APIs expose low-level details of the SSL protocol to application developers. As a consequence, developers often use SSL APIs incorrectly, misinterpreting and misunderstanding their manifold parameters, options, side effects, and return values."

57.5.4   Other uses

The Simple Mail Transfer Protocol (SMTP) can also be protected by TLS. These applications use public key certificates to verify the identity of endpoints.

TLS can also be used to tunnel an entire network stack to create a VPN, as is the case with OpenVPN and OpenConnect. Many vendors now marry TLS's encryption and authentication capabilities with authorization. There has also been substantial development since the late 1990s in creating client technology outside of the browser to enable support for client/server applications. When compared against traditional IPsec VPN technologies, TLS has some inherent advantages in firewall and NAT traversal that make it easier to administer for large remote-access populations.

TLS is also a standard method to protect Session Initiation Protocol (SIP) application signaling. TLS can be used to provide authentication and encryption of the SIP signaling associated with VoIP and other SIP-based applications.

57.6 Security

57.6.1 SSL 2.0

SSL 2.0 is flawed in a variety of ways:[166]

- Identical cryptographic keys are used for message authentication and encryption. (In SSL 3.0, MAC secrets may be larger than encryption keys, so messages can remain tamper resistant even if encryption keys are broken.[3])

- SSL 2.0 has a weak MAC construction that uses the MD5 hash function with a secret prefix, making it vulnerable to length extension attacks.

- SSL 2.0 does not have any protection for the handshake, meaning a man-in-the-middle downgrade attack can go undetected.

- SSL 2.0 uses the TCP connection close to indicate the end of data. This means that truncation attacks are possible: the attacker simply forges a TCP FIN, leaving the recipient unaware of an illegitimate end of data message (SSL 3.0 fixes this problem by having an explicit closure alert).

- SSL 2.0 assumes a single service and a fixed domain certificate, which clashes with the standard feature of virtual hosting in Web servers. This means that most websites are practically impaired from using SSL.

SSL 2.0 is disabled by default, beginning with Internet Explorer 7,[167] Mozilla Firefox 2,[168] Opera 9.5,[169] and Safari. After it sends a TLS "ClientHello", if Mozilla Firefox finds that the server is unable to complete the handshake, it will attempt to fall back to using SSL 3.0 with an SSL 3.0 "ClientHello" in SSL 2.0 format to maximize the likelihood of successfully handshaking with older servers.[170] Support for SSL 2.0 (and weak 40-bit and 56-bit ciphers) has been removed completely from Opera as of version 10.[171][172]

57.6.2 SSL 3.0

SSL 3.0 improved upon SSL 2.0 by adding SHA-1–based ciphers and support for certificate authentication.

From a security standpoint, SSL 3.0 should be considered less desirable than TLS 1.0. The SSL 3.0 cipher suites have a weaker key derivation process; half of the master key that is established is fully dependent on the MD5 hash function, which is not resistant to collisions and is, therefore, not considered secure. Under TLS 1.0, the master key that is established depends on both MD5 and SHA-1 so its derivation process is not currently considered weak. It is for this reason that SSL 3.0 implementations cannot be validated under FIPS 140-2.[173]

In October 2014, the vulnerability in the design of SSL 3.0 has been reported, which makes CBC mode of operation with SSL 3.0 vulnerable to the padding attack (see #POODLE attack).

57.6.3 TLS

TLS has a variety of security measures:

- Protection against a downgrade of the protocol to a previous (less secure) version or a weaker cipher suite.

- Numbering subsequent Application records with a sequence number and using this sequence number in the message authentication codes (MACs).

- Using a message digest enhanced with a key (so only a key-holder can check the MAC). The HMAC construction used by most TLS cipher suites is specified in RFC 2104 (SSL 3.0 used a different hash-based MAC).

- The message that ends the handshake ("Finished") sends a hash of all the exchanged handshake messages seen by both parties.

- The pseudorandom function splits the input data in half and processes each one with a different hashing algorithm (MD5 and SHA-1), then XORs them together to create the MAC. This provides protection even if one of these algorithms is found to be vulnerable.

57.6.4 Attacks against TLS/SSL

Significant attacks against TLS/SSL are listed below:

Note: On February 2015, IETF issued an informational RFC[174] summarizing the various known attacks against TLS/SSL.

Renegotiation attack

A vulnerability of the renegotiation procedure was discovered in August 2009 that can lead to plaintext injection attacks against SSL 3.0 and all current versions of TLS. For example, it allows an attacker who can hijack an https connection to splice their own requests into the beginning of the conversation the client has with the web server. The attacker can't actually decrypt the client-server communication, so it is different from a typical man-in-the-middle attack. A short-term fix is for web servers to stop allowing renegotiation, which typically will not require other changes unless client certificate authentication is used. To fix the vulnerability, a renegotiation indication extension was proposed for TLS. It will require the client and server to include and verify information about previous handshakes in any renegotiation handshakes.[175] This extension has become a proposed standard and has been assigned the number RFC 5746. The RFC has been implemented by several libraries.[176][177][178]

Version rollback attacks

Modifications to the original protocols, like **False Start**[179] (adopted and enabled by Google Chrome[180]) or Snap Start, have reportedly introduced limited TLS protocol version rollback attacks[181] or allowed modifications to the cipher suite list sent by the client to the server (an attacker may succeed in influencing the cipher suite selection in an attempt to downgrade the cipher suite strength, to use either a weaker symmetric encryption algorithm or a weaker key exchange[182]). A paper presented at an Association for Computing Machinery (ACM) conference on computer and communications security in 2012 demonstrates that the False Start extension is at risk: in certain circumstances it could allow an attacker to recover the encryption keys offline and to access the encrypted data.[183]

BEAST attack

On September 23, 2011 researchers Thai Duong and Juliano Rizzo demonstrated a proof of concept called **BEAST (Browser Exploit Against SSL/TLS)**[184] using a Java applet to violate same origin policy constraints, for a long-known cipher block chaining (CBC) vulnerability in TLS 1.0:[185][186] an attacker observing 2 consecutive ciphertext blocks C0, C1 can test if the plaintext block P1 is equal to x by choosing the next plaintext block $P2 = x \wedge C0 \wedge C1$; due to how CBC works C2 will be equal to C1 if x = P1. Practical exploits had not been previously demonstrated for this vulnerability, which was originally discovered by Phillip Rogaway[187] in 2002. The vulnerability of the attack had been fixed with TLS 1.1 in 2006, but TLS 1.1 had not seen wide adoption prior to this attack demonstration.

RC4 as a stream cipher is immune to BEAST attack. Therefore, RC4 was widely used as a way to mitigate BEAST attack on the server side. However, in 2013, researchers found more weaknesses in RC4. Thereafter enabling RC4 on server side was no longer recommended.[188]

Chrome and Firefox themselves are not vulnerable to BEAST attack,[46][189] however, Mozilla updated their NSS libraries to mitigate BEAST-like attacks. NSS is used by Mozilla Firefox and Google Chrome to implement SSL. Some web servers that have a broken implementation of the SSL specification may stop working as a result.[190]

Microsoft released Security Bulletin MS12-006 on January 10, 2012, which fixed the BEAST vulnerability by changing the way that the Windows Secure Channel (SChannel) component transmits encrypted network packets from the server end.[191]

Users of Internet Explorer on Windows 7, Windows 8 and Windows Server 2008 R2 can enable use of TLS 1.1 and 1.2, but this workaround will fail if it is not supported by the other end of the connection and will result in a fall-back to TLS 1.0.

Apple fixed BEAST vulnerability by implementing 1/n-1 split and turning it on by default in OS X Mavericks, released on October 22, 2013.[192]

CRIME and BREACH attacks

Main articles: CRIME (security exploit) and BREACH (security exploit)

The authors of the BEAST attack are also the creators of the later CRIME attack, which can allow an attacker to recover the content of web cookies when data compression is used along with TLS.[193][194] When used to recover the content of secret authentication cookies, it allows an attacker to perform session hijacking on an authenticated web session.

While the CRIME attack was presented as a general attack that could work effectively against a large number of protocols, including but not limited to TLS, and application-layer protocols such as SPDY or HTTP, only exploits against TLS and SPDY were demonstrated and largely mitigated in browsers and servers. The CRIME exploit against HTTP compression has not been mitigated at all, even though the authors of CRIME have warned that this vulnerability might be even more widespread than SPDY and TLS compression combined. In 2013 a new instance of the CRIME attack against HTTP compression, dubbed BREACH, was announced. Built based on the CRIME attack a BREACH attack can extract login tokens, email addresses or other sensitive information from TLS encrypted web traffic in as little as 30 seconds (depending on the number of bytes to be extracted), provided the attacker tricks the victim into visiting a malicious web link or is able to inject content into valid pages the user is visiting (ex: a wireless network under the control of the attacker).[195] All versions of TLS and SSL are at risk from BREACH regardless of the encryption algorithm or cipher used.[196] Unlike previous instances of CRIME, which can be successfully defended against by turning off TLS compression or SPDY header compression, BREACH exploits HTTP compression which cannot realistically be turned off, as virtually all web servers rely upon it to improve data transmission speeds for users.[195] This is a known limitation of TLS as it is susceptible to chosen-plaintext attack against the application-layer data it was meant to protect.

Timing attacks on padding

Earlier TLS versions were vulnerable against the padding oracle attack discovered in 2002. A novel variant, called the Lucky Thirteen attack, was published in 2013.

Some experts[30] also recommended avoiding Triple-DES CBC. Since the last supported ciphers developed to support any program using Windows XP's SSL/TLS library like Internet Explorer on Windows XP are RC4 and Triple-DES, and since RC4 is now deprecated (see discussion of RC4 attacks), this makes it difficult to support any version of SSL for any program using this library on XP.

A fix was released as the Encrypt-then-MAC extension to the TLS specification, released as RFC 7366.[197] The Lucky Thirteen attack can be mitigated in TLS 1.2 by using only AES_GCM ciphers; AES_CBC remains vulnerable.

POODLE attack

Main article: POODLE

On October 14, 2014, Google researchers published a vulnerability in the design of SSL 3.0, which makes CBC mode of operation with SSL 3.0 vulnerable to a padding attack (CVE-2014-3566). They named this attack **POODLE (Padding Oracle On Downgraded Legacy Encryption)**. On average, attackers only need to make 256 SSL 3.0 requests to reveal one byte of encrypted messages.[36][198]

Although this vulnerability only exists in SSL 3.0 and most clients and servers support TLS 1.0 and above, all major browsers voluntarily downgrade to SSL 3.0 if the handshakes with newer versions of TLS fail unless they provide the option for a user or administrator to disable SSL 3.0 and the user or administrator does so. Therefore, the man-in-the-middle can first conduct a version rollback attack and then exploit this vulnerability.[36][198]

In general, graceful security degradation for the sake of interoperability is difficult to carry out in a way that cannot be exploited. This is challenging especially in domains where fragmentation is high.[199]

On December 8, 2014, a variant of POODLE was announced that impacts TLS implementations that do not properly enforce padding byte requirements.[200]

RC4 attacks

Main article: RC4 § Security

Despite the existence of attacks on RC4 that broke its security, cipher suites in SSL and TLS that were based on RC4 were still considered secure prior to 2013 because the way in which they were used in SSL and TLS. In 2011, the RC4 suite was actually recommended as a work around for the BEAST attack.[201] New forms of attack disclosed in March 2013 conclusively demonstrated the feasibility of breaking RC4 in TLS, suggesting it was not a good workaround for BEAST.[35] An attack scenario was proposed by AlFardan, Bernstein, Paterson, Poettering and Schuldt that used newly discovered statistical biases in the RC4 key table[202] to recover parts of the plaintext with a large number of TLS encryptions.[203][204] An attack on RC4 in TLS and SSL that requires 13×2^{20} encryptions to break RC4 was unveiled on 8 July 2013 and later described as "feasible" in the accompanying presentation at a USENIX Security Symposium in August 2013.[205][206] In July 2015, subsequent improvements in the attack make it increasingly practical to defeat the security of RC4-encrypted TLS.[207]

As many modern browsers have been designed to defeat BEAST attacks (except Safari for Mac OS X 10.7 or earlier, for iOS 6 or earlier, and for Windows; see #Web browsers), RC4 is no longer a good choice for TLS 1.0. The CBC ciphers which were affected by the BEAST attack in the past have become a more popular choice for protection.[30] Mozilla and Microsoft recommend disabling RC4 where possible.[208][209] RFC 7465 prohibits the use of RC4 cipher suites in all versions of TLS.

On September 1, 2015, Microsoft, Google and Mozilla announced that RC4 cipher suites would be disabled by default in their browsers (Microsoft Edge, Internet Explorer 11 on Windows 7/8.1/10, Firefox, and Chrome) in early 2016.[210][211][212]

Truncation attack

A TLS truncation attack blocks a victim's account logout requests so that the user unknowingly remains logged into a web service. When the request to sign out is sent, the attacker injects an unencrypted TCP FIN message (no more data from sender) to close the connection. The server therefore doesn't receive the logout request and is unaware of the abnormal termination.[213]

Published in July 2013,[214] the attack causes web services such as Gmail and Hotmail to display a page that informs the user that they have successfully signed-out, while ensuring that the user's browser maintains authorization with the service, allowing an attacker with subsequent access to the browser to access and take over control of the user's logged-in account. The attack does not rely on installing malware on the victim's computer; attackers need only place themselves between the victim and the web server (e.g., by setting up a rogue wireless hotspot).[213] This vulnerability also requires access to the victim's computer.

Downgrade attacks: FREAK attack and Logjam attack

Main articles: FREAK and Logjam (computer security)

Downgrade attacks can force servers and clients to negotiate a connection using cryptographically weak keys. In 2014, a man-in-the-middle attack called FREAK was discovered affecting the OpenSSL stack, the default Android web browser, and some Safari browsers.[215] The attack involved tricking servers into negotiating a TLS connection using cryptographically weak 512 bit encryption keys.

Logjam is a security exploit discovered in May 2015 that exploits the option of using legacy "export-grade" 512-bit Diffie–Hellman keys dating back to the 1990s.[216] As with FREAK, it forces susceptible servers to downgrade to cryptographically weak 512 bit keys that an attacker can then deduce.

Implementation errors: Heartbleed bug, BERserk attack, Komodia root certificate

Main article: Heartbleed

The Heartbleed bug is a serious vulnerability specific to the implementation of SSL/TLS in the popular OpenSSL cryptographic software library, affecting versions 1.0.1 to 1.0.1f. This weakness, reported in April 2014, allows attackers to steal private keys from servers that should normally be protected.[217] The Heartbleed bug allows anyone on the Internet to read the memory of the systems protected by the vulnerable versions of the OpenSSL software. This compromises the secret private keys associated with the public certificates used to identify the service providers and to encrypt the traffic, the names and passwords of the users and the actual content. This allows attackers to eavesdrop on communications, steal data directly from the services and users and to impersonate services and users.[218] The vulnerability is caused by a buffer over-read bug in the OpenSSL software, rather than a defect in the SSL or TLS protocol specification.

On September 29, 2014 a variant of Daniel Bleichenbacher's PKCS#1 v1.5 RSA Signature Forgery vulnerability [219] was announced by Intel Security Advanced Threat Research. This attack, dubbed BERserk, is a result of incomplete ASN.1 length decoding of public key signatures in some SSL implementations, and allows a man-in-the-middle attack by forging a public key signature.[220]

In February 2015, after media reported the hidden pre-installation of Superfish adware on some Lenovo notebooks,[221] a researcher found a trusted root certificate on affected Lenovo machines to be insecure, as the keys could easily be accessed using the company name, Komodia, as a passphrase.[222] The Komodia library was designed to intercept client-side TLS/SSL traffic for parental control and surveillance, but it was also used in numerous adware programs, including Superfish, that were often surreptitiously installed unbeknownst to the computer user. In turn, these potentially unwanted programs installed the corrupt root certificate, allowing attackers to completely control web traffic and confirm false web sites as authentic.

Survey of websites vulnerable to attacks

As of September 2015, Trustworthy Internet Movement estimate the ratio of websites that are vulnerable to TLS attacks.[34]

57.6.5   Forward secrecy

Main article: Forward secrecy

Forward secrecy is a property of cryptographic systems which ensures that a session key derived from a set of public and private keys will not be compromised if one of the private keys is compromised in the future.[223] Without forward secrecy, if the server's private key is compromised, not only will all future TLS-encrypted sessions using that server certificate be compromised, but also any past sessions that used it as well (provided of course that these past sessions were intercepted and stored at the time of transmission).[224] An implementation of TLS can provide forward secrecy by requiring the use of ephemeral Diffie-Hellman key exchange to establish session keys, and some notable TLS implementations do so exclusively: e.g., Gmail and other Google HTTPS services that use OpenSSL.[225] However, many clients and servers supporting TLS (including browsers and web servers) are not configured to implement such restrictions.[226][227] In practice, unless a web service uses Diffie-Hellman key exchange to implement forward secrecy, all of the encrypted web traffic to and from that service can be decrypted by a third party if it obtains the server's master (private) key; e.g., by means of a court order.[228]

Even where Diffie-Hellman key exchange is implemented, server-side session management mechanisms can impact forward secrecy. The use of TLS session tickets (a TLS extension) causes the session to be protected by AES128-CBC-SHA256 regardless of any other negotiated TLS parameters, including forward secrecy ciphersuites, and the long-lived TLS session ticket keys defeat the attempt to implement forward secrecy.[229][230][231] Stanford University research in 2014 also found that of 473,802 TLS servers surveyed, 82.9% of the servers deploying ephemeral Diffie-Hellman (DHE) key exchange to support forward secrecy were using weak Diffie Hellman parameters. These weak paramater choices could potentially compromise the effectiveness of the forward secrecy that the servers sought to provide.[232]

Since late 2011, Google has provided forward secrecy with TLS by default to users of its Gmail service, along with Google Docs and encrypted search among other services.[233] Since November 2013, Twitter has provided forward secrecy with TLS to users of its service.[234] As of September 2015, 37.8% of TLS-enabled websites are configured to use cipher suites that provide forward secrecy to modern web browsers.[34]

57.6.6 Dealing with MITM attacks

Main article: Man-in-the-middle attack

How SSL Certificates Protect Against Man-in-the-Middle Attacks

This assumes that you know the basics of public key authentication and how a web browser communicates with a web server through a domain name. The setup process is between the company web server and the certificate authority. The normal use interaction is between the user's web browser and the company's web server.

Public Keys and Private Keys

The web server has a public key and a private key. The private key can decrypt a message encrypted by the public key. The public key can decrypt a message encrypted by the private key. The certificate authority has their own public key and private key.

The web server sends its company information, public key, and the domain name (to be associated with the SSL certificate) to the certificate authority. The certificate authority sends a confirmation message to the email address associated with the domain name, in order to prove that this request was made by the genuine owner of the domain name.

At this point, the certificate authority will wait until the domain name owner validates the request by email.

Certificate Signing

The certificate authority encrypts the web server's domain name, company info, and public key using their own private key. The certificate authority sends the encrypted result to the web server. This result is the SSL certificate, a text message containing the domain name, company info, and public key of the web server that has all been encrypted with the private key of the certificate authority. The web server sends this certificate to the user's browser.

Trusted Certificate Authorities

The web browser comes pre-loaded with a list of trusted certificate authorities and their public keys. The web browser decrypts the certificate using the public key of the corresponding certificate authority.

At this point, the web browser knows that the certificate and its contents are trustworthy because only a message encrypted with the certificate authority's private key could have been coherently decrypted by that certificate authority's public key.

The web browser now knows the trusted company info, public key, and domain name that is supposed to be associated with the web server, which is still suspicious. The web browser confirms that the domain name on the certificate matches the actual domain name of the web server.

At this point, if the domain names match, the web browser determines that the web server is trustworthy enough to send encrypted data to. Also at this point, the web browser determines that it can use the trusted public key of the certificate to encrypt its messages because only the private key of the genuine web server can decrypt that message.

Note that if an untrusted public key was used (not going through the certificate authority verification), the web browser may have encrypted and sent sensitive information only to be decrypted by the private key of a malicious server! In other words, it is imperative that the public key be trusted because encrypting a message with it is the only line of eavesdropping defense for the information that the web browser sends out.

Moving on, the web browser encrypts its message using the trusted public key and sends the encrypted message to the web server. The web server decrypts the message with the genuine private key, if it has one, then reads the decrypted message successfully.

Shared Secret Key

When the web server responds to the web browser, that message must be sent securely as well. The web browser could copy what the web server just did (excluding the certificate authority process) by generating a public key and private key for itself then sending its public key to the web server. This would establish a secure connection called 2-way asymmetric encryption. However, such communication is network-intensive and computationally taxing (relatively speaking).

So the standard approach is to use a shared secret key that can encrypt a message and also decrypt it. Using a temporary secret key (that will be deleted when the connection is terminated) for just this conversation ensures that messages are protected in the event of a compromised private key of the web server later on. Protecting encrypted messages for the future by having a temporary, conversation-specific secret key is called forward secrecy.

The web browser generates a secret key, encrypts it using the trusted public key, then sends it to the web server. If the web server is genuine, it will be able to decrypt the secret key successfully.

Note that the generation of the shared secret key is more secure and logically deep in practice than depicted above. In simple terms, the web browser sends the ingredients it wants to use to the web server. Both the web browser and web server add and mix with a private ingredient of their own, then exchange their mixtures. The way that the ingredients were mixed allows the web browser's mix and web server's private ingredient, or the web server's mix and web browser's private ingredient to combine into identical keys. Hence, even an outsider who decrypted and observed all communications up to this point would not know the private ingredients, hence would have difficulty finding out the final key. This process is called a Diffie–Hellman_key_exchange.

At this point, both the web browser and the web server have a shared secret key that they can use to encrypt and decrypt further communications henceforth.

Certificate pinning

Main article: HTTP Public Key Pinning

One way to detect and block many kinds of MITM attacks is "certificate pinning", sometimes called "SSL pinning" but more accurately called "public key pinning".[235] A client that does key pinning adds an extra step beyond the normal X.509 certificate validation: After obtaining the server's certificate in the standard way, the client checks the public key(s) in the server's certificate chain against a set of (hashes of) public keys for the server name. Typically the public key hashes are bundled with the application. For example, Google Chrome includes public key hashes for the *.google.com certificate that detected fraudulent certificates

in 2011. (Chromium does not enforce the hardcoded key pins.) Since then, Mozilla has introduced Public Key Pinning to its Firefox browser.[236]

In other systems the client hopes that the first time it obtains a server's certificate it is trustworthy and stores it; during later sessions with that server, the client checks the server's certificate against the stored certificate to guard against later MITM attacks.

Perspectives Project

The Perspectives Project[237] operates network notaries that clients can use to detect if a site's certificate has changed. By their nature, man-in-the-middle attacks place the attacker between the destination and a single specific target. As such, Perspectives would warn the target that the certificate delivered to the web browser does not match the certificate seen from other perspectives - the perspectives of other users in different times and places. Use of network notaries from a multitude of perspectives makes it possible for a target to detect an attack even if a certificate appears to be completely valid.

DNSChain

DNSChain[238] relies on the security that "block chains" provide to distribute public keys. It uses one pin to secure the connection to the DNSChain server itself, after which all other public keys (that are stored in a block chain) become accessible over a secure channel.

57.7   Protocol details

The TLS protocol exchanges *records*—which encapsulate the data to be exchanged in a specific format (see below). Each record can be compressed, padded, appended with a message authentication code (MAC), or encrypted, all depending on the state of the connection. Each record has a *content type* field that designates the type of data encapsulated, a length field and a TLS version field. The data encapsulated may be control or procedural messages of the TLS itself, or simply the application data needed to be transferred by TLS. The specifications (cipher suite, keys etc.) required to exchange application data by TLS, are agreed upon in the "TLS handshake" between the client requesting the data and the server responding to requests. The protocol therefore defines both the structure of payloads transferred in TLS and the procedure to establish and monitor the transfer.

57.7.1 TLS handshake

When the connection starts, the record encapsulates a "control" protocol—the handshake messaging protocol (*content type* 22). This protocol is used to exchange all the information required by both sides for the exchange of the actual application data by TLS. It defines the messages formatting or containing this information and the order of their exchange. These may vary according to the demands of the client and server—i.e., there are several possible procedures to set up the connection. This initial exchange results in a successful TLS connection (both parties ready to transfer application data with TLS) or an alert message (as specified below).

Basic TLS handshake

A simple connection example follows, illustrating a handshake where the server (but not the client) is authenticated by its certificate:

1. Negotiation phase:

   - A client sends a **ClientHello** message specifying the highest TLS protocol version it supports, a random number, a list of suggested cipher suites and suggested compression methods. If the client is attempting to perform a resumed handshake, it may send a *session ID*.

   - The server responds with a **ServerHello** message, containing the chosen protocol version, a random number, CipherSuite and compression method from the choices offered by the client. To confirm or allow resumed handshakes the server may send a *session ID*. The chosen protocol version should be the highest that both the client and server support. For example, if the client supports version 1.1 and the server supports version 1.2, version 1.1 should be selected; 1.0 should not be selected.

   - The server sends its **Certificate** message (depending on the selected cipher suite, this may be omitted by the server).[239]

   - The server sends its **ServerKeyExchange** message (depending on the selected cipher suite, this may be omitted by the server). This message is sent for all DHE and DH_anon ciphersuites.[1]

   - The server sends a **ServerHelloDone** message, indicating it is done with handshake negotiation.

   - The client responds with a **ClientKeyExchange** message, which may contain a *PreMasterSecret*, public key, or nothing. (Again, this depends on

the selected cipher.) This *PreMasterSecret* is encrypted using the public key of the server certificate.

   - The client and server then use the random numbers and *PreMasterSecret* to compute a common secret, called the "master secret". All other key data for this connection is derived from this master secret (and the client- and server-generated random values), which is passed through a carefully designed pseudorandom function.

2. The client now sends a **ChangeCipherSpec** record, essentially telling the server, "Everything I tell you from now on will be authenticated (and encrypted if encryption parameters were present in the server certificate)." The ChangeCipherSpec is itself a record-level protocol with content type of 20.

   - Finally, the client sends an authenticated and encrypted **Finished** message, containing a hash and MAC over the previous handshake messages.

   - The server will attempt to decrypt the client's *Finished* message and verify the hash and MAC. If the decryption or verification fails, the handshake is considered to have failed and the connection should be torn down.

3. Finally, the server sends a **ChangeCipherSpec**, telling the client, "Everything I tell you from now on will be authenticated (and encrypted, if encryption was negotiated)."

   - The server sends its authenticated and encrypted **Finished** message.

   - The client performs the same decryption and verification.

4. Application phase: at this point, the "handshake" is complete and the application protocol is enabled, with content type of 23. Application messages exchanged between client and server will also be authenticated and optionally encrypted exactly like in their *Finished* message. Otherwise, the content type will return 25 and the client will not authenticate.

Client-authenticated TLS handshake

The following *full* example shows a client being authenticated (in addition to the server like above) via TLS using certificates exchanged between both peers.

1. Negotiation Phase:

- A client sends a **ClientHello** message specifying the highest TLS protocol version it supports, a random number, a list of suggested cipher suites and compression methods.

- The server responds with a **ServerHello** message, containing the chosen protocol version, a random number, cipher suite and compression method from the choices offered by the client. The server may also send a *session id* as part of the message to perform a resumed handshake.

- The server sends its **Certificate** message (depending on the selected cipher suite, this may be omitted by the server).[239]

- The server sends its **ServerKeyExchange** message (depending on the selected cipher suite, this may be omitted by the server). This message is sent for all DHE and DH_anon ciphersuites.[1]

- The server requests a certificate from the client, so that the connection can be mutually authenticated, using a **CertificateRequest** message.

- The server sends a **ServerHelloDone** message, indicating it is done with handshake negotiation.

- The client responds with a **Certificate** message, which contains the client's certificate.

- The client sends a **ClientKeyExchange** message, which may contain a *PreMasterSecret*, public key, or nothing. (Again, this depends on the selected cipher.) This *PreMasterSecret* is encrypted using the public key of the server certificate.

- The client sends a **CertificateVerify** message, which is a signature over the previous handshake messages using the client's certificate's private key. This signature can be verified by using the client's certificate's public key. This lets the server know that the client has access to the private key of the certificate and thus owns the certificate.

- The client and server then use the random numbers and *PreMasterSecret* to compute a common secret, called the "master secret". All other key data for this connection is derived from this master secret (and the client- and server-generated random values), which is passed through a carefully designed pseudorandom function.

2. The client now sends a **ChangeCipherSpec** record, essentially telling the server, "Everything I tell you from now on will be authenticated (and encrypted if encryption was negotiated). " The ChangeCipherSpec is itself a record-level protocol and has type 20 and not 22.

- Finally, the client sends an encrypted **Finished** message, containing a hash and MAC over the previous handshake messages.

- The server will attempt to decrypt the client's *Finished* message and verify the hash and MAC. If the decryption or verification fails, the handshake is considered to have failed and the connection should be torn down.

3. Finally, the server sends a **ChangeCipherSpec**, telling the client, "Everything I tell you from now on will be authenticated (and encrypted if encryption was negotiated). "

- The server sends its own encrypted **Finished** message.

- The client performs the same decryption and verification.

4. Application phase: at this point, the "handshake" is complete and the application protocol is enabled, with content type of 23. Application messages exchanged between client and server will also be encrypted exactly like in their *Finished* message.

Resumed TLS handshake

Public key operations (e.g., RSA) are relatively expensive in terms of computational power. TLS provides a secure shortcut in the handshake mechanism to avoid these operations: resumed sessions. Resumed sessions are implemented using session IDs or session tickets.

Apart from the performance benefit, resumed sessions can also be used for Single sign-on as it is guaranteed that both the original session as well as any resumed session originate from the same client. This is of particular importance for the FTP over TLS/SSL protocol which would otherwise suffer from a man-in-the-middle attack in which an attacker could intercept the contents of the secondary data connections.[240]

Session IDs   In an ordinary *full* handshake, the server sends a *session id* as part of the **ServerHello** message. The client associates this *session id* with the server's IP address and TCP port, so that when the client connects again to that server, it can use the *session id* to shortcut the handshake. In the server, the *session id* maps to the cryptographic parameters previously negotiated, specifically the "master secret". Both sides must have the same "master secret" or the resumed handshake will fail (this prevents an eavesdropper from using a *session id*). The random data in the **ClientHello** and **ServerHello** messages virtually guarantee that the generated connection keys will be different from in the previous connection. In the RFCs, this type of handshake is called an *abbreviated* handshake. It is also described in the literature as a *restart* handshake.

1. Negotiation phase:

   - A client sends a **ClientHello** message specifying the highest TLS protocol version it supports, a random number, a list of suggested cipher suites and compression methods. Included in the message is the *session id* from the previous TLS connection.

   - The server responds with a **ServerHello** message, containing the chosen protocol version, a random number, cipher suite and compression method from the choices offered by the client. If the server recognizes the *session id* sent by the client, it responds with the same *session id*. The client uses this to recognize that a resumed handshake is being performed. If the server does not recognize the *session id* sent by the client, it sends a different value for its *session id*. This tells the client that a resumed handshake will not be performed. At this point, both the client and server have the "master secret" and random data to generate the key data to be used for this connection.

2. The server now sends a **ChangeCipherSpec** record, essentially telling the client, "Everything I tell you from now on will be encrypted." The ChangeCipherSpec is itself a record-level protocol and has type 20 and not 22.

   - Finally, the server sends an encrypted **Finished** message, containing a hash and MAC over the previous handshake messages.

   - The client will attempt to decrypt the server's *Finished* message and verify the hash and MAC. If the decryption or verification fails, the handshake is considered to have failed and the connection should be torn down.

3. Finally, the client sends a **ChangeCipherSpec**, telling the server, "Everything I tell you from now on will be encrypted. "

   - The client sends its own encrypted **Finished** message.

   - The server performs the same decryption and verification.

4. Application phase: at this point, the "handshake" is complete and the application protocol is enabled, with content type of 23. Application messages exchanged between client and server will also be encrypted exactly like in their *Finished* message.

Session tickets RFC 5077 extends TLS via use of session tickets, instead of session IDs. It defines a way to resume a TLS session without requiring that session-specific state is stored at the TLS server.

When using session tickets, the TLS server stores its session-specific state in a session ticket and sends the session ticket to the TLS client for storing. The client resumes a TLS session by sending the session ticket to the server, and the server resumes the TLS session according to the session-specific state in the ticket. The session ticket is encrypted and authenticated by the server, and the server verifies its validity before using its contents.

One particular weakness of this method with OpenSSL is that it always limits encryption and authentication security of the transmitted TLS session ticket to AES128-CBC-SHA256, no matter what other TLS parameters were negotiated for the actual TLS session.[230] This means that the state information (the TLS session ticket) is not as well protected as the TLS session itself. Of particular concern is OpenSSL's storage of the keys in an application-wide context (SSL_CTX), i.e. for the life of the application, and not allowing for re-keying of the AES128-CBC-SHA256 TLS session tickets without resetting the application-wide OpenSSL context (which is uncommon, error-prone and often requires manual administrative intervention).[231][229]

57.7.2 TLS record

This is the general format of all TLS records.

Content type This field identifies the Record Layer Protocol Type contained in this Record.

Version This field identifies the major and minor version of TLS for the contained message. For a ClientHello message, this need not be the *highest* version supported by the client.

Length

The length of Protocol message(s), MAC and Padding, not to exceed 2^{14} bytes (16 KiB).

Protocol message(s) One or more messages identified by the Protocol field. Note that this field may be encrypted depending on the state of the connection.

MAC and Padding A message authentication code computed over the Protocol message, with additional key material included. Note that this field may be encrypted, or not included entirely, depending on the state of the connection.

No MAC or Padding can be present at end of TLS records before all cipher algorithms and parameters have been negotiated and handshaked and then confirmed by sending a CipherStateChange record (see below) for signalling that these parameters will take effect in all further records sent by the same peer.

Handshake protocol

Most messages exchanged during the setup of the TLS session are based on this record, unless an error or warning occurs and needs to be signaled by an Alert protocol record (see below), or the encryption mode of the session is modified by another record (see ChangeCipherSpec protocol below).

Message type This field identifies the Handshake message type.

Handshake message data length This is a 3-byte field indicating the length of the handshake data, not including the header.

Note that multiple Handshake messages may be combined within one record.

Alert protocol

This record should normally not be sent during normal handshaking or application exchanges. However, this message can be sent at any time during the handshake and up to the closure of the session. If this is used to signal a fatal error, the session will be closed immediately after sending this record, so this record is used to give a reason for this closure. If the alert level is flagged as a warning, the remote can decide to close the session if it decides that the session is not reliable enough for its needs (before doing so, the remote may also send its own signal).

Level This field identifies the level of alert. If the level is fatal, the sender should close the session immediately. Otherwise, the recipient may decide to terminate the session itself, by sending its own fatal alert and closing the session itself immediately after sending it. The use of Alert records is optional, however if it is missing before the session closure, the session may be resumed automatically (with its handshakes).

Normal closure of a session after termination of the transported application should preferably be alerted with at least the *Close notify* Alert type (with a simple warning level) to prevent such automatic resume of a new session. Signalling explicitly the normal closure of a secure session before effectively closing its transport layer is useful to prevent or detect attacks (like attempts to truncate the securely transported data, if it intrinsically does not have a predetermined length or duration that the recipient of the secured data may expect).

Description This field identifies which type of alert is being sent.

ChangeCipherSpec protocol

CCS protocol type Currently only 1.

Application protocol

Length Length of application data (excluding the protocol header and including the MAC and padding trailers)

MAC 20 bytes for the SHA-1-based HMAC, 16 bytes for the MD5-based HMAC.

Padding Variable length; last byte contains the padding length.

57.8   Support for name-based virtual servers

From the application protocol point of view, TLS belongs to a lower layer, although the TCP/IP model is too coarse to show it. This means that the TLS handshake is usually (except in the STARTTLS case) performed before the application protocol can start. In the name-based virtual server feature being provided by the application layer, all co-hosted virtual servers share the same certificate because the server has to select and send a certificate immediately after the ClientHello message. This is a big problem in hosting environments because it means either sharing the same certificate among all customers or using a different IP address for each of them.

There are two known workarounds provided by X.509:

- If all virtual servers belong to the same domain, a wildcard certificate can be used.[241] Besides the loose host name selection that might be a problem or not, there is no common agreement about how to match wildcard certificates. Different rules are applied depending on the application protocol or software used.[242]

- Add every virtual host name in the subjectAltName extension. The major problem being that the certificate needs to be reissued whenever a new virtual server is added.

In order to provide the server name, RFC 4366 Transport Layer Security (TLS) Extensions allow clients to include a *Server Name Indication* extension (SNI) in the extended ClientHello message. This extension hints the server immediately which name the client wishes to connect to, so the server can select the appropriate certificate to send to the client.

57.9 Standards

57.9.1 Primary standards

The current approved version of TLS is version 1.2, which is specified in:

- RFC 5246: "The Transport Layer Security (TLS) Protocol Version 1.2".

The current standard replaces these former versions, which are now considered obsolete:

- RFC 2246: "The TLS Protocol Version 1.0".

- RFC 4346: "The Transport Layer Security (TLS) Protocol Version 1.1".

As well as the never standardized SSL 2.0 and 3.0, which are considered obsolete:

- Internet Draft (1995), SSL Version 2.0

- RFC 6101: "The Secure Sockets Layer (SSL) Protocol Version 3.0".

57.9.2 Extensions

Other RFCs subsequently extended TLS.

Extensions to TLS 1.0 include:

- RFC 2595: "Using TLS with IMAP, POP3 and ACAP". Specifies an extension to the IMAP, POP3 and ACAP services that allow the server and client to use transport-layer security to provide private, authenticated communication over the Internet.

- RFC 2712: "Addition of Kerberos Cipher Suites to Transport Layer Security (TLS)". The 40-bit cipher suites defined in this memo appear only for the purpose of documenting the fact that those cipher suite codes have already been assigned.

- RFC 2817: "Upgrading to TLS Within HTTP/1.1", explains how to use the Upgrade mechanism in HTTP/1.1 to initiate Transport Layer Security (TLS) over an existing TCP connection. This allows unsecured and secured HTTP traffic to share the same *well known* port (in this case, http: at 80 rather than https: at 443).

- RFC 2818: "HTTP Over TLS", distinguishes secured traffic from insecure traffic by the use of a different 'server port'.

- RFC 3207: "SMTP Service Extension for Secure SMTP over Transport Layer Security". Specifies an extension to the SMTP service that allows an SMTP server and client to use transport-layer security to provide private, authenticated communication over the Internet.

- RFC 3268: "AES Ciphersuites for TLS". Adds Advanced Encryption Standard (AES) cipher suites to the previously existing symmetric ciphers.

- RFC 3546: "Transport Layer Security (TLS) Extensions", adds a mechanism for negotiating protocol extensions during session initialisation and defines some extensions. Made obsolete by RFC 4366.

- RFC 3749: "Transport Layer Security Protocol Compression Methods", specifies the framework for compression methods and the DEFLATE compression method.

- RFC 3943: "Transport Layer Security (TLS) Protocol Compression Using Lempel-Ziv-Stac (LZS)".

- RFC 4132: "Addition of Camellia Cipher Suites to Transport Layer Security (TLS)".

- RFC 4162: "Addition of SEED Cipher Suites to Transport Layer Security (TLS)".

- RFC 4217: "Securing FTP with TLS".

- RFC 4279: "Pre-Shared Key Ciphersuites for Transport Layer Security (TLS)", adds three sets of new cipher suites for the TLS protocol to support authentication based on pre-shared keys.

Extensions to TLS 1.1 include:

- RFC 4347: "Datagram Transport Layer Security" specifies a TLS variant that works over datagram protocols (such as UDP).

- RFC 4366: "Transport Layer Security (TLS) Extensions" describes both a set of specific extensions and a generic extension mechanism.

- RFC 4492: "Elliptic Curve Cryptography (ECC) Cipher Suites for Transport Layer Security (TLS)".

- RFC 4680: "TLS Handshake Message for Supplemental Data".

- RFC 4681: "TLS User Mapping Extension".

- RFC 4785: "Pre-Shared Key (PSK) Ciphersuites with NULL Encryption for Transport Layer Security (TLS)".

- RFC 5054: "Using the Secure Remote Password (SRP) Protocol for TLS Authentication". Defines the TLS-SRP ciphersuites.

- RFC 5077: "Transport Layer Security (TLS) Session Resumption without Server-Side State".

- RFC 5081: "Using OpenPGP Keys for Transport Layer Security (TLS) Authentication", obsoleted by RFC 6091.

Extensions to TLS 1.2 include:

- RFC 5288: "AES Galois Counter Mode (GCM) Cipher Suites for TLS".

- RFC 5289: "TLS Elliptic Curve Cipher Suites with SHA-256/384 and AES Galois Counter Mode (GCM)".

- RFC 5746: "Transport Layer Security (TLS) Renegotiation Indication Extension".

- RFC 5878: "Transport Layer Security (TLS) Authorization Extensions".

- RFC 5932: "Camellia Cipher Suites for TLS"

- RFC 6066: "Transport Layer Security (TLS) Extensions: Extension Definitions", includes Server Name Indication and OCSP stapling.

- RFC 6091: "Using OpenPGP Keys for Transport Layer Security (TLS) Authentication".

- RFC 6176: "Prohibiting Secure Sockets Layer (SSL) Version 2.0".

- RFC 6209: "Addition of the ARIA Cipher Suites to Transport Layer Security (TLS)".

- RFC 6347: "Datagram Transport Layer Security Version 1.2".

- RFC 6367: "Addition of the Camellia Cipher Suites to Transport Layer Security (TLS)".

- RFC 6460: "Suite B Profile for Transport Layer Security (TLS)".

- RFC 6655: "AES-CCM Cipher Suites for Transport Layer Security (TLS)".

- RFC 7027: "Elliptic Curve Cryptography (ECC) Brainpool Curves for Transport Layer Security (TLS)".

- RFC 7251: "AES-CCM Elliptic Curve Cryptography (ECC) Cipher Suites for TLS".

- RFC 7301: "Transport Layer Security (TLS) Application-Layer Protocol Negotiation Extension".

- RFC 7366: "Encrypt-then-MAC for Transport Layer Security (TLS) and Datagram Transport Layer Security (DTLS)".

- RFC 7465: "Prohibiting RC4 Cipher Suites".

- RFC 7507: "TLS Fallback Signaling Cipher Suite Value (SCSV) for Preventing Protocol Downgrade Attacks".

- RFC 7568: "Deprecating Secure Sockets Layer Version 3.0".

- RFC 7627: "Transport Layer Security (TLS) Session Hash and Extended Master Secret Extension".

Encapsulations of TLS include:

- RFC 5216: "The EAP-TLS Authentication Protocol"

57.9.3 Informational RFCs

- RFC 7457: "Summarizing Known Attacks on Transport Layer Security (TLS) and Datagram TLS (DTLS)"

- RFC 7525: "Recommendations for Secure Use of Transport Layer Security (TLS) and Datagram Transport Layer Security (DTLS)"

57.10 See also

- Application-Layer Protocol Negotiation – a TLS extension used for SPDY and TLS False Start

- Bullrun (decryption program) – a secret anti-encryption program run by the U.S. National Security Agency

- Key ring file

- Multiplexed Transport Layer Security

- Obfuscated TCP

- RdRand

- Server gated cryptography

- SSL acceleration

- tcpcrypt

- Transport Layer Security Channel ID – a proposed protocol extension that improves web browser security via self-signed browser certificates

- Wireless Transport Layer Security

57.11 References

[1] T. Dierks, E. Rescorla (August 2008). "The Transport Layer Security (TLS) Protocol, Version 1.2".

[2] SSL: Intercepted today, decrypted tomorrow, Netcraft, 2013-06-25.

[3] A. Freier, P. Karlton, P. Kocher (August 2011). "The Secure Sockets Layer (SSL) Protocol Version 3.0".

[4] "SSL/TLS in Detail". *Microsoft TechNet*. Updated July 31, 2003.

[5] Thomas Y. C. Woo, Raghuram Bindignavle, Shaowen Su and Simon S. Lam, *SNP: An interface for secure network programming* Proceedings USENIX Summer Technical Conference, June **1994**

[6] "THE SSL PROTOCOL". Netscape Corporation. 2007. Archived from the original on 14 June 1997.

[7] Rescorla 2001

[8] Messmer, Ellen. "Father of SSL, Dr. Taher Elgamal, Finds Fast-Moving IT Projects in the Middle East". *Network World*. Retrieved 30 May 2014.

[9] Greene, Tim. "Father of SSL says despite attacks, the security linchpin has lots of life left". *Network World*. Retrieved 30 May 2014.

[10] "POODLE: SSLv3 vulnerability (CVE-2014-3566)". Retrieved 21 October 2014.

[11] Polk, Tim; McKay, Terry; Chokhani, Santosh (April 2014). "Guidelines for the Selection, Configuration, and Use of Transport Layer Security (TLS) Implementations" (PDF). National Institute of Standards and Technology. p. 67. Retrieved 2014-05-07.

[12] Dierks, T. and E. Rescorla (April 2006). "The Transport Layer Security (TLS) Protocol Version 1.1, RFC 4346".

[13] draft-ietf-tls-tls13-09 - The Transport Layer Security (TLS) Protocol Version 1.3

[14] draft-ietf-tls-tls13-latest

[15] Counting SSL certificates; netcraft; May 13, 2015.

[16] Law Enforcement Appliance Subverts SSL, Wired, 2010-04-03.

[17] New Research Suggests That Governments May Fake SSL Certificates, EFF, 2010-03-24.

[18] P. Eronen, Ed. "RFC 4279: Pre-Shared Key Ciphersuites for Transport Layer Security (TLS)". Internet Engineering Task Force. Retrieved 9 September 2013.

[19] D. Taylor, Ed. "RFC 5054: Using the Secure Remote Password (SRP) Protocol for TLS Authentication". Internet Engineering Task Force. Retrieved December 21, 2014.

[20] Gothard, Peter. "Google updates SSL certificates to 2048-bit encryption". *Computing*. Incisive Media. Retrieved 9 September 2013.

[21] Sean Turner (September 17, 2015). "Consensus: remove DSA from TLS 1.3".

[22] draft-chudov-cryptopro-cptls-04 - GOST 28147-89 Cipher Suites for Transport Layer Security (TLS)

[23] RFC 5288, RFC 5289

[24] RFC 6655, RFC 7251

[25] RFC 6367

[26] RFC 5932, RFC 6367

[27] RFC 6209

[28] RFC 4162

[29] "NIST Special Publication 800-57 *Recommendation for Key Management — Part 1: General (Revised)*" (PDF). 2007-03-08. Retrieved 2014-07-03.

[30] Qualys SSL Labs. "SSL/TLS Deployment Best Practices". Retrieved 2 June 2015.

[31] RFC 5469

[32] draft-ietf-tls-chacha20-poly1305 The ChaCha20-Poly1305 AEAD Cipher for Transport Layer Security

[33] "HTTP VS HTTPS". Retrieved 2015-02-12.

[34] As of September 3, 2015. "SSL Pulse: Survey of the SSL Implementation of the Most Popular Web Sites". Retrieved 2015-09-10.

[35] ivanr. "RC4 in TLS is Broken: Now What?". Qualsys Security Labs. Retrieved 2013-07-30.

[36] Bodo Möller, Thai Duong and Krzysztof Kotowicz. "This POODLE Bites: Exploiting The SSL 3.0 Fallback" (PDF). Retrieved 2014-10-15.

[37] "What browsers support Extended Validation (EV) and display an EV indicator?". Symantec. Retrieved 2014-07-28.

[38] "SHA-256 Compatibility". Retrieved 2015-06-12.

[39] "ECC Compatibility". Retrieved 2015-06-13.

[40] "Tracking the FREAK Attack". Retrieved 2015-03-08.

[41] "FREAK: Factoring RSA Export Keys". Retrieved 2015-03-08.

[42] Google (2012-05-29). "Dev Channel Update". Retrieved 2011-06-01.

[43] Google (2012-08-21). "Stable Channel Update". Retrieved 2012-08-22.

[44] Chromium Project (2013-05-30). "Chromium TLS 1.2 Implementation".

[45] "The Chromium Project: BoringSSL". Retrieved 2015-09-05.

[46] "Chrome Stable Release". Chrome Releases. Google. 2011-10-25. Retrieved 2015-02-01.

[47] "SVN revision log on Chrome 10.0.648.127 release". Retrieved 2014-06-19.

[48] "ImperialViolet - CRIME". 2012-09-22. Retrieved 2014-10-18.

[49] "SSL/TLS Overview". 2008-08-06. Retrieved 2013-03-29.

[50] "Chromium Issue 90392". 2008-08-06. Retrieved 2013-06-28.

[51] "Issue 23503030 Merge 219882". 2013-09-03. Retrieved 2013-09-19.

[52] "Issue 278370: Unable to submit client certificates over TLS 1.2 from Windows". 2013-08-23. Retrieved 2013-10-03.

[53] Möller, Bodo (October 14, 2014). "This POODLE bites: exploiting the SSL 3.0 fallback". Google Online Security blog. Google (via Blogspot). Retrieved 2014-10-28.

[54] "An update on SSLv3 in Chrome.". Security-dev. Google. 2014-10-31. Retrieved 2014-11-04.

[55] "Stable Channel Update". Mozilla Developer Network. Google. 2014-02-20. Retrieved 2014-11-14.

[56] "Changelog for Chrome 33.0.1750.117". Google. Google. Retrieved 2014-11-14.

[57] "Issue 318442: Update to NSS 3.15.3 and NSPR 4.10.2". Retrieved 2014-11-14.

[58] "Issue 693963003: Add minimum TLS version control to about:flags and Finch gate it. - Code Review". Retrieved 2015-01-22.

[59] "Issue 375342: Drop RC4 Support". Retrieved 2015-05-22.

[60] "Issue 436391: Add info on end of life of SSLVersionFallbackMin & SSLVersionMin policy in documentation". Retrieved 2015-04-19.

[61] "Issue 490240: Increase minimum DH size to 1024 bits (tracking bug)". Retrieved 2015-05-29.

[62] "SSLSocket | Android Developers". Retrieved 2015-03-11.

[63] "What browsers work with Universal SSL". Retrieved 2015-06-15.

[64] "Android 5.0 Behavior Changes | Android Developers". Retrieved 2015-03-11.

[65] "Security in Firefox 2". 2008-08-06. Retrieved 2009-03-31.

[66] "Attack against TLS-protected communications". Mozilla Security Blog. Mozilla. 2011-09-27. Retrieved 2015-02-01.

[67] "Introduction to SSL". MDN. Retrieved 2014-06-19.

[68] "NSS 3.15.3 Release Notes". Mozilla Developer Network. Mozilla. Retrieved 2014-07-13.

[69] "MFSA 2013-103: Miscellaneous Network Security Services (NSS) vulnerabilities". Mozilla. Mozilla. Retrieved 2014-07-13.

[70] "Bug 565047 – (RFC4346) Implement TLS 1.1 (RFC 4346)". Retrieved 2013-10-29.

[71] "Bug 480514 – Implement support for TLS 1.2 (RFC 5246)". Retrieved 2013-10-29.

[72] "Bug 733647 – Implement TLS 1.1 (RFC 4346) in Gecko (Firefox, Thunderbird), on by default". Retrieved 2013-12-04.

[73] "Firefox Notes – Desktop". 2014-02-04. Retrieved 2014-02-04.

[74] "Bug 861266 – Implement TLS 1.2 (RFC 5246) in Gecko (Firefox, Thunderbird), on by default". Retrieved 2013-11-18.

[75] "The POODLE Attack and the End of SSL 3.0". Mozilla blog. Mozilla. 2014-10-14. Retrieved 2014-10-28.

[76] "Firefox — Notes (34.0) — Mozilla". mozilla.org. 2014-12-01. Retrieved 2015-04-03.

[77] "Bug 1083058 - A pref to control TLS version fallback". bugzilla.mozilla.org. Retrieved 2014-11-06.

[78] "Bug 1036737 - Add support for draft-ietf-tls-downgrade-scsv to Gecko/Firefox". bugzilla.mozilla.org. Retrieved 2014-10-29.

[79] "Bug 1166031 - Update to NSS 3.19.1". bugzilla.mozilla.org. Retrieved 2015-05-29.

[80] "Bug 1088915 - Stop offering RC4 in the first handshakes". bugzilla.mozilla.org. Retrieved 2014-11-04.

[81] "Firefox — Notes (39.0) — Mozilla". mozilla.org. 2015-06-30. Retrieved 2015-07-03.

[82] "Google, Microsoft, and Mozilla will drop RC4 encryption in Chrome, Edge, IE, and Firefox next year". VentureBeat. 2015-09-01. Retrieved 2015-09-05.

[83]

[84] Microsoft (2012-09-05). "Secure Channel". Retrieved 2012-10-18.

[85] Microsoft (2009-02-27). "MS-TLSP Appendix A". Retrieved 2009-03-19.

[86] "What browsers only support SSLv2?". Retrieved 2014-06-19.

[87] "SHA2 and Windows - Windows PKI blog - Site Home - TechNet Blogs". 2010-09-30. Retrieved 2014-07-29.

[88] http://msdn.microsoft.com/en-us/library/windows/desktop/aa380512(v=vs.85).aspx

[89] http://support.microsoft.com/kb/948963

[90] "Vulnerability in Schannel Could Allow Security Feature Bypass (3046049)". 2015-03-10. Retrieved 2015-03-11.

[91] "Vulnerability in Schannel Could Allow Information Disclosure (3061518)". 2015-05-12. Retrieved 2015-05-22.

[92] "HTTPS Security Improvements in Internet Explorer 7". Retrieved 2013-10-29.

[93] [http://support.microsoft.com/gp/msl-ie-dotnet-an

[94] "Windows 7 adds support for TLSv1.1 and TLSv1.2 - IEInternals - Site Home - MSDN Blogs". Retrieved 2013-10-29.

[95] Thomlinson, Matt (2014-11-11). "Hundreds of Millions of Microsoft Customers Now Benefit from Best-in-Class Encryption". Microsoft Security. Retrieved 2014-11-14.

[96] Microsoft security advisory: Update for disabling RC4

[97] Microsoft (2013-09-24). "IE11 Changes". Retrieved 2013-11-01.

[98] "February 2015 security updates for Internet Explorer". 2015-02-11. Retrieved 2015-02-11.

[99] "Update turns on the setting to disable SSL 3.0 fallback for protected mode sites by default in Internet Explorer 11". Retrieved 2015-02-11.

[100] "Vulnerability in SSL 3.0 Could Allow Information Disclosure". 2015-04-14. Retrieved 2015-04-14.

[101] "Release Notes: Important Issues in Windows 8.1 Preview". Microsoft. 2013-06-24. Retrieved 2014-11-04.

[102] "W8.1(IE11) vs RC4 | Qualys Community". Retrieved 2014-11-04.

[103] [http://www.zdnet.com/article/some-windows-10-enterprise-users-wont-get-microsofts-edge-browser

[104] http://forum.xda-developers.com/windows-phone-8/development/poodle-ssl-vulnerability-secure-windows-t2906203

[105] "What TLS version is used in Windows Phone 8 for secure HTTP connections?". Microsoft. Retrieved 2014-11-07.

[106] https://www.ssllabs.com/ssltest/viewClient.html?name=IE%20Mobile&version=10&platform=Win%20Phone%208.0

[107] "Platform Security". Microsoft. 2014-06-25. Retrieved 2014-11-07.

[108] "Changelog for Opera 9.0 for Windows" at Opera.com

[109] "Opera 2 series". Retrieved 2014-09-20.

[110] "Opera 3 series". Retrieved 2014-09-20.

[111] "Opera 4 series". Retrieved 2014-09-20.

[112] "Changelog for Opera 5.x for Windows". Retrieved 2014-06-19.

[113] "Changelog for Opera [8] Beta 2 for Windows". Retrieved 2014-06-19.

[114] "Web Specifications Supported in Opera 9". Retrieved 2014-06-19.

[115] "Opera: Opera 10 beta for Windows changelog". Retrieved 2014-06-19.

[116] "About Opera 11.60 and new problems with some secure servers". 2011-12-11. Archived from the original on 2012-01-18.

[117] "Security changes in Opera 25; the poodle attacks". 2014-10-15. Retrieved 2014-10-28.

[118] "Unjam the logjam". 2015-06-09. Retrieved 2015-06-11.

[119] "Advisory: RC4 encryption protocol is vulnerable to certain brute force attacks". 2013-04-04. Retrieved 2014-11-14.

[120] "On the Precariousness of RC4". 2013-03-20. Archived from the original on 2013-11-12. Retrieved 2014-11-17.

[121] "Dev.Opera — Opera 14 for Android Is Out!". 2013-05-21. Retrieved 2014-09-23.

[122] "Dev.Opera — Introducing Opera 15 for Computers, and a Fast Release Cycle". 2013-07-02. Retrieved 2014-09-23.

[123] same as Chrome 26–29

[124] same as Chrome 30 and later

[125] same as Chrome 33 and later

[126] "Apple Secures Mac OS X with Mavericks Release - eSecurity Planet". 2013-10-25. Retrieved 2014-06-23.

[127] Ristic, Ivan. "Is BEAST Still a Threat?". *qualys.com*.

[128] Ivan Ristić (2013-10-31). "Apple enabled BEAST mitigations in OS X 10.9 Mavericks". Retrieved 2013-11-07.

[129] Ivan Ristić (2014-02-26). "Apple finally releases patch for BEAST". Retrieved 2014-07-01.

[130] http://support.apple.com/kb/HT6531

[131] http://support.apple.com/kb/HT6541

[132] "About Security Update 2015-002". Retrieved 2015-03-09.

[133] "About the security content of OS X Mavericks v10.9". Retrieved 2014-06-20.

[134] "User Agent Capabilities: Safari 8 / OS X 10.10". Qualsys SSL Labs. Retrieved 2015-03-07.

[135] "About the security content of OS X Yosemite v10.10.4 and Security Update 2015-005". Retrieved 2015-07-03.

[136] Apple (2011-10-14). "Technical Note TN2287 – iOS 5 and TLS 1.2 Interoperability Issues". Retrieved 2012-12-10.

[137] Liebowitz, Matt (2011-10-13). "Apple issues huge software security patches". *NBCNews.com*. Retrieved 2012-12-10.

[138] MWR Info Security (2012-04-16). "Adventures with iOS UIWebviews". Retrieved 2012-12-10., section "HTTPS (SSL/TLS)"

[139] "Secure Transport Reference". Retrieved 2014-06-23. kSSLProtocol2 is deprecated in iOS

[140] "iPhone 3.0: Mobile Safari Gets Enhanced Security Certificate Visualization | The iPhone Blog". 2009-03-31. Archived from the original on 2009-04-03.

[141] https://www.ssllabs.com/ssltest/viewClient.html?name=Safari&version=7&platform=iOS%207.1

[142] schurtertom (October 11, 2013). "SOAP Request fails randomly on one Server but works on an other on iOS7". Retrieved January 5, 2014.

[143] "User Agent Capabilities: Safari 8 / iOS 8.1.2". Qualsys SSL Labs. Retrieved 2015-03-07.

[144] "About the security content of iOS 8.2". Retrieved 2015-03-09.

[145] "About the security content of iOS 8.4". Retrieved 2015-07-03.

[146] Oracle. "Java Cryptography Architecture Oracle Providers Documentation". Retrieved 2012-08-16.

[147] Oracle. "JDK 8 Security Enhancements". Retrieved 2015-02-25.

[148] "Version 1.11.13, 2015-01-11 — Botan". 2015-01-11. Retrieved 2015-01-16.

[149] "[gnutls-devel] GnuTLS 3.4.0 released". 2015-04-08. Retrieved 2015-04-16.

[150] "OpenBSD 5.6 Released". 2014-11-01. Retrieved 2015-01-20.

[151] "LibreSSL 2.3.0 Released". 2015-09-23. Retrieved 2015-09-24.

[152] "MatrixSSL - News". Retrieved 2014-11-09.

[153] "NSS 3.19 release notes". *Mozilla Developer Network*. Mozilla. Retrieved 2015-05-06.

[154] "NSS 3.14 release notes". *Mozilla Developer Network*. Mozilla. Retrieved 2012-10-27.

[155] "NSS 3.15.1 release notes". *Mozilla Developer Network*. Mozilla. Retrieved 2013-08-10.

[156] "Major changes between OpenSSL 1.0.0h and OpenSSL 1.0.1 [14 Mar 2012]". 2012-03-14. Retrieved 2015-01-20.

[157] "RSA BSAFE Technical Specification Comparison Tables" (PDF).

[158] TLS cipher suites in Microsoft Windows XP and 2003

[159] SChannel Cipher Suites in Microsoft Windows Vista

[160] TLS Cipher Suites in SChannel for Windows 7, 2008R2, 8, 2012

[161] "[wolfssl] wolfSSL 3.6.6 Released". 2015-08-20. Retrieved 2015-08-25.

[162] "Java™ SE Development Kit 8, Update 31 Release Notes". Retrieved 2015-01-22.

[163] "Technical Note TN2287: iOS 5 and TLS 1.2 Interoperability Issues". *iOS Developer Library*. Apple Inc. Retrieved 2012-05-03.

[164] https://dev.ssllabs.com/ssltest/clients.html

[165] Georgiev, Martin and Iyengar, Subodh and Jana, Suman and Anubhai, Rishita and Boneh, Dan and Shmatikov, Vitaly (2012). *The most dangerous code in the world: validating SSL certificates in non-browser software. Proceedings of the 2012 ACM conference on Computer and communications security* (PDF). pp. 38–49. ISBN 978-1-4503-1651-4.

[166] Joris Claessens, Valentin Dem, Danny De Cock, Bart Preneel, Joos Vandewalle (2002). "On the Security of Today's Online Electronic Banking Systems". *Computers & Security* **21** (3): 253–265. doi:10.1016/S0167-4048(02)00312-7.

[167] Lawrence, Eric (2005-10-22). "IEBlog: Upcoming HTTPS Improvements in Internet Explorer 7 Beta 2". MSDN Blogs. Retrieved 2007-11-25.

[168] "Bugzilla@Mozilla — Bug 236933 – Disable SSL2 and other weak ciphers". Mozilla Corporation. Retrieved 2007-11-25.

[169] "Opera 9.5 for Windows Changelog" at Opera.com: "Disabled SSL v2 and weak ciphers."

[170] "Firefox still sends SSLv2 handshake even though the protocol is disabled". 2008-09-11.

[171] "Opera 10 for Windows changelog" at Opera.com: "Removed support for SSL v2 and weak ciphers"

[172] Pettersen, Yngve (2007-04-30). "10 years of SSL in Opera — Implementer's notes". Opera Software. Archived from the original on October 12, 2007. Retrieved 2007-11-25.

[173] National Institute of Standards and Technology (December 2010). "Implementation Guidance for FIPS PUB 140-2 and the Cryptographic Module Validation Program" (PDF).

[174] "RFC 7457 : Summarizing Known Attacks on Transport Layer Security (TLS) and Datagram TLS (DTLS)".

[175] Eric Rescorla (2009-11-05). "Understanding the TLS Renegotiation Attack". *Educated Guesswork*. Retrieved 2009-11-27.

[176] "SSL_CTX_set_options SECURE_RENEGOTIATION". *OpenSSL Docs*. 2010-02-25. Retrieved 2010-11-18.

[177] "GnuTLS 2.10.0 released". *GnuTLS release notes*. 2010-06-25. Retrieved 2011-07-24.

[178] "NSS 3.12.6 release notes". *NSS release notes*. 2010-03-03. Retrieved 2011-07-24.

[179] A. Langley; N. Modadugu; B. Moeller (2 June 2010). "Transport Layer Security (TLS) False Start". *Internet Engineering Task Force*. IETF. Retrieved 31 July 2013.

[180] Wolfgang, Gruener. "False Start: Google Proposes Faster Web, Chrome Supports It Already". Archived from the original on October 7, 2010. Retrieved 9 March 2011.

[181] Brian, Smith. "Limited rollback attacks in False Start and Snap Start". Retrieved 9 March 2011.

[182] Adrian, Dimcev. "False Start". *Random SSL/TLS 101*. Retrieved 9 March 2011.

[183] Mavrogiannopoulos, Nikos and Vercautern, Frederik and Velichkov, Vesselin and Preneel, Bart (2012). *A cross-protocol attack on the TLS protocol. Proceedings of the 2012 ACM conference on Computer and communications security* (PDF). pp. 62–72. ISBN 978-1-4503-1651-4.

[184] Thai Duong and Juliano Rizzo (2011-05-13). "Here Come The ⊕ Ninjas".

[185] Dan Goodin (2011-09-19). "Hackers break SSL encryption used by millions of sites".

[186] "Y Combinator comments on the issue". 2011-09-20.

[187] "Security of CBC Ciphersuites in SSL/TLS: Problems and Countermeasures". 2004-05-20. Archived from the original on 2012-06-30.

[188] Ristic, Ivan (Sep 10, 2013). "Is BEAST Still a Threat?". Retrieved 8 October 2014.

[189] "Attack against TLS-protected communications". *Mozilla Security Blog*. Mozilla. 2011-09-27. Retrieved 2015-02-01.

[190] Brian Smith (2011-09-30). "(CVE-2011-3389) Rizzo/Duong chosen plaintext attack (BEAST) on SSL/TLS 1.0 (facilitated by websockets −76)".

[191] "Vulnerability in SSL/TLS Could Allow Information Disclosure (2643584)". 2012-01-10.

[192] Ristic, Ivan (Oct 31, 2013). "Apple Enabled BEAST Mitigations in OS X 10.9 Mavericks". Retrieved 8 October 2014.

[193] Dan Goodin (2012-09-13). "Crack in Internet's foundation of trust allows HTTPS session hijacking". Ars Technica. Retrieved 2013-07-31.

[194] Dennis Fisher (September 13, 2012). "CRIME Attack Uses Compression Ratio of TLS Requests as Side Channel to Hijack Secure Sessions". ThreatPost. Archived from the original on September 15, 2012. Retrieved 2012-09-13.

[195] Goodin, Dan (1 August 2013). "Gone in 30 seconds: New attack plucks secrets from HTTPS-protected pages". *Ars Technica*. Condé Nast. Retrieved 2 August 2013.

[196] Leyden, John (2 August 2013). "Step into the BREACH: New attack developed to read encrypted web data". *The Register*. Retrieved 2 August 2013.

[197] P. Gutmann (September 2014). "Encrypt-then-MAC for Transport Layer Security (TLS) and Datagram Transport Layer Security (DTLS)".

[198] Bodo Möller (October 14, 2014). "This POODLE bites: exploiting the SSL 3.0 fallback". Retrieved 2014-10-15.

[199] Hagai Bar-El. "Poodle flaw and IoT". Retrieved 15 October 2014.

[200] Langley, Adam (December 8, 2014). "The POODLE bites again". Retrieved 2014-12-08.

[201] security – Safest ciphers to use with the BEAST? (TLS 1.0 exploit) I've read that RC4 is immune – Server Fault

[202] Pouyan Sepehrdad, Serge Vaudenay, Martin Vuagnoux (2011). "Discovery and Exploitation of New Biases in RC4". *Lecture Notes in Computer Science* **6544**: 74–91. doi:10.1007/978-3-642-19574-7_5.

[203] Green, Matthew. "Attack of the week: RC4 is kind of broken in TLS". *Cryptography Engineering*. Retrieved March 12, 2013.

[204] Nadhem AlFardan, Dan Bernstein, Kenny Paterson, Bertram Poettering and Jacob Schuldt. "On the Security of RC4 in TLS". Royal Holloway University of London. Retrieved March 13, 2013.

[205] AlFardan, Nadhem J.; Bernstein, Daniel J.; Paterson, Kenneth G.; Poettering, Bertram; Schuldt, Jacob C. N. (8 July 2013). "On the Security of RC4 in TLS and WPA" (PDF). Retrieved 2 September 2013.

[206] AlFardan, Nadhem J.; Bernstein, Daniel J.; Paterson, Kenneth G.; Poettering, Bertram; Schuldt, Jacob C. N. (15 August 2013). *On the Security of RC4 in TLS* (PDF). 22nd USENIX Security Symposium. p. 51. Retrieved 2 September 2013. Plaintext recovery attacks against RC4 in TLS are feasible although not truly practical

[207] Goodin, Dan. "Once-theoretical crypto attack against HTTPS now verges on practicality". *Ars Technical*. Conde Nast. Retrieved 16 July 2015.

[208] "Mozilla Security Server Side TLS Recommended Configurations". Mozilla. Retrieved 2015-01-03.

[209] "Security Advisory 2868725: Recommendation to disable RC4". Microsoft. 2013-11-12. Retrieved 2013-12-04.

[210] "Ending support for the RC4 cipher in Microsoft Edge and Internet Explorer 11". Microsoft Edge Team. September 1, 2015.

[211] Langley, Adam (Sep 1, 2015). "Intent to deprecate: RC4".

[212] Barnes, Richard (Sep 1, 2015). "Intent to ship: RC4 disabled by default in Firefox 44".

[213] John Leyden (1 August 2013). "Gmail, Outlook.com and e-voting 'pwned' on stage in crypto-dodge hack". *The Register*. Retrieved 1 August 2013.

[214] "BlackHat USA Briefings". *Black Hat 2013*. Retrieved 1 August 2013.

[215] "SMACK: State Machine AttaCKs".

[216] Dan Goodin (2015-05-20). "HTTPS-crippling attack threatens tens of thousands of Web and mail servers". Ars Technica.

[217] "Why is it called the 'Heartbleed Bug'?".

[218] "Heartbleed Bug vulnerability [9 April 2014]".

[219] Daniel Bleichenbacher (August 2006). "Bleichenbacher's RSA signature forgery based on implementation error".

[220] Intel Security: Advanced Threat Research (September 2014). "BERserk".

[221] Lenovo PCs ship with Man-In-The-Middle adware that breaks HTTPS connections, Dan Goodin, Arstechnica, February 19, 2015

[222] Komodia Superfish SSL validation is broken

[223] Diffie, Whitfield; van Oorschot, Paul C.; Wiener, Michael J. (June 1992). "Authentication and Authenticated Key Exchanges". *Designs, Codes and Cryptography* **2** (2): 107–125. doi:10.1007/BF00124891. Retrieved 2008-02-11.

[224] Discussion on the TLS mailing list in October 2007

[225] "Protecting data for the long term with forward secrecy". Retrieved 2012-11-05.

[226] Vincent Bernat. "SSL/TLS & Perfect Forward Secrecy". Retrieved 2012-11-05.

[227] "SSL Labs: Deploying Forward Secrecy". Qualys.com. 25 June 2013. Retrieved 10 July 2013.

[228] Ristic, Ivan (5 August 2013). "SSL Labs: Deploying Forward Secrecy". Qualsys. Retrieved 31 August 2013.

[229] Langley, Adam (27 June 2013). "How to botch TLS forward secrecy". *imperialviolet.org*.

[230] Daignière, Florent. "TLS "Secrets": Whitepaper presenting the security implications of the deployment of session tickets (RFC 5077) as implemented in OpenSSL" (PDF). Matta Consulting Limited. Retrieved 7 August 2013.

[231] Daignière, Florent. "TLS "Secrets": What everyone forgot to tell you..." (PDF). Matta Consulting Limited. Retrieved 7 August 2013.

[232] L.S. Huang; S. Adhikarla; D. Boneh; C. Jackson (2014). "An Experimental Study of TLS Forward Secrecy Deployments". *IEEE Internet Computing* (IEEE) **18** (6): 43–51. Retrieved 16 October 2015.

[233] "Protecting data for the long term with forward secrecy". Retrieved 2014-03-07.

[234] Hoffman-Andrews, Jacob. "Forward Secrecy at Twitter". *Twitter*. Twitter. Retrieved 2014-03-07.

[235] "Certificate Pinning".

[236] "Public key pinning released in Firefox"

[237] Perspectives Project

[238] DNSChain

[239] These certificates are currently X.509, but RFC 6091 also specifies the use of OpenPGP-based certificates.

[240] Chris (2009-02-18). "vsftpd-2.1.0 released – Using TLS session resume for FTPS data connection authentication". Scarybeastsecurity. blogspot.com. Retrieved 2012-05-17.

[241] *Wildcard SSL Certificate overview*, retrieved 2015-07-02

[242] *Named-based SSL virtual hosts: how to tackle the problem* (PDF), retrieved 2012-05-17

57.12 Further reading

- Wagner, David; Schneier, Bruce (November 1996). "Analysis of the SSL 3.0 Protocol" (PDF). *The Second USENIX Workshop on Electronic Commerce Proceedings*. USENIX Press. pp. 29–40.

- Eric Rescorla (2001). *SSL and TLS: Designing and Building Secure Systems*. United States: Addison-Wesley Pub Co. ISBN 0-201-61598-3.

- Stephen A. Thomas (2000). *SSL and TLS essentials securing the Web*. New York: Wiley. ISBN 0-471-38354-6.

- Bard, Gregory (2006). "A Challenging But Feasible Blockwise-Adaptive Chosen-Plaintext Attack On Ssl". *International Association for Cryptologic Research* (136). Retrieved 2011-09-23.

- Canvel, Brice. "Password Interception in a SSL/TLS Channel". Retrieved 2007-04-20.

- IETF Multiple Authors. "RFC of change for TLS Renegotiation". Retrieved 2009-12-11.

- Creating VPNs with IPsec and SSL/TLS Linux Journal article by Rami Rosen

- Polk, Tim; McKay, Kerry; Chokhani, Santosh (April 2014). "Guidelines for the Selection, Configuration, and Use of Transport Layer Security (TLS) Implementations" (PDF). National Institute of Standards and Technology. Retrieved 2014-05-07.

57.13 External links

Specifications (see Standards section for older SSL 2.0, SSL 3.0, TLS 1.0, TLS 1.1 links)

- RFC 5246 - The Transport Layer Security (TLS) Protocol Version 1.2

- IETF (Internet Engineering Task Force) TLS Workgroup

Other

- OWASP: Transport Layer Protection Cheat Sheet

- A talk on SSL/TLS that tries to explain things in terms that people might understand.

- SSL: Foundation for Web Security

- TLS Renegotiation Vulnerability – IETF Tools

- Trustworthy Internet Movement – SSL Pulse – Survey of TLS/SSL implementation of the most popular web sites

- How to Generate CSR for SSL

This article is based on material taken from the Free On-line Dictionary of Computing prior to 1 November 2008 and incorporated under the "relicensing" terms of the GFDL, version 1.3 or later.

Chapter 58

Twinkle (software)

For the Wikipedia software, see WP:TWINKLE.

Twinkle is a free and open source software application for Voice over Internet Protocol (VoIP) voice communications in IP networks, such as the Internet.[1][2] It is designed for GNU/Linux operating systems and uses the Qt toolkit for its graphical user interface. For call signaling it employs the Session Initiation Protocol (SIP). It also features direct IP-to-IP calls. Media streams are transmitted via the Real-time Transport Protocol (RTP) which may be encrypted with the Secure Real-time Transport Protocol (SRTP) and the ZRTP security protocols.

Since version 1.3.2 (September 2008), Twinkle supports message exchange and a *buddy-list* feature for presence notification, showing the online-status of predefined communications partners (provider-support needed).

58.1 Supported audio formats

- G.711 A-law: 64 kbit/s payload, 8 kHz sampling rate

- G.711 μ-law: 64 kbit/s payload, 8 kHz sampling rate

- G.726: 16, 24, 32 or 40 kbit/s payload, 8 kHz sampling rate

- GSM: 13 kbit/s payload, 8 kHz sampling rate

- G.729: 8 kbit/s payload, 8 kHz sampling rate

- iLBC: 13.3 or 15.2 kbit/s payload, 8 kHz sampling rate

- Speex narrow band: 15.2 kbit/s payload, 8 kHz sampling rate

- Speex wide band: 28 kbit/s payload, 16 kHz sampling rate

- Speex ultra wide band: 36 kbit/s payload, 32 kHz sampling rate

58.2 See also

- Comparison of VoIP software

- List of SIP software

- Opportunistic encryption

58.3 References

[1] Twinklephone new upstream website

[2] Twinklephone original website

Chapter 59

Typhoid adware

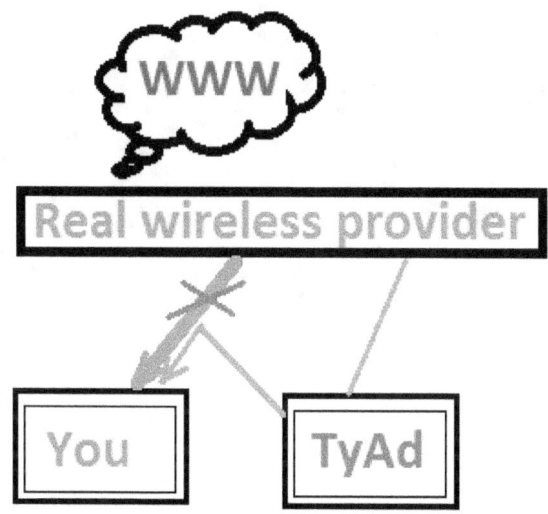

How typhoid adware works

Typhoid adware is a type of computer security threat that uses a Man-in-the-middle attack to inject advertising into web pages a user visits when using a public network, like a WiFi hotspot. Researchers from the University of Calgary identified the issue, which does not require the affected computer to have adware installed in order to display advertisements on this computer. The researchers said that the threat was not yet observed, but described its mechanism and potential countermeasures.[1][2]

59.1 Description

The environment for the threat to work is an area of non-encrypted wireless connection, such as a wireless internet cafe or other WiFi hotspots. Typhoid adware would trick a laptop to recognize it as the wireless provider and inserts itself into the route of the wireless connection between the computer and the actual provider. After that the adware may insert various advertisements into the data stream to appear on the computer during the browsing session. In this way even a video stream, e.g., from YouTube may be modified. What is more, the adware may run from an infested computer whose owner would not see any manifestations, yet will affect neighboring ones. For the latter peculiarity it was named in an analogy with Typhoid Mary (Mary Mallon), the first identified person who never experienced any symptoms yet spread infection.[1][3] At the same time running antivirus on the affected computer is useless, since it has no adware installed.

The implemented proof of concept was described in an article written in March 2010, by Daniel Medeiros Nunes de Castro, Eric Lin, John Aycock, and Mea Wang.[3]

While typhoid adware is a variant of the well-known man-in-the-middle attack, the researchers point out a number of new important issues, such as protection of video content and growing availability of public wireless internet access which are not well-monitored.[3][4]

Researchers say that annoying advertisements are only the tip of the iceberg. A serious danger may come from, e.g., promotions of rogue antivirus software seemingly coming from a trusted source.[1]

59.2 Defenses

Suggested countermeasures include:

- Various approaches to detection of ARP spoofing, rogue DHCP servers and other "man-in-the-middle" tricks in the network by network administrators[3]

- Detection of content modification[3]

- Detection of timing anomalies[3]

All these approaches have been investigated earlier in other contexts.[3]

59.3   See also

- Countermeasure (computer)

- Mobile virus

- Piggybacking (Internet access)

- Threat (computer)

- Vulnerability (computing)

- Wireless LAN security

- Wireless intrusion prevention system

59.4   References

[1] "Will Typhoid adware become an epidemic?"

[2] Beware Typhoid Adware

[3] "Typhoid Adware"

[4] "New Threat For Wireless Networks: Typhoid Adware".

Chapter 60

User Activity Monitoring

In the field of information security, **User Activity Monitoring** (UAM) is the monitoring and recording of user actions. UAM captures user actions, including the use of applications, windows opened, system commands executed, check boxes clicked, text entered/edited, URLs visited and nearly every other on-screen event to protect data by ensuring that employees and contractors are staying within their assigned tasks, and posing no risk to the organization.

UAM software can deliver video-like playback of user activity and process the videos into user activity logs that keep step-by-step records of user actions that can be searched and analyzed to investigate any out-of-scope activities.

60.1 Issues

The need for UAM has risen in the last decade due to the increase in security incidents that directly or indirectly involve user credentials, exposing company information or sensitive files. In 2014, there were 761 data breaches in the United States, resulting in over 83 million exposed customer and employee records.[1] With 76% of these breaches resulting from weak or exploited user credentials, UAM has become a significant component of IT infrastructure.[2] The main populations of users that UAM aims to mitigate risks with are:

60.1.1 Contractors

Contractors are used in organizations to complete various IT and operational tasks. Remote vendors that have access to company data are risks to company infrastructure. Even with no malicious intent, an external user like a contractor is a major security liability.

60.1.2 Everyday Business Users

70% of regular business users admitted to having access to more data than necessary. Generalized accounts give regular business users access to classified company data.[3] This makes insider threats a reality for any business that uses generalized accounts.

60.1.3 IT Users

Administrator accounts are heavily monitored due to the high profile nature of their access. However, current log tools can generate "log fatigue" on these admin accounts. Log fatigue is the overwhelming sensation of trying to handle a vast amount of logs on an account as a result of too many user actions. Harmful user actions can easily be overlooked with thousands of user actions being compiled every day.

60.1.4 Overall User Risk

According to the Verizon Data Breach Incident Report, "The first step in protecting your data is in knowing where it is and who has access to it."[1] In today's IT environment, "there is a lack of oversight and control over how and who among employees has access to confidential, sensitive information." [4] This apparent gap is one of many factors that have resulted in a major number of security issues for companies.

60.2 The Three Major Components of UAM

Most companies that use UAM usually separate the necessary aspects of UAM into three major components.

60.2.1   Visual Forensics

Visual Forensics involves creating a visual summary of potentially hazardous user activity. Each user action is logged, and recorded. Once a user session is completed, UAM has created both a written record and a visual record, whether it be screen-captures or video of exactly what a user has done. This written record differs from that of a SIEM or logging tool, because it captures data at a user-level not at a system level –providing plain English logs rather than SysLogs (originally created for debugging purposes). These textual logs are paired with the corresponding screen-captures or video summaries. Using these corresponding logs and images, the visual forensics component of UAM allows for organizations to search for exact user actions in case of a security incident. In the case of a security threat, i.e. a data breach, Visual Forensics are used to show exactly what a user did, and everything leading up to the incident. Visual Forensics can also be used to provide evidence to any law enforcement that investigate the intrusion.

60.2.2   User Activity Alerting

User Activity Alerting serves the purpose of notifying whoever operates the UAM solution to a mishap or misstep concerning company information. Real-time alerting enables the console administrator to be notified the moment an error or intrusion occurs. Alerts are aggregated for each user to provide a user risk profile and threat ranking. Alerting is customizable based on combinations of users, actions, time, location, and access method. Alerts can be triggered simply such as opening an application, or entering a certain keyword or web address. Alerts can also be customized based on user actions within an application, such as deleting or creating a user and executing specific commands.

60.2.3   User Behavior Analytics

User behavior analytics add an additional layer of protection that will help security professionals keep an eye on the weakest link in the chain. By monitoring user behavior, with the help of dedicated software that analyzes exactly what the user does during their session, security professionals can attach a risk factor to the specific users and/or groups, and immediately be alerted with a red flag warning when a high-risk user does something that can be interpreted as a high-risk action such as exporting confidential customer information, performing large database queries that are out of the scope of their role, accessing resources that they shouldn't be accessing and so forth.

60.3   Major Features of UAM

60.3.1   Capturing User Activity

UAM collects user data by recording activity by every user on applications, web pages and internal systems and databases. UAM spans all access levels and access strategies (RDP, SSH, Telnet, ICA, direct console login, etc....). Some UAM solutions pair with Citrix and VMware environments.[5]

60.3.2   User Activity Logs

UAM solutions transcribe all documented activities into user activity logs. UAM logs match up with video-playbacks of concurrent actions. Some examples of items logged are names of applications run, titles of pages opened, URLs, text (typed, edited, copied/pasted), commands, and scripts.

60.3.3   Video-like Playback

UAM uses screen-recording technology that captures individual user actions. Each video-like playback is saved and accompanied by a user activity log. Playbacks differ from traditional video playback to screen scraping, which is the compiling of sequential screen shots into a video like replay. The user activity logs combined with the video-like playback provides a searchable summary of all user actions. This enables companies to not only read, but also view exactly what a particular user did on company systems.

60.4   User Privacy Concerns

Some companies and employees have raised issue with the user privacy aspect of UAM. They believe employees will resist the idea of having their actions monitored, even if it is being done for security purposes. In reality, most UAM strategies address these concerns.

While it is possible to monitor every single user action, the purpose of UAM systems is not to snoop on employee browsing history. UAM solutions use policy-based activity recording, which enables the console administrator to program exactly what is and isn't monitored.

60.5   Audit and Compliance

Many regulations require a certain level of UAM while others only require logs of activity for audit purposes.

UAM meets a variety of regulatory compliance requirements (HIPAA, ISO 27001, SOX, PCI etc....). UAM is typically implemented for the purpose of audits and compliance, to serve as a way for companies to make their audits easier and more efficient. An audit information request for information on user activity can be met with UAM. Unlike normal log or SIEM tools, UAM can help speed up an audit process by building the controls necessary to navigate an increasingly complex regulatory environment. The ability to replay user actions provides support for determining the impact on regulated information during security incident response.

60.6 Appliance-Based vs. Software-Based

UAM solutions have two deployment models. Appliance-based monitoring approaches that use dedicated hardware to conduct monitoring by looking at network traffic. Software-based monitoring approaches that use software agents installed on the nodes accessed by users.

More commonly, UAM solutions are software-based and require the installation of an agent on systems (servers, desktops, VDI servers, Terminal servers) across which users you want to monitor. These agents capture user activity and reports information back to a central console for storage and analysis. These solutions may be quickly deployed in a phased manner by targeting high-risk users and systems with sensitive information first, allowing the organization to get up and running quickly and expand to new user populations as the business requires.

60.7 References

[1] "Data Breach Reports" (PDF). *Identity Theft Resource Center*. 31 December 2014. Retrieved 19 January 2015.

[2] "2014 Data Breach Investigation Report". *Verizon*. 14 April 2014. Retrieved 19 January 2015.

[3] "Virtualisation: Exposing the Intangible Enterprise". *Enterprise Management Associates*. 14 August 2014. Retrieved 19 January 2015.

[4] "Corporate Data: A Protected Asset or a Ticking Time Bomb?" (PDF). *Ponemon Institute*. December 2014. Retrieved 19 January 2015.

[5] "ObserveIT for Citrix: The Smart Alternative to SmartAuditor!". *ObserveIT*. Retrieved 19 January 2015.

Chapter 61

Vanish (computer science)

Vanish is a project at the University of Washington which endeavors to "give users control over the lifetime of personal data stored on the web."[1] The project proposes to allow a user to enter information that he or she will send out across the internet, thereby relinquishing control of it. However, the user is able to include an "expiration date," after which the information is no longer usable by anyone who may have a copy of it, even the creator.[2] The Vanish approach was found to be vulnerable to a Sybil attack, and thus insecure, by a team called Unvanish from the University of Texas, University of Michigan, and Princeton.[3]

61.1 Theory

Vanish acts by automating the encryption of information entered by the user with an encryption key that is unknown to the user. Along with the actual information the user enters, he or she also enters metadata concerning how long the information should remain available. The system then encrypts the information, but does not store either the encryption key or the, instead breaking up the decryption key into smaller components that are disseminated across distributed hash tables, or DHTs via the internet. The DHTs refresh information within their nodes on a set schedule unless told to persist the information. The time-delay entered by the user in the metadata controls how long the DHTs should allow the information to persist, but once that time period is over, the DHTs will reuse those nodes, making the information about the decryption stored irretrievable. As long as the decryption key may be reassembled from the DHTs, the information is retrievable. However, the time initially entered by the user has lapsed, the information is not recoverable, as the user was never informed of the decryption key.[4]

61.2 Implementation

Vanish currently exists as a Firefox plug-in which allows a user to enter text into either a standard Gmail email or Facebook message, and choose to send the message via Vanish. The message is then encrypted and sent via the normal networking pathways through the cloud to the recipient. The recipient must have the same Firefox plug-in to decrypt the message. The plugin accesses BitTorrent DHTs, which have 8-hour lifespans. This means the user may select an expiration date for the message in increments of 8 hours. After the expiration of the user-defined time span, the information in the DHT is overwritten, thereby eliminating the key. While both the user and recipient may have copies of the original encrypted message, the key used to turn it back into plain text is now gone.

Although this particular instance of the data has become inaccessible, it's important to note that the information can always be saved by other means before expiration (copied, or even via screen shots) and published again.

61.3 See also

- Cryptography
- Internet privacy
- Proactive Cyber Defence

61.4 References

[1] "' This article will self-destruct: A tool to make online personal data vanish". http://www.washington.edu/. Retrieved 2009-07-21.

[2] "'Privacy Tool Makes Internet Postings Vanish '". InformationWeek. Retrieved 2009-07-24.

[3] "'Unvanish: Reconstructing Self-Destructing Data'".

[4] "' Vanish: Increasing Data Privacy with Self-Destructing Data" (PDF). http://vanish.cs.washington.edu. Retrieved 2010-12-07.

Chapter 62

Verifiable computing

Verifiable computing (or verified computation or verified computing) is enabling a computer to offload the computation of some function, to other perhaps untrusted clients, while maintaining verifiable results. The other clients evaluate the function and return the result with a proof that the computation of the function was carried out correctly. The introduction of this notion came as a result of the increasingly common phenomenon of "outsourcing" computation to untrusted users in projects such as SETI@home and also to the growing desire of weak clients to outsource computational tasks to a more powerful computation service like in Cloud computing. The term *verifiable computing* was formalized by Rosario Gennaro, Craig Gentry, and Bryan Parno.[1]

62.1   Motivation and overview

The growing desire to outsource computational tasks from a relatively weak computational device (client) to a more powerful computation services (worker), and the problem of dishonest workers who modify their client's software to return plausible results without performing the actual work [2] motivated the formalization of the notion of Verifiable Computation.[1]

Verifiable computing is not only concerned with getting the result of the outsourced function on the client's input and the proof of its correctness, but also with the client being able to verify the proof with significantly less computational effort than computing the function from scratch.

Considerable attention has been devoted to verifying the computation of functions performed by untrusted workers including the use of secure coprocessors,[3][4] Trusted Platform Modules (TPMs),[5] interactive proofs,[6][7] probabilistically checkable proofs,[8][9] efficient arguments,[10][11] and Micali's CS proofs.[12] These verifications are either interactive which require the client to interact with the worker to verify the correctness proof,[10][11] or are non-interactive protocols which can be proven in the random oracle model.[12]

62.2   Verification by replication

The largest verified computation (SETI@home) uses verification by replication.

The SETI@home verification process involves one client machine and many worker machines. The client machine sends identical workunits to multiple computers (at least 2).

When not enough results are returned in a reasonable amount of time—due to machines accidentally turned off, communication breakdowns, etc. -- or the results do not agree—due to computation errors, cheating by submitting false data without actually doing the work, etc. -- then the client machine sends more identical workunits to other worker machines. Once a minimum quorum (often 2) of the results agree, then the client assumes those results (and other identical results for that workunit) are correct. The client grants credit to all machines that returned the correct results.

62.3   Verifiable computation

Gennaro et al.[1] defined the notion of Verifiable Computation Scheme as a protocol between two polynomial time parties to collaborate on the computation of a function F: $\{0,1\}^n \rightarrow \{0,1\}^m$. This scheme consists of three main phases:

62.3.1   Preprocessing

This stage is performed once by the client in order to calculate some auxiliary information associated with F. Part of this information is public to be shared with the worker while the rest is private and kept with the client.

62.3.2 Input preparation

In this stage, the client calculates some auxiliary information about the input of the function. Part of this information is public while the rest is private and kept with the client. The public information is sent to the worker to compute F on the input data.

62.3.3 Output computation and verification

In this stage, the worker uses the public information associated with the function F and the input, which are calculated in the previous two phases, to compute an encoded output of the function F on the provided input.

This result is then returned to the client to verify its correctness by computing the actual value of the output by decoding the result returned by the worker using the private information calculated in the previous phases.

The defined notion of verifiable computation scheme minimizes the interaction between the client and the worker into exactly two messages, where a single message sent from each party to the other party during the different phases of the protocol.[1]

62.3.4 An example scheme based on Fully homomorphic encryption

Gennaro et al.[1] defined a verifiable computation scheme for any function *F* using Yao's Garbled Circuit[13][14] combined with a fully homomorphic encryption system.

This verifiable computation scheme **VC** is defined as follows:[1]

VC = (**KeyGen, ProbGen, Compute, Verify**) consists of four algorithms as follows:

1. **KeyGen(F, λ) $\rightarrow$ (PK, SK)**: The randomized key generation algorithm generates two keys, public and private, based on the security parameter λ. The public key encodes the target function F and is sent to the worker to compute F. On the other hand, the secret key is kept private by the client.

2. **ProbGenSK(x) $\rightarrow$ (σx, τx)**: The problem generation algorithm encodes the function input x into two values, public and private, using the secret key SK. The public value σx is given to the worker to compute F(x) with, while the secret value τx is kept private by the client.

3. **ComputePK(σx) $\rightarrow$ σy**: The worker computes an encoded value σy of the function's output y = F(x) using the client's public key PK and the encoded input σx.

4. **VerifySK(τx,σy) $\rightarrow$ y $\cup$ $\perp$**: The verification algorithm converts the worker's encoded output σy into the actual output of the function F using both the secret key SK and the secret "decoding" τx. It outputs y = F(x) if the σy represents a valid output of F on x, or outputs $\perp$ otherwise.

The protocol of the verifiable computations scheme defined by Gennaro et al.[1] works as follows:

The function F should be represented as a Boolean circuit on which the key generation algorithm would be applied. The key generation algorithm runs Yao's garbling procedure over this Boolean circuit to compute the public and secret keys. The public key (PK) is composed of all the ciphertexts that represent the garbled circuit, and the secret key (SK) is composed of all the random wire labels. The generated secret key is then used in the problem generation algorithm. This algorithm first generates a new pair of public and secret keys for the homomorphic encryption scheme, and then uses these keys with the homomorphic scheme to encrypt the correct input wires, represented as the secret key of the garbled circuit. The produced ciphertexts represent the public encoding of the input (σx) that is given to the worker, while the secret key (τx) is kept private by the client. After that, the worker applies the computation steps of the Yao's protocol over the ciphertexts generated by the problem generation algorithm. This is done by recursively decrypting the gate ciphertexts until arriving to the final output wire values (σy). The homomorphic properties of the encryption scheme enable the worker to obtain an encryption of the correct output wire. Finally, the worker returns the ciphertexts of the output to the client who decrypts them to compute the actual output y = F(x) or $\perp$.

The definition of the verifiable computation scheme states that the scheme should be both correct and secure. **Scheme Correctness** is achieved if the problem generation algorithm produces values that enable an honest worker to compute encoded output values that will verify successfully and correspond to the evaluation of F on those inputs. On the other hand, a verifiable computation scheme is *secure* if a malicious worker cannot convince the verification algorithm to accept an incorrect output for a given function F and input x.

62.4 Practical verifiable computing

Although it was shown that verifiable computing is possible in theory (using fully homomorphic encryption or via probabilistically checkable proofs), most of the known constructions are very expensive in practice. Recently, some researchers have looked at making verifiable computation practical. One such effort is the work of UT Austin researchers .[15] The authors start with an argument system based on probabilistically checkable proofs and reduce its costs by a factor of 10^{20}. They also implemented their techniques in the Pepper system. The authors note that "Our conclusion so far is that, as a tool for building secure systems, PCPs and argument systems are not a lost cause."

62.5   References

[1] Gennaro, Rosario; Gentry, Craig; Parno, Bryan (31 August 2010). *Non-Interactive Verifiable Computing: Outsourcing Computation to Untrusted Workers* (PDF). CRYPTO 2010.

[2] Molnar, D. (2000). "The SETI@Home problem". *ACM Crossroads* **7** (1).

[3] Smith, S.; Weingart, S. (1999). "Building a high-performance, programmable secure coprocessor". *Computer Networks* **31**: 831–960. doi:10.1016/S1389-1286(98)00019-X.

[4] Yee, B. (1994). *Using Secure Coprocessors* (PhD thesis). Carnegie Mellon University.

[5] Trusted Computing Group (July 2007). *Trusted platform module main specification.* 1.2, Revision 103.

[6] L. Babai (1985). "Trading group theory for randomness." In *Proceedings of the ACM Symposium on Theory of Computing (STOC)*, New York, NY, USA, pp. 421–429

[7] S. Goldwasser, S. Micali, and C. Rackoff (1989). "The knowledge complexity of interactive proof-systems." *SIAM Journal on Computing*, **18**(1), pp. 186–208

[8] S. Arora and S. Safra (1998). "Probabilistically checkable proofs: a new characterization of NP." *Journal of the ACM*, **45**, pp.501-555

[9] O. Goldreich (1998). *Modern Cryptography, Probabilistic Proofs and Pseudorandomness.* Algorithms and Combinatorics series, **17**, Springer

[10] J. Kilian (1992). "A note on efficient zero-knowledge proofs and arguments (extended abstract)." In *Proceedings of the ACM Symposium on Theory of Computing (STOC)*

[11] J. Kilian (1995). "Improved efficient arguments (preliminary version)." In *Proceedings of Crypto*, London, UK, pp. 311–324. Springer-Verlag

[12] S. Micali (1994). "CS proofs (extended abstract)." In *Proceedings of the IEEE Symposium on Foundations of Computer Science*, pp. 436-453.

[13] A. Yao (1982). "Protocols for secure computations." In *Proceedings of the IEEE Symposium on Foundations of Computer Science*, pp. 160-164

[14] A. Yao (1986). "How to generate and exchange secrets." In *Proceedings of the IEEE Symposium on Foundations of Computer Science*, pp. 162-167

[15] Setty, Srinath; McPherson, Richard; Blumberg, Andrew J.; Walfish, Michael (February 2012).   *Making Argument Systems for Outsourced Computation Practical (Sometimes).* Network & Distributed System Security Symposium (NDSS) 2012.

Chapter 63

Voice over IP Security Alliance

The **Voice over IP Security Alliance** (**VOIPSA**) was launched in early 2005 to bring together Voice over IP and information security vendors, providers, and thought leaders to address current and emerging security threats to VoIP. The stated mission statement is:

> VOIPSA's mission is to promote the current state of VoIP security research, VoIP security education and awareness, and free VoIP testing methodologies and tools.

63.1   External links

- Official site

Chapter 64

Whisper Systems

This article is about a subsidiary of Twitter. For the open-source software project, see Open Whisper Systems.

Whisper Systems was an enterprise mobile security company that was acquired by Twitter in November 2011.[1][2] The company was co-founded by security researcher Moxie Marlinspike and roboticist Stuart Anderson in 2010.[3] Some of Whisper Systems' software was made available under free software licenses after the acquisition,[4] which led to the creation of Open Whisper Systems.[5]

64.1 History

Security researcher Moxie Marlinspike and roboticist Stuart Anderson co-founded Whisper Systems in 2010.[6][3] In addition to launching TextSecure in May 2010, Whisper Systems produced RedPhone, an application that provides encrypted voice calls.[7] They also developed a firewall and tools for encrypting other forms of data.[3][8]

On November 28, 2011, Whisper Systems announced that it had been acquired by Twitter. The financial terms of the deal were not disclosed by either company.[9] The acquisition was done "primarily so that Mr. Marlinspike could help the then-startup improve its security".[10] Shortly after the acquisition, Whisper Systems' RedPhone service was made unavailable.[11] Some criticized the removal, arguing that the software was "specifically targeted [to help] people under repressive regimes" and that it left people like the Egyptians in "a dangerous position" during the events of the 2011 Egyptian revolution.[12]

Twitter released TextSecure as free and open-source software under the GPLv3 license in December 2011.[13][14][3][15] RedPhone was also released under the same license in July 2012.[16]

Marlinspike later left Twitter and founded Open Whisper Systems as a collaborative Open Source project for the continued development of TextSecure and RedPhone.[17][5]

As of May 2015, Open Whisper Systems consists of a large community of volunteer Open Source contributors, as well as a small team of dedicated grant-funded developers. Their active projects include TextSecure, RedPhone, and Signal.[18]

64.2 See also

- List of mergers and acquisitions by Twitter

64.3 References

[1] Tom Cheredar (November 28, 2011). "Twitter acquires Android security startup Whisper Systems". VentureBeat. Retrieved 2011-12-21.

[2] Brad McCarty (28 November 2011). "Twitter acquires mobile data security gurus Whisper Systems". Retrieved 16 January 2015.

[3] Garling, Caleb (2011-12-20). "Twitter Open Sources Its Android Moxie | Wired Enterprise". Wired.com. Retrieved 2011-12-21.

[4] Pete Pachal (2011-12-20). "Twitter Takes TextSecure, Texting App for Dissidents, Open Source". Mashable. Retrieved 2014-03-01.

[5] "A New Home". Open Whisper Systems. 2013-01-21. Retrieved 2014-03-01.

[6] "Company Overview of Whisper Systems Inc.". Bloomberg Businessweek. Retrieved 2014-03-04.

[7] Andy Greenberg (2010-05-25). "Android App Aims to Allow Wiretap-Proof Cell Phone Calls". Forbes. Retrieved 2014-02-28.

[8] "Secure your Android mobile – Use Whisper Systems free security app Whispercore « Technology updates by Techburrp". Techburrp.com. Retrieved 2011-12-21.

[9] Tom Cheredar (November 28, 2011). "Twitter acquires Android security startup Whisper Systems". VentureBeat. Retrieved 2011-12-21.

[10] Yadron, Danny (9 July 2015). "Moxie Marlinspike: The Coder Who Encrypted Your Texts". *The Wall Street Journal*. Retrieved 10 July 2015.

[11] Andy Greenberg (2011-11-28). "Twitter Acquires Moxie Marlinspike's Encryption Startup Whisper Systems". Forbes. Retrieved 2011-12-21.

[12] Garling, Caleb (2011-11-28). "Twitter Buys Some Middle East Moxie | Wired Enterprise". Wired.com. Retrieved 2011-12-21.

[13] Chris Aniszczyk (20 December 2011). "The Whispers Are True". *The Twitter Developer Blog*. Twitter. Archived from the original on 24 October 2014. Retrieved 22 January 2015.

[14] "TextSecure is now Open Source!". Whisper Systems. 20 December 2011. Archived from the original on 6 January 2012. Retrieved 22 January 2015.

[15] Pete Pachal (2011-12-20). "Twitter Takes TextSecure, Texting App for Dissidents, Open Source". Mashable. Retrieved 2014-03-01.

[16] "RedPhone is now Open Source!". Whisper Systems. 18 July 2012. Archived from the original on 31 July 2012. Retrieved 22 January 2015.

[17] Andy Greenberg (29 July 2014). "Your iPhone Can Finally Make Free, Encrypted Calls". Wired. Retrieved 18 January 2015.

[18] "Open Whisper Systems". *Github*. Retrieved 17 May 2015.

Chapter 65

Wireless intrusion prevention system

In computing, a **wireless intrusion prevention system** (WIPS) is a network device that monitors the radio spectrum for the presence of unauthorized access points (*intrusion detection*), and can automatically take countermeasures (*intrusion prevention*).

65.1 Purpose

The primary purpose of a WIPS is to prevent unauthorized network access to local area networks and other information assets by wireless devices. These systems are typically implemented as an overlay to an existing Wireless LAN infrastructure, although they may be deployed standalone to enforce no-wireless policies within an organization. Some advanced wireless infrastructure has integrated WIPS capabilities.

Large organizations with many employees are particularly vulnerable to security breaches[1] caused by rogue access points. If an employee (trusted entity) in a location brings in an easily available wireless router, the entire network can be exposed to anyone within range of the signals.

In July 2009, the PCI Security Standards Council published wireless guidelines[2] for PCI DSS recommending the use of WIPS to automate wireless scanning for large organizations.

65.2 Intrusion detection

A **wireless intrusion detection system** (WIDS) monitors the radio spectrum for the presence of unauthorized, rogue access points and the use of wireless attack tools. The system monitors the radio spectrum used by wireless LANs, and immediately alerts a systems administrator whenever a rogue access point is detected. Conventionally it is achieved by comparing the MAC address of the participating wireless devices.

Rogue devices can spoof MAC address of an autho-

rized network device as their own. New research uses fingerprinting approach to weed out devices with spoofed MAC addresses. The idea is to compare the unique signatures exhibited by the signals emitted by each wireless device against the known signatures of pre-authorized, known wireless devices.[3]

65.3 Intrusion prevention

In addition to intrusion detection, a WIPS also includes features that prevent against the threat *automatically*. For automatic prevention, it is required that the WIPS is able to accurately detect and automatically classify a threat.

The following types of threats can be prevented by a good WIPS:

- Rogue AP – WIPS should understand the difference between Rogue AP and External (neighbor's) AP

- Mis-configured AP

- Client Mis-association

- Unauthorized association

- Man in the Middle Attack

- *Ad hoc* Networks

- MAC-Spoofing

- Honeypot / Evil Twin Attack

- Denial of Service (DoS) Attack

65.4 Implementation

WIPS configurations consist of three components:

- **Sensors** — These devices contain antennas and radios that scan the wireless spectrum for packets and are installed throughout areas to be protected

- **Server** — The WIPS server centrally analyzes packets captured by sensors

- **Console** — The console provides the primary user interface into the system for administration and reporting

A simple intrusion detection system can be a single computer, connected to a wireless signal processing device, and antennas placed throughout the facility. For huge organizations, a Multi Network Controller provides central control of multiple WIPS servers, while for SOHO or SMB customers, all the functionality of WIPS is available in single box.

In a WIPS implementation, users first define the operating wireless policies in the WIPS. The WIPS sensors then analyze the traffic in the air and send this information to WIPS server. The WIPS server correlates the information validates it against the defined policies and classifies if it is a threat. The administrator of the WIPS is then notified of the threat, or, if a policy has been set accordingly, the WIPS takes automatic protection measures.

WIPS is configured as either a network implementation or a hosted implementation.

65.4.1 Network Implementation

In a network WIPS implementation, Server, Sensors and the Console are all placed inside a private network and are not accessible from the internet.

Sensors communicate with the Server over a private network using a private port. Since the Server resides on the private network, users can access the Console only from within the private network.

A network implementation is suitable for organizations where all locations are within the private network.

65.4.2 Hosted Implementation

In a hosted WIPS implementation, Sensors are installed inside a private network. However, the Server is hosted in secure data center and is accessible on the Internet. Users can access the WIPS Console from anywhere on the Internet. A hosted WIPS implementation is as secure as a network implementation because the data flow is encrypted between Sensors and Server, as well as between Server and Console.

A hosted WIPS implementation requires very little configuration because the Sensors are programmed to automatically look for the Server on the Internet over a secure SSL connection.

For a large organization with locations that are not a part of a private network, a hosted WIPS implementation simplifies deployment significantly because Sensors connect to the Server over the internet without requiring any special configuration. Additionally, the Console can be accessed securely from anywhere on the Internet.

Hosted WIPS implementations are available in an on-demand, subscription-based software as a service model.[4] Hosted implementations may be appropriate for organizations looking to fulfill the minimum scanning requirements of PCI DSS.

65.5 See also

- Wardriving

- Wireless LAN security

- Typhoid adware

65.6 References

[1] "Fitting the WLAN Security pieces together". pcworld.com. Retrieved 2008-10-30.

[2] "PCI DSS Wireless Guidelines" (PDF). Retrieved 2009-07-16.

[3] "University research aims at more secure Wi-Fi".

[4] "Security SaaS hits WLAN community". networkworld.com. Retrieved 2008-04-07.

Chapter 66

X.1035

ITU-T Recommendation **X.1035** specifies a password-authenticated key agreement protocol that ensures mutual authentication of two parties by using a Diffie–Hellman key exchange to establish a symmetric cryptographic key. The use of Diffie-Hellman exchange ensures perfect forward secrecy—a property of a key establishment protocol that guarantees that compromise of a session key or long-term private key after a given session does not cause the compromise of any earlier session.

In X.1035, the exchange is protected from the man-in-the-middle attack. The authentication relies on a pre-shared secret (e.g., password), which is protected (i.e., remains unrevealed) to an eavesdropper preventing an off-line dictionary attack.

The protocol can be used in a wide variety of applications including those with pre-shared secrets based on possibly weak passwords.

X.1035 was approved on 13 February 2007 by ITU-T Study Group 17.[1]

66.1 Applications

G.hn, an ITU-T standard that specifies high-speed (up to 1 Gbit/s) local area networking over existing home wires (power lines, phone lines and coaxial cables), uses X.1035 for authentication and key exchange.

66.2 References

[1] ITU-T Recommendation X.1035, Password-authenticated key exchange (PAK) protocol

Chapter 67

Zooko's triangle

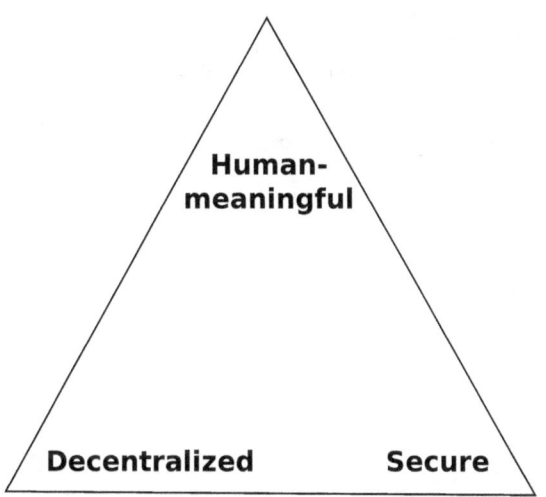

Zooko's triangle defines the three desirable traits of a network protocol identifier as Human-meaningful,, Decentralized *and* Secure.

Zooko's triangle is a diagram named after Zooko Wilcox-O'Hearn which sets out a conjecture for any system for giving names to participants in a network protocol. At the vertices of the triangle are three properties that are generally considered desirable for such names:[1]

- **Human-meaningful**: The quality of meaningfulness and memorability to the users of the naming system. Domain names and nicknaming are naming systems that are highly memorable.

- **Decentralized**: The lack of a centralized authority for determining the meaning of a name. Instead, measures such as a Web of trust are used.

- **Secure**: The quality that there is one, unique and specific entity to which the name maps. For instance, domain names are unique because there is just one party able to prove that they are the owner of each domain name.

Of these three properties, Zooko's conjecture states that no single kind of name can achieve more than two. So the edges of the triangles represent the three possible choices for a naming scheme:

- **Decentralized and human-meaningful**: This is true of nicknames people choose for themselves.

- **Secure and human-meaningful**: This is the property that domain names and URLs aim for.

- **Secure and decentralized**: This is a property of OpenPGP public key fingerprints.

Zooko's conjecture was disproved through creating systems that exhibit all three properties.

67.1 Solutions

The contribution of Zooko's triangle is that it encouraged systems designers to explore how to attain all three properties.

The original name systems designed featured two out of three properties, following Zooko's triangle:

- **Compromise decentralization**: DNSSec offers a secure, human-meaningful naming scheme, but is not decentralized.

- **Compromise human-readability**: .onion addresses and bitcoin addresses are secure and decentralized and are not human-meaningful, to most.

- **Compromise security**: I2P uses name translation services which are decentralized and provide human-meaningful names, but which relies on trusting third parties. Petname systems demonstrate that a naming system can be built by dynamically translating between different possible kinds of names.[2]

After such systems were explored, Zooko's conjecture was disproved by implementing systems that exhibit all three properties. Computer scientist Nick Szabo illustrated that all three properties can be achieved up to the limits of Byzantine fault tolerance.[3]

The internet activist Aaron Swartz described a naming system based on Bitcoin which tries to square Zooko's triangle by employing Bitcoin's distributed blockchain as a proof-of-work to establish consensus of domain name ownership.[4] These systems remain vulnerable to sybil attacks,[5] but are secure under Byzantine assumptions. A few months after the proposal, Namecoin was released which implements the concept.

Following Namecoin, other platforms were developed which defy Zooko's conjecture, such as Twister. Recently, Monero team released OpenAlias,[6] a DNS aliasing technology to ensure human-readability whilst keeping the already-existing decentralization and security properties. OpenAlias works with any cryptocurrency, including Bitcoin. Clients that support OpenAlias include Electrum (from version 2.0 onwards) and the Coin.Space web-based wallet.

67.2   See also

- OpenAlias

- Namecoin

- Petname

67.3   References

[1] Zooko Wilcox-O'Hearn. "Names: Decentralized, Secure, Human-Meaningful: Choose Two". Archived from the original on 2001-10-20.

[2] Mark Steigler, Zooko, *An Introduction to Petname Systems*, Feb 2005

[3] Nick Szabo, Secure Property Titles, 1998

[4] Aaron Swartz, Squaring the Triangle: Secure, Decentralized, Human-Readable Names, Aaron Swartz, January 6, 2011

[5] Dan Kaminsky, Spelunking the Triangle: Exploring Aaron Swartz's Take On Zooko's Triangle, January 13, 2011

[6] Monero core team (2014-09-19). "OpenAlias". Retrieved 2015-02-03.

67.4   External links

- Zooko Wilcox-O'Hearn, Names: Decentralized, Secure, Human-Meaningful: Choose Two – the essay highlighting this difficulty

- Marc Stiegler, An Introduction to Petname Systems – a clear introduction

- Nick Szabo, Secure Property Titles – argues that all three properties can be achieved up to the limits of Byzantine fault tolerance.

- Bob Wyman, The Persistence of Identity: Updating Zooko's Pyramid

- Paul Crowley, Squaring Zooko's Triangle

- Aaron Swartz, Squaring the Triangle using a technique from Bitcoin

67.5 Text and image sources, contributors, and licenses

67.5.1 Text

- **Secure communication** *Source:* https://en.wikipedia.org/wiki/Secure_communication?oldid=682087085 *Contributors:* Edward, Michael Hardy, Haakon, Reddi, Greenrd, Itai, Lowellian, Niteowlneils, Grm wnr, FT2, Bart133, Danhash, Mindmatrix, Pol098, Stefanomione, Blisco, BD2412, Rjwilmsi, Vegaswikian, JYOuyang, Intgr, Jmorgan, Chaser, Shaddack, Fram, Rearden9, SmackBot, KelleyCook, Bluebot, Oli Filth, Lexlex, Kasyapa, Scwlong, Frap, Squigish, Gbinal, Clicketyclack, Ckatz, Jec, Bonkerzbanks, CmdrObot, Wafulz, AgentPeppermint, Eeyore-Tim, Isilanes, CosineKitty, Codegrinder, Calltech, Jim.henderson, R'n'B, CommonsDelinker, TWCarlson, Hellno2, Palat, Kinhead, One half 3544, Andy Dingley, WJetChao, Jerryobject, Capitalismojo, Stan1972, Wnt, DumZiBoT, Misop, Sweeper tamonten, Addbot, Tothwolf, Parkertrail, Porkolt60, Yobot, AnomieBOT, Jim1138, LilHelpa, PaulHammond2, Woken Wanderer, Dave Al S, Dewritech, Bamyers99, Helpful Pixie Bot, Wbm1058, KillPR, Lugia2453, P2Peter, Lemnaminor, WCS100, PHYencryption, Sanhan555, Alexandr Galvita, Dodi 8238, Vipole, VkhMpLzWUbYFcWAT, Jacllondon, J13m7 and Anonymous: 54

- **Automated Certificate Management Environment** *Source:* https://en.wikipedia.org/wiki/Automated_Certificate_Management_Environment?oldid=670215798 *Contributors:* The Anome

- **Automatic Secure Voice Communications Network** *Source:* https://en.wikipedia.org/wiki/Automatic_Secure_Voice_Communications_Network?oldid=576752051 *Contributors:* Woohookitty, Zotel, SmackBot, Alaibot, Jim.henderson, Dthomsen8, Addbot, Yobot, Erik9bot, Xplosionist, Hmainsbot1, P2Peter and Anonymous: 1

- **BID 150** *Source:* https://en.wikipedia.org/wiki/BID_150?oldid=636370745 *Contributors:* Ceyockey, TexasAndroid, Bob Moffatt, Mysid, Alaibot, Trackorack, Will Beback Auto and Anonymous: 1

- **Blacker (security)** *Source:* https://en.wikipedia.org/wiki/Blacker_(security)?oldid=623504072 *Contributors:* Bearcat, Alan Liefting, Merenta, Cjcollier, RussBot, Xaosflux, Andy M. Wang, Cydebot, Buckshot06, McSly, Funandtrvl, Arendedwinter, Trivialist, Alvin Seville, W Nowicki, DrilBot, Rabbabodrool and Anonymous: 5

- **CA/Browser Forum** *Source:* https://en.wikipedia.org/wiki/CA/Browser_Forum?oldid=682094254 *Contributors:* Visorstuff, Kevjudge, Lakshmin, Chris the speller, Nagle, Cydebot, Cryptoki, Rror, Katharine908, Yobot, AnomieBOT, MLauba, Erik9bot, John of Reading, Dewritech, Chmarkine, Codename Lisa, Palmbeachguy, ArmbrustBot, Ben DigiCert, Aureomarginata, Slrawlins, WikiGopi and Anonymous: 9

- **CCMP** *Source:* https://en.wikipedia.org/wiki/CCMP?oldid=676216016 *Contributors:* Mac, Indefatigable, Inkling, Matt Crypto, DRE, Hellisp, TheObtuseAngleOfDoom, Rich Farmbrough, Dajhorn, L33th4x0rguy, Kfitzner, Palthainon, FlaBot, Dake~enwiki, Ospalh, Abune, SmackBot, Eskimbot, KelleyCook, Mgeorg~enwiki, Jbarcelo, Fat64~enwiki, Jesse Viviano, Dawnseeker2000, Widefox, Dimawik, Magioladitis, Seanscian, Jesant13, VolkovBot, Urbanrenewal, Mopskatze, Alexbot, Lartoven, RobLP1, Addbot, Yobot, Jackie, DSisyphBot, Trappist the monk, Thelittlemouse, EmausBot, ClueBot NG, Rkm28-NJITWILL, Helpful Pixie Bot and Anonymous: 43

- **Cipher suite** *Source:* https://en.wikipedia.org/wiki/Cipher_suite?oldid=686065670 *Contributors:* Lowellian, ZeroOne, RussBot, SmackBot, A5b, The Letter J, Widefox, Yintan, Niceguyedc, Anon lynx, Addbot, DE logics, Codename Lisa, Metamorforme42, WLeibbrandt and Anonymous: 10

- **Common Data Link** *Source:* https://en.wikipedia.org/wiki/Common_Data_Link?oldid=621247840 *Contributors:* Joeljkp, Beland, Drbreznjev, Kenyon, Woohookitty, RussBot, Malcolma, SmackBot, Hmains, Gobonobo, Alaibot, X96lee15, Fabrictramp, Adavidb, Landarski, Skier Dude, The.famous.adventurer, DeknMike, Sun Creator, Leofric1, DumZiBoT, Pichilingi, Addbot, Lightbot, AnomieBOT, Slacker2585, BattyBot, ChrisGualtieri and Anonymous: 2

- **Computer security** *Source:* https://en.wikipedia.org/wiki/Computer_security?oldid=685842509 *Contributors:* Tobias Hoevekamp, Derek Ross, Tuxisuau, Brion VIBBER, Eloquence, Zardoz, Mav, Robert Merkel, The Anome, Stephen Gilbert, Taw, Arcade~enwiki, Graham Chapman, Dachshund, Arvindn, PierreAbbat, Fubar Obfusco, SimonP, Ben-Zin~enwiki, Ant, Ark~enwiki, Heron, Dwheeler, Chuq, Iorek~enwiki, Frecklefoot, Edward, Michael Hardy, Pnm, Kku, Ixfd64, Dcljr, Dori, Arpingstone, CesarB, Haakon, Ronz, Snoyes, Yaronf, Nikai, Smaffy, Qwert, Mydogategodshat, Jengod, JidGom, Aarontay, Gingekerr, Taxman, Joy, Vaceituno, Khym Chanur, Pakaran, Robbot, Yas~enwiki, Fredrik, ZimZalaBim, Rursus, Texture, KellyCoinGuy, 2501~enwiki, Hadal, Tobias Bergemann, David Gerard, Honta, Wolf530, Tom harrison, Dratman, Mike40033, Siroxo, C17GMaster, Matt Crypto, SWAdair, Bobblewik, Wmahan, Mu, Geni, Antandrus, Beland, Mako098765, CSTAR, GeoGreg, Marc Mongenet, Gscshoyru, Joyous!, Bluefoxicy, Squash, Strbenjr, Mike Rosoft, Kmccoy, Monkeyman, Pyrop, Rich Farmbrough, Rhobite, Leibniz, FT2, Jesper Laisen, ArnoldReinhold, YUL89YYZ, Jwalden, Zarutian, MeltBanana, Sperling, Bender235, ZeroOne, Moa3333, JoeSmack, Danakil, Omnifarious, Jensbn, El C, Joanjoc~enwiki, Marcok, Perspective, Spearhead, EurekaLott, Nigelj, Stesmo, Smalljim, Rvera~enwiki, Myria, Adrian~enwiki, Boredzo, ClementSeveillac, JohnyDog, Poweroid, Alansohn, Quiggles, Arthena, Lightdarkness, Cdc, Mrholybrain, Caesura, Gbeeker, Raraoul, Filx, Proton, M3tainfo, Suruena, HenkvD, 2mcm, Wikicaz, H2g2bob, Condor33~enwiki, Bsdlogical, Johntex, Dan100, Woohookitty, Daira Hopwood, Al E., Prashanthns, Zhen-Xjell, Palica, Kesla, Vininim, Graham87, Clapaucius, BD2412, Icey, Sjakkalle, Rjwilmsi, Seidenstud, Koavf, Guyd, DeadlyAssassin, Dookie~enwiki, Edggar, Oblivious, QuickFox, Kazrak, Ddawson, Ligulem, Smtully, Aapo Laitinen, Ground Zero, RexNL, Alvin-cs, BMF81, JonathanFreed, Jmorgan, J.Ammon, Bgwhite, Hall Monitor, Digitalme, Gwernol, FrankTobia, Elfguy, Wavelength, NTBot~enwiki, Alan216, StuffOfInterest, Foxxygirltamara, Stephenb, Gaius Cornelius, Ptomes, Morphh, Salsb, Wimt, Bachrach44, AlMac, Irishguy, Albedo, Rmky87, Amcfreely, Deku-shrub, Romal, Peter Schmiedeskamp, Zzuuzz, Gorgonzilla, Papergrl, Arthur Rubin, Ka-Ping Yee, Juliano, GraemeL, Rlove, JoanneB, Whouk, NeilN, SkerHawx, SmackBot, Mmernex, Tripletmot, Reedy, KnowledgeOfSelf, TestPilot, Kosik, McGeddon, Stretch 135, Ccalvin, Manjunathbhatt, Gilliam, Ohnoitsjamie, Skizzik, Lakshmin, Kurykh, Autarch, Snori, Miquonranger03, Deli nk, Jenny MacKinnon, Kungming2, Jonasyorg, Timothy Clemans, Frap, Ponnampalam, Nixeagle, KevM, JonHarder, Wine Guy, Cpt~enwiki, Krich, Bslede, Richard001, Stor stark7, Newtonlee, Doug Bell, Harryboyles, Kuru, Geoinline, Disavian, Robofish, Joffeloff, Kwestin, Mr. Lefty, Beetstra, Jadams76, Ehheh, Boxflux, Kvng, Chadnibal, Wfgiuliano, Dthvt, IvanLanin, DavidHOzAu, Lcamtuf, CmdrObot, Tional, ShelfSkewed, Michael B. Trausch, Phatom87, Cydebot, Mblumber, Future Perfect at Sunrise, Blackjackmagic, UncleBubba, Gogo Dodo, Anonymi, Anthonyhcole, GRevolution824, Clovis Sangrail, SpK, Njan, Ebyabe, Thijs!bot, Epbr123, The Punk, Kpavery, Wistless, Oarchimondeo, RichardVeryard, EdJohnston, Druiloor, SusanLesch, I already forgot, Sheridbm, AntiVandalBot, Obiwankenobi,

Shirt58, Marokwitz, Khhodges, Ellenaz, Manionc, Chill doubt, Dmerrill, SecurityGuy, JAnDbot, Jimothytrotter, Barek, MER-C, The Transhumanist, Technologyvoices, Tqbf, Dave Nelson, Acroterion, Raanoo, VoABot II, Ukuser, JNW, Michi.bo, Szh~enwiki, Hubbardaie, Arctific, Froid, JXS, AlephGamma, Rohasnagpal, Catgut, WhatamIdoing, Marzooq, Gerrardperrett, Thireus, Devmem, Dharmadhyaksha, DerHexer, JaGa, Rcseacord, XandroZ, Gwern, SolitaryWolf, CliffC, =JeffH, Sjjupadhyay~enwiki, Bertix, Booker.ercu, J.delanoy, Gam2121, Maurice Carbonaro, Public Menace, Jesant13, Jreferee, JA.Davidson, Katalaveno, Touisiau, Ansh1979, Toon05, Mufka, Largoplazo, Dubhe.sk, YoavD, Bonadea, Red Thrush, RJASE1, Cralar, Javeed Safai, ABF, Wiki-ay, Davidwr, Zifert, Crazypete101, Dictouray, Shanata, Haseo9999, Falcon8765, Pctechbytes, Sapphic, Donnymo, FutureDomain, Smith bruce, Kbrose, JonnyJD, Lxicm, Whitehatnetizen, Jargonexpert, SecurInfos~enwiki, Ml-crest, Immzw4, Sephiroth storm, Graceup, Yuxin19, Agilmore, JohnManuel, Flyer22, Jojalozzo, Riya.agarwal, Corp Vision, Lightmouse, KathrynLybarger, Mscwriter, Soloxide, StaticGull, Capitalismojo, PabloStraub, Rinconsoleao, Denisarona, White Stealth, Ishisaka, WikipedianMarlith, Sfan00 IMG, Elassint, ClueBot, Shonharris, PipepBot, TransporterMan, Supertouch, Add32, Emantras, Tanglewood4, Mild Bill Hiccup, Niceguyedc, Dkontyko, Trivialist, Gordon Ecker, DragonBot, Dwcmsc, Excirial, Socrates2008, Dcampbell30, Moomoo987, Dr-Mx, Rbilesky, DanielPharos, Versus22, HarrivBOT, Fathisules, Raysecurity, XLinkBot, BodhisattvaBot, Solinym, Skarebo, Wingfamily, WikiDao, MystBot, Dsimic, JimWalker67, Addbot, Non-dropframe, Cst17, MrOllie, Passport90, Favonian, Torla42, AgadaUrbanit, Tassedethe, Jarble, Ben Ben, Tartarus, Luckas-bot, Yobot, OrgasGirl, The Grumpy Hacker, Librsh, Cyanoa Crylate, Grammaton, THEN WHO WAS PHONE?, Dr Roots, Sweerek, AnomieBOT, JDavis680, Jim1138, Galoubet, Dwayne, Piano non troppo, AdjustShift, Rwhalb, Quantumseven, HRV, Vijay Varadharajan, Materialscientist, Aneah, Stationcall, ArthurBot, Cameron Scott, Intelati, Securitywiki, Hi878, Coolkidmoa, Zarcillo, Mark Schierbecker, Pradameinhoff, Amaury, George1997, Architectchamp, =Josh.Harris, Shadowjams, President of hittin' that ass, FrescoBot, Bingo-101a, Nageh, Ionutzmovie, Cudwin, Expertour, Intelligentsium, Pinethicket, I dream of horses, Edderso, Access-bb, Yahia.barie, RedBot, MastiBot, Wlalng123, Mentmic, Dac04, Banej, Codemaster32, Tjmannos, Nitesh13579, Lotje, Sumone10154, Arkelweis, Ntlhui, Aoidh, Endpointsecurity, Tbhotch, Jesse V., DARTH SIDIOUS 2, Ripchip Bot, Panda Madrid, DASHBot, Julie188, EmausBot, Timtempleton, Dewritech, Active Banana, P@ddington, Susfele, Dolovis, Cosmoskramer, Alxndrpaz, AvicAWB, Bar-abban, Ocaasi, Solipsys, Tolly4bolly, Sharpie66, DennisIsMe, Veryfoolish, Geohac, ChuispastonBot, Pastore Italy, Tentontunic, Sepersann, Gadgad1973, Rocketrod1960, Jramio, ClueBot NG, AAriel42, Name Omitted, Enfcer, Iliketurtlesmeow, Widr, Helpful Pixie Bot, TechGeek70, Curb Chain, Calabe1992, BG19bot, Mollsiebee, M0rphzone, Rubmum, Mohilekedar, Karlomagnus, IraChesterfield, Sburkeel, Zune0112, Venera Seyranyan, Wondervoll, Mihai.scridonesi, Jtlopez, Nfirdosian, Alessandra Napolitano, Wannabemodel, Keeper03, BattyBot, Popescucalin, Arr4, Mrt3366, Cyberbot II, Khazar2, Peter A. Wolff, Soulparadox, Ilker Savas, BIG ISSUE LADY, Saturdayswiki, Dexbot, Jmitola, Mogism, Pete Mahen, Lugia2453, Doopbridge, Sbhalotra, SFK2, Arjungiri, Jamesx12345, ElinaSy, Patna01, Dr Dinosaur IV, Pdecalculus, Mbmexpress, Idavies007, RaheemaHussain, Cyberlawjustin, Rkocher, MoHafesji, ResearcherQ, Westonbowden, Peter303x, Karinera, OccultZone, LukeJeremy, Robevans123, Chima4mani, ClyderRakker46, Jonathan lampe, Jppcap, Leejjung86, Azulfiqar, IrvingCarR, Nyashinski, Monkbot, Nitzy99, Carpalclip3, RicardoBanchez, Owais Khursheed, Oushee, 405Duke, BrettofMoore, Gr3yHatf00l, Thetechgirl, Fimatic, Hchaudh3, AndrewKin, JRPolicy, Pacguy, HVanIderstine, Leeemily, FormerPatchEditor, Pixelized frog, 3 of Diamonds, Johngot, Bmore84, Informationsystemgeeks, PardonTheComma, Nemoanon, JohnEvans79, Hatebott, Ajay Yankee, Nbyrd2000, Soderbounce, Tejalpatel, Tonybdistil, Shedletskyy, Prshnktiitk and Anonymous: 697

- **CONDOR secure cell phone** *Source:* https://en.wikipedia.org/wiki/CONDOR_secure_cell_phone?oldid=525834910 *Contributors:* Apoc2400, RJFJR, Moreati, Nate1481, SmackBot, SomeBody, ALR, Jim.henderson, Shoessss, Kiranchandran, Yobot, FrescoBot and Anonymous: 2

- **Crypto phone** *Source:* https://en.wikipedia.org/wiki/Crypto_phone?oldid=663916243 *Contributors:* Ronz, Davidgothberg, Apokrif, Intgr, SmackBot, Harryboyles, Alaibot, Voyaging, Jim.henderson, Anaxial, Liveste, Elitre, JL-Bot, Into The Fray, Ziduwang, ClueBot, Addbot, Yobot, Talion86, Joaquin008, FrescoBot, Adaliaholding, EmausBot, HollywoodsEnd, Faij, KH-1, Gauntman1, Vureb, Phantomencrypt and Anonymous: 19

- **CSipSimple** *Source:* https://en.wikipedia.org/wiki/CSipSimple?oldid=670324396 *Contributors:* Bender235, Stesmo, Dl2000, Cydebot, Widefox, Magioladitis, Jerryobject, Yobot, FrescoBot, Fæ, Rolandschweiger, R3gis3r, Qetuth, Dexbot and Anonymous: 8

- **Data breach** *Source:* https://en.wikipedia.org/wiki/Data_breach?oldid=678991331 *Contributors:* Gzuckier, Discospinster, ArnoldReinhold, YUL89YYZ, Skyring, Woohookitty, BD2412, Intgr, Bachrach44, DouglasHeld, SmackBot, Xaosflux, Kuru, Langhorner, EdC~enwiki, MaxEnt, CliffC, R'n'B, Wikimandia, Billinghurst, Ehibbard, Rahulchic, Enviroboy, Dlrohrer2003, ClueBot, Ottawahitech, Trivialist, DumZiBoT, Sulaimanelladan, Frietta, Addbot, Download, Tassedethe, Lightbot, LuK3, AnomieBOT, Bluerasberry, Vgopinathan, Jgeorge60, Coconutstudio, Nbdubya, Erik9, SD5, Canberranone, Full-date unlinking bot, RjwilmsiBot, Rabbabodrool, RaqiwasSushi, DataHemorrhage, ChuispastonBot, ClueBot NG, Jack Greenmaven, Infinitethreats, Compfreak7, BrianWo, Ccwesley, BattyBot, GerardBalthasar, Dobie80, EggensPaar, Fomeister, BurritoBazooka, Stuckgreenpixel, Philroc, SamanthaPuckettIndo, Felygus, Professornova, Oiyarbepsy, Indopap, Akstotz and Anonymous: 37

- **Defense Red Switch Network** *Source:* https://en.wikipedia.org/wiki/Defense_Red_Switch_Network?oldid=677776723 *Contributors:* The Anome, Hellbus, Jim.henderson, FrescoBot, Neøn and P2Peter

- **Fishbowl (secure phone)** *Source:* https://en.wikipedia.org/wiki/Fishbowl_(secure_phone)?oldid=663114623 *Contributors:* ArnoldReinhold, Funandtrvl, Dawynn, Tim-mnm, Dewritech, GoingBatty, L.J. Tibbs and Anonymous: 4

- **Greynet** *Source:* https://en.wikipedia.org/wiki/Greynet?oldid=597986794 *Contributors:* Derek Ross, Uncle G, Justin Hirsh, Wongm, Splash, IslandHopper973, Pragmatist, SmackBot, Frap, JonHarder, Clicketyclack, Breno, CmdrObot, Cowpriest2, MarshBot, CosineKitty, Donnymo, Securegreynets, MrOllie, LilHelpa, TheAMmollusc, FrescoBot, Gildedtiger, ClueBot NG, SoledadKabocha and Anonymous: 8

- **The Guardian Project (software)** *Source:* https://en.wikipedia.org/wiki/The_Guardian_Project_(software)?oldid=681024273 *Contributors:* Zigger, Bender235, TLSuda, Grye, Widefox, Yawnbox, Grshiplett, Funandtrvl, Frmorrison, Yobot, Bluerasberry, Dolovis, Sharhues, BG19bot, RadicalRedRaccoon, Dodi 8238, Nullnullthree, EditGuardianProject and Anonymous: 7

- **HTTPS** *Source:* https://en.wikipedia.org/wiki/HTTPS?oldid=686005887 *Contributors:* Damian Yerrick, WojPob, The Anome, Andre Engels, Arcade~enwiki, Fnielsen, Arvindn, Aldie, Fubar Obfusco, William Avery, Ben-Zin~enwiki, LapoLuchini, Lorenzarius, Patrick, Kwertii, Zeno Gantner, Pde, Card~enwiki, Mac, Angela, Yaronf, Brettz9, Jebba, Scott, IMSoP, Ww, Andrewman327, Furrykef, Bevo, Robbot, Jredmond, Nurg, Polonius, Ashley Y, Rrjanbiah, Wereon, Mjscud, Profoss, Raeky, HaeB, Martinwguy, Dbenbenn, Msiebuhr, Everyking, Varlaam, Rick Block, Johan Burati, Canon, AlistairMcMillan, Matt Crypto, Wmahan, Chowbok, Pgan002, Calm, SURIV, Thomas Springer, Onco p53, Allaryin, Bumm13, Kevin B12, Mysidia, Zfr, TonyW, Topaz, Subsume, Demiurge, Adashiel, Safety Cap, Mike Rosoft, Mormegil, JTN, RossPatterson, أحمد, Brianhe, YUL89YYZ, S.K., Plugwash, -jkb-, Shanes, Bobo192, Smalljim, LeonardoGregianin, SpeedyGonsales, Cherlin, Zachlipton, Tlaresch, Alansohn, Arthena, Jeltz, Jeffhos, Lord Pistachio, Water Bottle, Malo, Scott5114, Snowolf, Jheald, RainbowOfLight, Randy Johnston,

BDD, Versageek, HenryLi, Zntrip, OwenX, Mindmatrix, Ruud Koot, Zzyzx11, 🔲🔲🔲🔲🔲, Toussaint, Mandarax, Spezied, Maros, Reisio, Jshadias, Rjwilmsi, Koavf, Dazmax, PinchasC, Pako, Vegaswikian, Gabipetrovay, Fred Bradstadt, Yamamoto Ichiro, Degeberg, Intgr, GringoCroco, Chobot, NiceGuyAlberto~enwiki, Peterl, YurikBot, Wavelength, Angus Lepper, Stephenb, Dequid, Irishguy, Mortein, Buss, Bozoid, Jeremy Visser, Hrvoje Simic, SimonMorgan, Zzuuzz, Closedmouth, Shirishag75, GraemeL, Neilka, Kevin, Anclation~enwiki, DVD R W, SmackBot, Mmernex, Royalguard11, McGeddon, Lord Snoeckx, WookieInHeat, Billdanbury, BiT, Grawity, Lakshmin, Keegan, SlimJim, DStoykov, Stubblyhead, EncMstr, Clumpidy, OrangeDog, JGXenite, Can't sleep, clown will eat me, Chainz, JonHarder, EditorsChoice, Rrburke, Cybercobra, Ddas, RaCha'ar, Thomasyen, Rnicoll, Esb, Slach~enwiki, Acdx, Daniel.Cardenas, ThurnerRupert, Ksn, Minna Sora no Shita, Noishe, Slakr, Stwalkerster, Beetstra, Optakeover, Kvng, Tawkerbot, Ariel Pontes, J Di, StephenBuxton, Elgaroo, Pimlottc, Courcelles, MrBoo, Tawkerbot2, CmdrObot, SupaStarGirl, LCarl, GHe, Jesse Viviano, Edwin, Logical2u, Cydebot, Gogo Dodo, Tawkerbot4, Codetiger, DumbBOT, Dark-Link, Lachlan Hunt, Omicronpersei8, Allamiro, Thijs!bot, Epbr123, Acodring, Electron9, TheJosh, Srose, AlefZet, AntiVandalBot, Yonatan, Widefox, Guy Macon, Prolog, Esmond.pitt, Rbb1181, Tigga, MER-C, Netzen, Andonic, Hut 8.5, Xoneca, Gaurav1424, Magioladitis, Gekedo, Parsecboy, Bongwarrior, VoABot II, Steven Walling, Nyttend, JPG-GR, Destynova, BrianGV, Ragimiri, LorenzoB, Jma89, DerHexer, Thompson.matthew, Cocytus, Hdt83, MartinBot, Felipe1982, RockMFR, Trusilver, Ps ttf, Jesant13, Mahewa, Inspire22, DarwinPeacock, Produke, Ziggymarley01, Jon335, Chriswiki, TomasBat, NewEnglandYankee, MetsFan76, BrunoHarbulot, Remember the dot, DorganBot, Chaos5023, AlastairIrvine, Lexein, Mrqcho, LeilaniLad, Daniel347x, Anonymous Dissident, Otto42, Qxz, Someguy1221, MICKLEK, Wenli, Meters, Dnicolaou, Picojeff, Andrew4u, Insanity Incarnate, AlleborgoBot, Blood sliver, Matagascar, SieBot, Gerakibot, Caltas, Khvalamde, Jruderman, DougWare, Dabomb87, Denisarona, Loren.wilton, ClueBot, The Thing That Should Not Be, Quinxorin, Kst2005, Halvorsen brian, Excirial, Kain Nihil, Alexbot, Socrates2008, GrahamJAT, Thingg, Versus22, Porchcorpter, Johnuniq, NERIC-Security, DumZiBoT, XLinkBot, Jordan Gray, C. A. Russell, WikHead, SilvonenBot, RP459, Sweeper tamonten, Addbot, Ghettoblaster, Kne1p, Some jerk on the Internet, FrankAndProust, CanadianLinuxUser, Jim10701, WorldlyWebster, Af.pf, Glane23, Bassbonerocks, Favonian, Planecrash111, ﻣﺎﻧﻲ, Legobot, Luckas-bot, Yobot, Veraladeramanera, Edoe, Metamorph123, AnomieBOT, Helixer, Mauro Lanari, JackieBot, Piano non troppo, Kingpin13, Materialscientist, Barakw, Oiudfgogsdf, GrouchoBot, Chrismiceli, RibotBOT, Zzzplayer, N419BH, Smallman12q, OldPeculier, Erik9, Nageh, Lonaowna, StaticVision, Nojan, Sixmeters, Redrose64, Gautier lebon, Tóraí, ContinueWithCaution, TobeBot, DixonDBot, Lotje, Nickyus, Vrenator, Allen4names, Reach Out to the Truth, DARTH SIDIOUS 2, TjBot, Dibblethewrecker, NerdyScienceDude, Peanut.pookie, Rollins83, EmausBot, John of Reading, Immunize, Zollerriia, Heymid, Meskalitos, RenamedUser01302013, Wikipelli, Zizomis, Lucas Thoms, Thecheesykid, Tambarge, Xdxter, KuduIO, TRB143, Sbmeirow, Palosirkka, Senator2029, Neil P. Quinn, Mjbmrbot, ClueBot NG, This lousy T-shirt, Dfarrell07, OpenInfoForAll, O.Koslowski, Glen 3B, Ke din, Mayhaymate, Wikingtubby, Annysmith001, Contentasaurus, Legovador, Dougvj, Tony Tan, Ahora, Chmarkine, Brainsuccess, Eidab, Filing Flunky, Iggytko, IRedRat, BattyBot, Justincheng12345-bot, Jack No1, TcDylan, RoxyFlox, Arcandam, Dexbot, Juppyw, Lezabees, SoledadKabocha, Listroiderbob, Mogism, Will Sandberg, Makecat-bot, Onemach, Frosty, SFK2, Leemon2010, BurritoBazooka, Faizan, Melonkelon, DavidLeighEllis, ArmbrustBot, Ajmal Ahammed, Haminoon, Meteor sandwich yum, Acebarry, 32RB17, PnT12, Whisper of the heart, ObiWanKenobi11, Monkbot, BethNaught, Tjlester, Qwertyxp2000, Lolwikicansuckit, WikiGopi, Infinite0694, Shahnawaz khan5, Brendanashworth, Brchelmo, Shubham joshi bhopali, Wesko, Matthews david and Anonymous: 579

- **HTTPS Everywhere** *Source:* https://en.wikipedia.org/wiki/HTTPS_Everywhere?oldid=680309816 *Contributors:* DocWatson42, Zigger, OwenBlacker, Bender235, Ruud Koot, GünniX, Intgr, King of Hearts, ViperSnake151, Timtrent, My Gussie, DGG, Neil Smithline, Citation bot, KommX, Phette23, Winterst, Jonesey95, Tom.Reding, John of Reading, Mz7, AManWithNoPlan, Palosirkka, Senator2029, RadicalRedRaccoon, BattyBot, Robin van der Vliet, Dexbot, Acebarry, Patrios, Risc64 and Anonymous: 7

- **Infinit** *Source:* https://en.wikipedia.org/wiki/Infinit?oldid=680268002 *Contributors:* Rwalker, Wikiisawesome, Jerryobject, Mycure, Yobot, W Nowicki, Cnwilliams, Stalwart111, Dewritech, GoingBatty, Palosirkka, Moritz37, BG19bot, Mark Arsten, ArmbrustBot, Tractor Tyres, Dsprc, Jesuufamtobie, Leftfoot leftfoot and Anonymous: 7

- **Inter-protocol exploitation** *Source:* https://en.wikipedia.org/wiki/Inter-protocol_exploitation?oldid=662154707 *Contributors:* The Anome, Alan Liefting, Kravietz, Woohookitty, Daira Hopwood, SmackBot, Hmains, Frap, Hoof Hearted, NickPenguin, Cydebot, Alaibot, Jacobko, Atomsdoubt~enwiki, Antiaxis, Chzz, Yobot, Erik9 and Anonymous: 4

- **Internet Security Research Group** *Source:* https://en.wikipedia.org/wiki/Internet_Security_Research_Group?oldid=662318146 *Contributors:* Bearcat, Giraffedata, MZMcBride, Mehmetaergun and Anonymous: 1

- **Key ring file** *Source:* https://en.wikipedia.org/wiki/Key_ring_file?oldid=632576392 *Contributors:* Spinningspark, The Earwig, BG19bot, Rohitbhave and Thetechgirl

- **Let's Encrypt** *Source:* https://en.wikipedia.org/wiki/Let{}s_Encrypt?oldid=684883740 *Contributors:* The Anome, Lousyd, Thue, Khalid hassani, Rich Farmbrough, MZMcBride, Intgr, Kashmiri, Raysonho, Magioladitis, Mehmetaergun, Konsumkind, Editorofthewiki, AnomieBOT, Actiuinformatica, Mjdtjm, Tony Tan, MeanMotherJr, Zhaofeng Li, Sim3213, Impsswoon, Seatsea, Sarr Cat, Ndevor and Anonymous: 9

- **Linoma Software** *Source:* https://en.wikipedia.org/wiki/Linoma_Software?oldid=674357798 *Contributors:* Inkling, Chowbok, Mark McLoughlin, Rjwilmsi, Bgwhite, SmackBot, Tim@, CmdrObot, Cydebot, CommonsDelinker, Nono64, Wuhwuzdat, Beeblebrox, AnomieBOT, Erik9bot, FrescoBot, Dirkzwart, Ammodramus, Mean as custard, GoingBatty, BG19bot, SharpeIT, HandyHanthorn, JoyceatHome and Anonymous: 10

- **Nautilus (secure telephone)** *Source:* https://en.wikipedia.org/wiki/Nautilus_(secure_telephone)?oldid=637952504 *Contributors:* BenM, Srleffler, Rearden9, Ratarsed, Bluebot, Frap, OrphanBot, Arctic-Editor, Alaibot, Martha.weinberg, Geraldo Perez, Banclark24 and Anonymous: 5

- **Network Security Services** *Source:* https://en.wikipedia.org/wiki/Network_Security_Services?oldid=685847818 *Contributors:* AxelBoldt, The Anome, Sanxiyn, Jeffq, Mattflaschen, Matt Crypto, Beland, Abdull, Kaie, Pmsyyz, Zachlipton, Ceyockey, Armando, Ruud Koot, Boblord, FeldBum, Hawaiian717, RussBot, Ospalh, Shirishag75, Hmains, Gutworth, Frap, Johnchiu, Cydebot, Widefox, Wilee, Bouktin, Nyttend, STBot, Hschulze, Crashie, Int21h, Yobot, MisterSSL, AnomieBOT, Phistuck, Winterst, Jandalhandler, Alvolkov, Hannob, Fedjmike, BG19bot, Ptancredi, X64Architecture, ArmbrustBot, Comp.arch, Xue Fuqiao, Claw of Slime, Backerupper and Anonymous: 33

- **Numbers relay page** *Source:* https://en.wikipedia.org/wiki/Numbers_relay_page?oldid=655418466 *Contributors:* Bearcat, Nakon, War wizard90, Amon16, CorenSearchBot, BG19bot, Smileguy91 and Iaritmioawp

- **Observeit** *Source:* https://en.wikipedia.org/wiki/Observeit?oldid=679790092 *Contributors:* BG19bot, Hmainsbot1, Filedelinkerbot, Rachel123s and Timludy9

- **Open Whisper Systems** *Source:* https://en.wikipedia.org/wiki/Open_Whisper_Systems?oldid=685914265 *Contributors:* Muzzle, Southwest, Bgwhite, Rearden9, Nbauman, Varnent, Unbuttered Parsnip, Yobot, AnomieBOT, Lotje, SporkBot, Hallows AG, Exercisephys, Mogism, Kulandru mor, Dodi 8238, Suelru, BethNaught, TheBeastdot, Doug0707 and Anonymous: 7

- **Petname** *Source:* https://en.wikipedia.org/wiki/Petname?oldid=602803390 *Contributors:* TonyW, Woohookitty, Koavf, RussBot, SmackBot, Einemnet, JonHarder, Iridescent, Cydebot, PamD, Widefox, PGibbons, Malik Shabazz, Michaeldsuarez, Niemeyerstein en, Kathleen.wright5, Mhockey, Miami33139, JakobVoss, Yobot, Erik9, Oneforfortytwo, Jonesey95, BattyBot, Someone not using his real name and Anonymous: 7

- **PGPfone** *Source:* https://en.wikipedia.org/wiki/PGPfone?oldid=635336685 *Contributors:* Prz, Ronz, Securiger, Matt Crypto, ArnoldReinhold, Interiot, Alvis, Armando, Rjwilmsi, FlaBot, Shaddack, Brandon, Rearden9, SmackBot, Ratarsed, Alaibot, Weakness, Martha.weinberg, TooTall-Sid, A box of sticks, Dlrohrer2003, MystBot, Addbot, Dawynn, AnomieBOT, Fpietrosanti, Surv1v4l1st, HollywoodsEnd and Anonymous: 11

- **Phoner** *Source:* https://en.wikipedia.org/wiki/Phoner?oldid=670324243 *Contributors:* Stesmo, Arnomane, Beagel, HeikoS, Funandtrvl, Addbot, DASHBot, Tholme, Patrick87 and Anonymous: 5

- **Presumed security** *Source:* https://en.wikipedia.org/wiki/Presumed_security?oldid=542981926 *Contributors:* Cydebot, DanielPharos, Addbot, Tartarus, Erik9bot, Codemaster32, BG19bot and Anonymous: 1

- **Quicknet** *Source:* https://en.wikipedia.org/wiki/Quicknet?oldid=576752140 *Contributors:* SmackBot, Cander0000, Nomad2u2001, Dthomsen8, Erik9bot, Xianrenb and Anonymous: 1

- **Red/black concept** *Source:* https://en.wikipedia.org/wiki/Red/black_concept?oldid=665867823 *Contributors:* Charles Matthews, DavidCary, Matt Crypto, ArnoldReinhold, Cjcollier, BDD, Kenyon, Eyreland, Josh Parris, RussBot, PhilKnight, AlephGamma, Molly-in-md, Dmwheel1, Sfan00 IMG, Trivialist, DanielPharos, Yobot, Citation bot, Lotje, H3llBot, Helpful Pixie Bot, CitationCleanerBot, Codename Lisa, Tony Mach and Anonymous: 7

- **RetroShare** *Source:* https://en.wikipedia.org/wiki/RetroShare?oldid=679249801 *Contributors:* Olathe, Tosha, Stesmo, Billymac00, Hns~enwiki, Woohookitty, Dadu~enwiki, Jimp, RussBot, Dialectric, Deku-shrub, Frap, Lambiam, ShelfSkewed, Widefox, Smartse, Dktz, Tiagofassoni, Nw man, 83d40m, Black Kite, Tigerix, Trustable, Escape Orbit, PixelBot, Rhododendrites, DanielPharos, Duffbeerforme, Addbot, Legobot, Luckas-bot, Yobot, Flose, Xqbot, GrouchoBot, TheSameGuy, FrescoBot, LittleWink, Suggestednickname, Smatrese, Willy Weazley, Bikepunk2, Liamzebedee, ZéroBot, Arpabone, Sphks, Rezabot, Jhdfj, Nightwalker77, Csoler, BG19bot, Northamerica1000, Cliff12345, ChrisGualtieri, Codename Lisa, Mogism, Jamesmcmahon0, OynetTheCoinissary, Dodi 8238, Tjandamurra, Dadu, VirtuOZ and Anonymous: 41

- **Secret broadcast** *Source:* https://en.wikipedia.org/wiki/Secret_broadcast?oldid=625922638 *Contributors:* David Newton, Matt Crypto, Bobblewik, ArnoldReinhold, Jnestorius, Aard, Woohookitty, RHaworth, Tetraminoe, Stefanomione, Edison, Nneonneo, Cliffb, Junglecat, SmackBot, Brossow, Imzadi1979, CrypticBacon, Sloman, Bluebot, Otto4711, Catsmoke, Wa3frp, Fyrael, Lightbot, AnomieBOT, Xqbot, Bravo Foxtrot, PhnomPencil, Neøn, BattyBot, TortoiseWrath, 32RB17 and Anonymous: 13

- **Secure Communications Interoperability Protocol** *Source:* https://en.wikipedia.org/wiki/Secure_Communications_Interoperability_Protocol?oldid=686003723 *Contributors:* Ronz, Aarchiba, Inkling, Matt Crypto, Bobblewik, TheObtuseAngleOfDoom, Jcm, Pmsyyz, ArnoldReinhold, Kbh3rd, Davidgothberg, Cburnett, Pol098, ApprenticeFan, Wongm, RussBot, Brandon, Terrybader, SmackBot, Mmernex, GBL, Bluebot, DantheCowMan, MartinRe, Cydebot, AntiVandalBot, Widefox, CosineKitty, Jay haines, Dekimasu, Julianbashford, Jim.henderson, SoxBot, WikHead, Keir.tomasso, Lightbot, Yobot, Unsourced, Fpietrosanti, W Nowicki, HamburgerRadio, Velowiki, Liechtensteiner~enwiki, Acer FAMS, Cequica and Anonymous: 24

- **Secure Electronic Transaction** *Source:* https://en.wikipedia.org/wiki/Secure_Electronic_Transaction?oldid=678825258 *Contributors:* The Anome, Ijon, Sbisolo, David Gerard, Matt Crypto, SWAdair, Wk muriithi, Sietse Snel, Alansohn, Argilo, SouthernNights, WouterBot, Juanpdp, Deville, SmackBot, Ignacioerrico, Gilliam, WMXX, SkyWalker, Becheung~enwiki, Cydebot, Mblumber, Urdna, Dulciana, Axel.fois, RJaguar3, Flyer22, Grimey109, Jarjoura, ClueBot, Luismgarcia, Addbot, Tcncv, Tide rolls, Lightbot, Ptbotgourou, Xqbot, Capricorn42, Tad Lincoln, Yun-Yammka, MuffledThud, Shadowjams, BenzolBot, RedBot, Wiking, EmausBot, Zollerriia, MainFrame, ClueBot NG, Eventomer, Phamnhatkhanh, Fjhaq, Faizan, Capt decibel, Monkbot and Anonymous: 60

- **Secure Terminal Equipment** *Source:* https://en.wikipedia.org/wiki/Secure_Terminal_Equipment?oldid=680399082 *Contributors:* Julesd, Inkling, Matt Crypto, Bumm13, Pmsyyz, ArnoldReinhold, Slambo, Davidgothberg, Woohookitty, Wongm, Wavelength, Terrybader, SmackBot, Mauls, Chris the speller, Dmolavi, Frap, Chazchaz101, Thud495, Luna Santin, Jim.henderson, ABF, Eve Hall, Desertfox592, Lightbot, Yobot, Hellogod, Surv1v4l1st, Ida Shaw, P2Peter and Anonymous: 16

- **Secure voice** *Source:* https://en.wikipedia.org/wiki/Secure_voice?oldid=647255007 *Contributors:* Jnc, Matt Crypto, WhiteDragon, Rama, ArnoldReinhold, Bender235, Davidgothberg, Oleg Alexandrov, Woohookitty, Eyreland, Joel7687, Terrybader, SmackBot, Oli Filth, Stattouk, Woodshed, Luna Santin, Albany NY, Jim.henderson, Maurice Carbonaro, TreasuryTag, Lightmouse, Ciphony, Niceguyedc, WikHead, G0rn, Yobot, Rubinbot, Fpietrosanti, Hkuykend, FrescoBot, Rodolfor, MichaelStreetNLD, EmausBot, EJ257T40R, Snotbot, Helpful Pixie Bot, Neøn, Khazar2, Arlene47, Bunkerfunker, Prodx3, Gauntman1 and Anonymous: 24

- **Security Protocols Open Repository** *Source:* https://en.wikipedia.org/wiki/Security_Protocols_Open_Repository?oldid=532225382 *Contributors:* Rich Farmbrough, ArnoldReinhold, Stevenayre, Ddcc, Ideogram, Addbot and Yobot

- **Server Name Indication** *Source:* https://en.wikipedia.org/wiki/Server_Name_Indication?oldid=682351003 *Contributors:* Damian Yerrick, The Anome, Msiebuhr, Pascal666, Rchandra, Pne, Oneiros, Two Bananas, Pwaring, Abdull, Dimmer, Pmsyyz, Bender235, Plugwash, Edward Z. Yang, BenediktWildenhain, Philantrop, Giraffedata, LCamel, Psz, LFaraone, Kenyon, Muhgcee, ChrisNoe, Daira Hopwood, Ruud Koot, Graham87, Elvey, Rjwilmsi, Raztus, Nneonneo, Syced, Aveekbh, Pmc, TheAnarcat, Virtualblackfox, Jengelh, Pelago, Syrthiss, Rwalker, Jeremy Visser, Janfrode, Gslin, JRey, JLaTondre, Urkle0, Vellmont, SmackBot, Gigglesworth, Bmearns, Gilliam, DomQ, Eug, Thumperward, Frap, Tronicum, Duckbill, Zvis~enwiki, Spartanfan10, Tobiasly, Trou, Ceplm, Jec, Majora4, HDCase, Philipp Kern, AndrewHowse, Marc W. Abel, ClarkMills, Nick Number, Gioto, Widefox, Marokwitz, Fredden, JAnDbot, Struthious Bandersnatch, Zevnik, Falkvinge, Philippe23, EXCEPTION NOT HANDLED, Onkelringelhuth, TXiKiBoT, Porjo, ShlomoS, Tjorven, MJaggard, Ildmur, MidSpeck, Rosuav, Niceguyedc, Magamiako, NuclearWarfare, RDailey79, SF007, DumZiBoT, Cristiklein, MystBot, Addbot, Jc7k, Tothwolf, MrOllie, Technologov, Yobot, Bunnyhop11, TheWishy, Iangfc, Dmarquard, AnomieBOT, SexyTyranno, Rat2, Purpledikkweed, Erik9bot, FrescoBot, Winterst, Torkel.bjornson,

Brentl99, Daxim, Nickyus, Drmonocle, Noloader, Dewritech, ZéroBot, FlippyFlink, Cf. Hay, Matthewgarysmith, Ldimartino, Flokater, Cathyjf, BG19bot, Maartenvanrooijen, Chmarkine, Eyuval, BjarniRunar, Hiddenray, Tmiiluva, Kariantti, Kaushal.dj, Wz4gz, Uri.may, Martijnotto, Shanemadden, TanzDSE, Eserte12, Swypych, Mstancombe, Claw of Slime, Dai Pritchard, Bertrand lupart, Jdleider, Pppingme, JavaProphet, Jose.nobile and Anonymous: 153

- **Signal (software)** *Source:* https://en.wikipedia.org/wiki/Signal_(software)?oldid=682274151 *Contributors:* Beeblebrox, Niceguyedc, Yobot, Jim1138, Nodove, Dodi 8238, Spiderjerky and Anonymous: 2

- **SMTPS** *Source:* https://en.wikipedia.org/wiki/SMTPS?oldid=640343932 *Contributors:* ChristopherA, Feezo, Apokrif, Ebertek, Mattsim, GrahamHardy, Logan, Den Hieperboree, Addbot, Informationtheory, CXCV, FrescoBot, Gqqnb, Rixs, Bluefist, Jht001, Felixkasza, BG19bot, Majx, Ajducks and Anonymous: 6

- **Sqlnet.ora** *Source:* https://en.wikipedia.org/wiki/Sqlnet.ora?oldid=674993955 *Contributors:* Apokrif, Jsb~enwiki, Cedar101, CorenSearchBot, Jim1138, Iurii.Fedyshyn, Jandalhandler, VernoWhitney, BG19bot, ChrisGualtieri, Éire FireDragon, Jkstill and Anonymous: 1

- **STU-I** *Source:* https://en.wikipedia.org/wiki/STU-I?oldid=686001686 *Contributors:* Bumm13, RevRagnarok, ArnoldReinhold, Davidgothberg, Redvers, Woohookitty, MarkPos, RussBot, SmackBot, Scarletsmith, Cydebot, Widefox, Valerius Tygart, Gridiron Scholar, Jim.henderson, Rosiestep, Excirial, SoxBot, Miami33139, Yobot, Capricorn42 and Anonymous: 2

- **STU-II** *Source:* https://en.wikipedia.org/wiki/STU-II?oldid=595905977 *Contributors:* Matt Crypto, WhiteDragon, Bumm13, RevRagnarok, Pmsyyz, ArnoldReinhold, Davidgothberg, Woohookitty, Apokrif, Scarletsmith, Cydebot, Jim.henderson, Whitebox, Dillard421, SoxBot, Yobot, P2Peter and Anonymous: 4

- **Temporal Key Integrity Protocol** *Source:* https://en.wikipedia.org/wiki/Temporal_Key_Integrity_Protocol?oldid=679096801 *Contributors:* Jimfbleak, JidGom, Ferdinand Pienaar, Matt Crypto, Soman, JoshG, Pmsyyz, ArnoldReinhold, YUL89YYZ, Bender235, Guy Harris, L33th4x0rguy, Stephan Leeds, Woohookitty, Lofor, Nandakumarg, Bratch, Allen Moore, FlaBot, Intgr, YurikBot, Hairy Dude, Yuhong, NFH, Derelk, SmackBot, KelleyCook, Ryan Roos, Luís Felipe Braga, Euchiasmus, Morten, Peyre, Hu12, Jesse Viviano, T23c, Quibik, Rmicallef, Thijs!bot, Dawnseeker2000, Deflective, Austin512, VolkovBot, TXiKiBoT, Arleyl, Jamelan, AlleborgoBot, Rilkas, Atif.t2, ClueBot, Addbot, Luckas-bot, Yobot, Royote, Xizhi.zhu, ArthurBot, Xqbot, Giddeon Fox, Frysalebald, FrescoBot, LucienBOT, Darr247, RedBot, Dcostello2, EmausBot, Westley Turner, ChocolateBall, BattyBot, TheJJJunk, Userings, Monkbot, Thenetsec and Anonymous: 44

- **TLS termination proxy** *Source:* https://en.wikipedia.org/wiki/TLS_termination_proxy?oldid=676671736 *Contributors:* Bearcat, Iantresman, Thumperward, Dbu, PamD, Drwilco, Silas S. Brown, Glane23, Lotje and Anonymous: 4

- **TLS-PSK** *Source:* https://en.wikipedia.org/wiki/TLS-PSK?oldid=680354071 *Contributors:* YUL89YYZ, GBL, Widefox, CommonsDelinker, WOSlinker, Xeno8, Tuntable, Liorkaplan, Manuel (mpg), ArmbrustBot and Anonymous: 2

- **TLS-SRP** *Source:* https://en.wikipedia.org/wiki/TLS-SRP?oldid=635311543 *Contributors:* Qslack, FauxFaux, BG19bot, ArmbrustBot and Anonymous: 5

- **Tor (anonymity network)** *Source:* https://en.wikipedia.org/wiki/Tor_(anonymity_network)?oldid=685692056 *Contributors:* AxelBoldt, Edward, Nealmcb, Michael Hardy, Lquilter, Karada, Pde, SebastianHelm, MichaelJanich, CesarB, Haakon, 5ko, Den fjättrade ankan~enwiki, Julesd, Andres, Mxn, Feedmecereal, Guaka, WhisperToMe, Ewald, K1Bond007, Thue, AnonMoos, Mordomo, Phil Boswell, Chrism, Lowellian, Chris Roy, Jeronim, Auric, Commonchaos, David Edgar, HaeB, Connelly, Centrx, DocWatson42, Yama, Zigger, Romanpoet, Alison, Niteowlneils, Kravietz, Pascal666, Gzornenplatz, Tangerine Cossack, Edcolins, Delta G, SoWhy, OverlordQ, Quarl, CaribDigita, Oneiros, Kubieziel, Lev, Indolering, Gscshoyru, Creidieki, Shiftchange, Discospinster, Helohe, Rich Farmbrough, FT2, Samboy, Gronky, Cariaso, Bender235, ESkog, Twiek, El C, Nile, Dennis Brown, John Vandenberg, B0at, Nhandler, Sleske, Krellis, Qedragon, Orangemarlin, Orzetto, Alansohn, Gary, The RedBurn, PatrickFisher, Graingert, Bios~enwiki, Jehannette, Isaac, Jrleighton, Stephan Leeds, Geraldshields11, H2g2bob, Kusma, Dan100, SJMurdoch, Dismas, Dtobias, Gmaxwell, Rolloffle, Thryduulf, Hoziron, Simetrical, Lemi4, Woohookitty, Kupojsin, AirBa~enwiki, Sburke, StradivariusTV, Lofor, Deeahbz, Armando, Jhartmann, Sam916, LogicalDash, Gerbrant, Marudubshinki, Volland, Mendaliv, Rjwilmsi, Koavf, Tangotango, Bruce1ee, Bensin, Nandesuka, Lotu, Tommy Kronkvist, FlaBot, SchuminWeb, Jsheehy, Pathoschild, Gurch, Alexjohnc3, Jrtayloriv, Riki, OpenToppedBus, Silivrenion, Sherab~enwiki, JiVE, Chobot, Bgwhite, Peterl, YurikBot, Jimp, Wolfmankurd, MMuzammils, RussBot, Ansell, Hydrargyrum, David Woodward, CambridgeBayWeather, Shaddack, Schoen, Varnav, Thbarnes, NawlinWiki, Nowa, Anomo, Drsayis2, Burntpsilocybin, Ezeu, Voidxor, EEMIV, Deku-shrub, Vlad, Brisvegas, User27091, Graciella, Ninly, Icedog, Closedmouth, Arthur Rubin, SMcCandlish, 430072, Nlitement, Synergyplease, Snaxe920, NeilN, Jimerb, Tom Morris, That Guy, From That Show!, SmackBot, ManaUser, Herostratus, InverseHypercube, C.Fred, Grye, Wegesrand, Mazaczek, BiT, Rotemliss, PeterSymonds, Ohnoitsjamie, Lighthill, Skizzik, Remohammadi, Chris the speller, StephenH, Thumperward, Tweak232, Mithaca, DHN-bot~enwiki, Frap, Neo139, OSborn, Theprez98, Adamantios, Albertalbs, Flyguy649, Ianmacm, Cybercobra, MichaelBillington, 4hodmt, Philpraxis~enwiki, Henning Makholm, Rhkramer, Salamurai, Kukini, Oneangrydwarf, AThing, John, Littleman TAMU, Gobonobo, Robofish, Joffeloff, MonstaPro, IronGargoyle, PseudoSudo, 16@r, Drdevil44, Notwist, PRRfan, Afecks, Sgutkind, Dr.K., Vashtihorvat, Darry2385, Dl2000, Kencf0618, Sixstone~enwiki, The7thmagus, Elryacko, JHP, Twas Now, Timrem, StephenFalken, Wafulz, Sir Vicious, Zarex, Einsensteiner, Nczempin, Endareth, IntrigueBlue, WeggeBot, Phatom87, AnthonyCheng, A876, UncleBubba, Christian75, Foo bars, Gimmetrow, Blackm0re, Thijs!bot, Thomas Skogestad, Qwyrxian, Pstanton, Reil, Ljean, Hcobb, Esbullin, Dawnseeker2000, AntiVandalBot, Gioto, Luna Santin, Guy Macon, Bladestorm, Rainonwood, Isilanes, L0b0t, Qwerty Binary, Daytona2, Barek, MER-C, Skomorokh, Doctorhawkes, Joneyf1, Wkenzie, Dvehrs, .anacondabot, Think outside the box, Wseltzer, Nyttend, Comradecommisar, Perebot~enwiki, LorenzoB, Roberth Edberg, Philg88, Sue Gardner, STCL, Yawnbox, Yiwen017, Gwern, Oren0, AVRS, CommonsDelinker, Tickerhead, J.delanoy, Cyanolinguophile, Jreferee, OohBunnies!, Aryisfat, Dispenser, Mahewa, Jeepday, Kapparo, 83d40m, Петър Петров, Remember the dot, Gelisiyo, Lakersforce, Cralar, Black Kite, LeeColleton, VolkovBot, Davidwr, Aesopos, Rei-bot, Wikidemon, Liko81, Broadbot, Silent52, Rps5, Rhinux, AlleborgoBot, Kbrose, SieBot, Nubiatech, Gerakibot, Dawn Bard, Feoray, Belorn, Flyer22, Jimthing, Free Software Knight, PhilMacD, Callidior, Cyb3rdemon, StaticGull, Anchor Link Bot, HighInBC, WikiLaurent, Digisus, ImageRemovalBot, Mwenge, Animeronin, Czarkoff, Kl4m-AWB, Der Golem, VQuakr, P.T.isfirst, Mondo6, Airwot4, Jalanpalmer, Ashashyou, Vonkaiser44, Excirial, Alexbot, Murod, Straightpress, Rhododendrites, Arjayay, TheRedPenOfDoom, JasonAQuest, Jack-A-Roe, Scalhotrod, Omody, Apparition11, SF007, Expertjohn, Chris1834, Helixweb, XLinkBot, AnotherSolipsist, I'm On The Rise, WillOakland, Avoided, Snurre86, ErkinBatu, Galzigler, Realworth, Zinger0, Lamadrid~enwiki, Kajabla, Addbot, Mortense, Yoenit, Elenaignatova, Ismouton, LaaknorBot, CarsracBot, Debresser, Jasper Deng, Bigzteve, Lukejamesoconnor, Ibrahim-saeed, Bluebusy, Gyzome, Jarble, Swarm, Prof7bit, Arrakis-cDc, Yobot, CFeyecare, The Grumpy Hacker, Legobot II, Bhagwad, Kmolnar, Grrrrrtehgrrrrman, 4th-otaku, Resubew, AnomieBOT, DemocraticLuntz, 1exec1,

Jim1138, Kingpin13, Materialscientist, Aff123a, Citation bot, JohnnyB256, LilHelpa, Xqbot, Editing for clarity, TracyMcClark, Gilo1969, Omnipaedista, Carrite, Henk.muller, Fixentries, Howard McCay, Samwb123, Superhoneybee, FrescoBot, Surv1v4l1st, Mfwitten, Igna, 猫ノ森, Jersey92, Appleton1324, Gnepets, Onlyjob, RedBot, Puppier, Dannyx28, Lotje, Callanecc, Wilton gorske, Athaba, Aoidh, David Hedlund, Elite-hobo, C4K3, Tbhotch, Jesse V., Diablo-fan, AXRL, Sinistlor, RjwilmsiBot, Tomchen1989, Star-Syrup, DaveyHanks101, Rollins83, EmausBot, WikitanvirBot, Gat101, Japs 88, GoingBatty, RA0808, Taylor 808, JohnValeron, Peaceray, Dalegudmunsen, HugC12, Rkononenko, Mz7, Pro translator, Cogiati, Checkingfax, Ida Shaw, IThinkTheseUsernamesAreStupid, H3llBot, SkyBon, Rcsprinter123, OrpheusSang, Happy-DudeThe MadTim, Palosirkka, Donner60, MainFrame, Maucelli, JaredThornbridge, Mjbmrbot, Mikhail Ryazanov, Helpsome, ClueBot NG, Strugee, Robin Mathew Rajan, Hyiltiz, Atagar, WIERDGREENMAN, Jenova20, James65.pike, Wibble666, Vacation9, Twillisjr, Chisme, Frosty2204, Maddee moo, Rezabot, Riveravaldez, MerlIwBot, Helpful Pixie Bot, HMSSolent, Lowercase sigmabot, BG19bot, Pine, Roberticus, WikiTryHardDieHard, KShiger, MusikAnimal, Frze, Ekotren420, Dlampton, Paganinip, Dan653, Piguy101, Boxerpop82, Eman2129, Ter-rel Shumway, Mysticete, Chmarkine, Cliff12345, Power2794, Vennila1983, Tianjiao, Metsfreak2121, BricksReallyExist, LlamaAl, Cyberbot II, ChrisGualtieri, Anthonyb123, Khazar2, Rezonansowy, Codename Lisa, Tsuruya, DinnerDude, Nuclear awesome, Adbenedictlewis, Jemap-pelleungarcon, Marblesoda, Wassup2190, Hillbillyholiday, BobThePlatypus, 128b-net, Jamesmcmahon0, Jaaayzzzzzee, Tentinator, Wuerzele, Nodove, Dustin V. S., Fpigerre, Pahty, Caitlin.swartz, Alexd1010101, Haminoon, Akh81, Makhoondi, Spacenut42, Oranjelo100, Wikifan2744, اوورایشگر-1, Mody633, Dodi 8238, Grjgt893u34, Acebarry, Officialdrgamer, Monkbot, Stefano.desabbata, Hkphe, Randomrandomrandom, KaosMuppet, BethNaught, Thompsonswiki, Chainsaw-penguin, Sblog13, Gamergod321, Danielsun174, U9y0x46md247bg5ivb7z, WordSeven-teen, ChamithN, Sheckyrubenstein, Kethrus, Pagesclo, EmailExperto, John Smithssssssssss, Wolesslap, Computerpc0710, SnowdenFan, Aidan-Mulv, Billy d kidd, Posterboi1, MichaelGReed and Anonymous: 502

- **Tox (protocol)** *Source:* https://en.wikipedia.org/wiki/Tox_(protocol)?oldid=682078637 *Contributors:* Topbanana, Alexanderino, Stesmo, Diego Moya, Yurivict, Benlisquare, RussBot, Rwalker, Deadbeef, Dktz, Magioladitis, Pikolas, ImageRemovalBot, Czarkoff, Niceguyedc, Curath, Geo g guy, Yobot, Fleshgrinder, AnomieBOT, Cedesguin, ContinueWithCaution, Digital.Maniac, Jesse V., BillyPreset, Mikhail Ryazanov, Vacation9, BG19bot, Cjrobe, BattyBot, Rarkenin, ScotXW, Dodi 8238, Filedelinkerbot, Alexskc, Stqism, LaKapitano, Fjreegman, JFreegman, AdamRecende, Promctagonist, Astonex, PropIex, Ekkima, Mlkj, CandyGumdrop, LumenTeun, Henry Newman, Sagem1917, Dvor5525 and Anonymous: 60

- **Transport Layer Security** *Source:* https://en.wikipedia.org/wiki/Transport_Layer_Security?oldid=686387081 *Contributors:* Damian Yerrick, AxelBoldt, WojPob, Bryan Derksen, Zundark, The Anome, Etu, Andre Engels, Aldie, William Avery, Mjb, PeterB, Branko, Nyco~enwiki, Ed-ward, Nealmcb, Michael Hardy, GABaker, Qslack, Anders Feder, Haakon, Mac, Strebe, 5ko, Marumari, Yaronf, Nikai, Ed Brey, Adam Conover, Mydogategodshat, Emperorbma, Ww, The Anomebot, Sanxiyn, Dougjih, Tpbradbury, Furrykef, Olathe, Robbot, Pfortuny, RichiH, Chealer, Nurg, Tim Ivorson, Ashdurbat, Rholton, Rasmus Faber, Davodd, Tbutzon, HaeB, Mattflaschen, Giftlite, Lunkwill, Cfp, Thorne, DavidCary, Zigkill, Levin, MadmanNova, Everyking, Rick Block, Gracefool, Kravietz, AlistairMcMillan, Matt Crypto, Wmahan, Thomas Springer, Beland, ShakataGaNai, Nils~enwiki, Hgfernan, OwenBlacker, Rlcantwell, Biot, Kelson, Olivier Debre, Abdull, Mike Rosoft, Ta bu shi da yu, Imroy, JTN, GoodStuff~enwiki, Discospinster, Rich Farmbrough, Pmsyyz, Smyth, Mani1, Bender235, Plugwash, Evice, CanisRufus, Mwanner, Kgaughan, Rhomboid, Ray Dassen, Ninels, Mpvdm, SpeedyGonsales, Nk, Zr40, Sleske, Blodulv, Haham hanuka, Krellis, CKlunck, ChristopherA, Clement-Seveillac, Jigen III, Alansohn, Interiot, Uogl, Neuhaus~enwiki, Nealcardwell, Abaybas, JoaoRicardo, Calton, Psz, Nitrogenx, Paul1337, Rey-Brujo, Stephan Leeds, Suruena, Star General, Freyr, Versageek, Voxadam, Feezo, Jef-Infojef, Weyes, Simetrical, Mindmatrix, Daira Hop-wood, Borb, Niqueco, Ruud Koot, Trödel, Davidfstr, Eyreland, Eruionnnyron, Jasonpearce, FBarber, Gerbrant, Marudubshinki, Msiddalingaiah, Graham87, Seneces, Elvey, Jclemens, Phoenix-forgotten, Sjö, Rjwilmsi, Wiarthurhu, Maxim Razin, Ghalas, Lukegilman, Andrzej P. Woz-niak, Boblord, FlaBot, Thomasgud, Margosbot~enwiki, Ysangkok, FireballDWF2, Intgr, Fresheneesz, Jmaister~enwiki, Fritzophrenic, GreyCat, VishalJBhatt, Typhoonhurricane, Colenso, Jmorgan, Chobot, Moocha, Bgwhite, Zimbabweed, Shaggyjacobs, Siddhant, YurikBot, Wavelength, RobotE, Hairy Dude, Yuhong, Shaddack, Bovineone, Vanished user kjdioejh329io3rksdkj, Toehead2001, Schlafly, Albedo, Davemck, Mgcsinc, Leotohill, Bota47, Oscardt, Lzyiii, Zzuuzz, Wilfrednilsen, Rushyo, JoanneB, Alias Flood, Anclation~enwiki, Mardus, TDM, Wizofaus, Vell-mont, Lundse, SmackBot, Mmernex, Branlon, 0x6adb015, Xaosflux, Gilliam, Youremyjuliet, Skizzik, Lakshmin, Chris the speller, Apankrat, Jprg1966, Thumperward, Armour Hotdog, Snori, Ber, Omniplex, Colonies Chris, Jdthood, Meetabu, N.MacInnes, Frap, Alphathon, JonHarder, Michel SALES, Ianmacm, Plustgarten, Martijn Hoekstra, Tony esopi patra, Ametheus, Daniel.Cardenas, LeoNomis, Pilotguy, Morten Brørup, TJJFV, ThurnerRupert, Jmanico, Nmav, Ksn, Breno, Scetoaux, NongBot~enwiki, IronGargoyle, Nagle, RomanSpa, BlindWanderer, Tacke, Stupid Corn, Beetstra, MTSbot~enwiki, Ant honey, DouglasCalvert, Olegos, Andrew Hampe, Paul Foxworthy, Beno1000, Alec it, TwelveBaud, AbsolutDan, Felixcatuk, Jesse Viviano, Ternto333, Cydebot, Ntsimp, Groovy12, UncleBubba, Robinalden, Verdy p, Burke Libbey, M. B., Jr., Swagatata, Gionnico, Ysimonson, Vinayr rao, Hanche, Wdspann, Thijs!bot, Koeplinger, Acodring, Wmasterj, Dawnseeker2000, AntiVandal-Bot, Davidoff, Widefox, Guy Macon, QuiteUnusual, Itistoday, Isilanes, Julie Deanna, VictorAnyakin, Esmond.pitt, Deadbeef, Arsenikk, JAnD-bot, Ppelleti, Barek, PaleAqua, Martinkunev, Tqbf, Raanoo, Enjoi4586, GreatEgret, Magioladitis, Bongwarrior, VoABot II, Joblack~enwiki, Antientropic, Aka042, Papadopa, JaGa, Kgfleischmann, Molf, Ericnay, WLU, Gwern, Jc monk, Michaelfowler, Mårten Berglund, Rettetast, SP-Cartman, CommonsDelinker, Mange01, Aethedor, Startcom, Titiri, VAcharon, Jesant13, Yonidebot, Iida-yosiaki, Arronax50, Produke, C1010, Simon.may.007, 83d40m, Mundocani, 806f0F, Remember the dot, Ross Fraser, Idioma-bot, Juhovh~enwiki, FloydRTurbo, VolkovBot, Speaker to Lampposts, Michaelkrauklis, Philip Trueman, Oconnor663, Mischling, Dictouray, Mickraus, Thunderbritches, Lradrama, Stefonic, Arkoon, Nerwal, Flyingw, Jamelan, Dkgdkg, WinTakeAll, Schmalls, PieterDeBruijn, Erth64net, Falcon8765, Anna512, Spinningspark, Marrowmonkey, Hottdee, LittleBenW, Matthew V Ball, Kbrose, Nubiatech, Dreamafter, Gerakibot, J-p krelli, Crossland~enwiki, Mayevski, Dogbyter, Ravi-aulakh, MinorContributor, ObscurO, Int21h, Svick, Spartan-James, Cajunbill, Emk, Cellmate707, Martarius, ClueBot, Mrbbking, Traveler100, Deedub1983, JWilk, Pasi Eronen, Czhower, Mild Bill Hiccup, Twkd, Uncle Milty, Timberframe, Loftenter, Ripsss, Excirial, Alexbot, Anon lynx, Kpsmithuk, Sun Creator, Legacypac, Complicated1, Aprock, Vijay.kotari, Blackbearded, Devon Sean McCullough, Digi-cs, Mitch Ames, RP459, Sweeper tamonten, Addbot, Ghettoblaster, Writermonique, Mabdul, Madigral, KitchM, CanadianLinuxUser, MrOllie, CarsracBot, Debresser, MagnetiK, Nonno88, Fryed-peach, Luckas-bot, Yobot, Themfromspace, Bunnyhop11, TaBOT-zerem, TheWishy, THEN WHO WAS PHONE?, Iangfc, Chaliy, Sachuraju, Fleshgrinder, MisterSSL, Jas4711, AnomieBOT, Noq, Ciphers, Rubinbot, Jlehen, Jim1138, Ma-terialscientist, Xizhi.zhu, Citation bot, 天使のくれた花, Xqbot, Koektrommel, Amenel, Gidoca, Barakw, MichaelCoates, KommX, Webguynik, Kyng, PHansen, SCΛRECROW, Aprogrammer, Shadowjams, Toyotabedzrock, Sesu Prime, Andrei.wap, FrescoBot, Doedoejohn, ExportRadical, Avbentem, CanadianNine, HamburgerRadio, Hawk-Eagle, Winterst, Stevie-sen, I dream of horses, Mr Heine, Jandalhandler, Yonatan.graber, Robvanvee, FoxBot, Conseguenza, Trappist the monk, Tracef2112, Wutherings, Lotje, Nickyus, Polurupraveen, Comet Tuttle, Gzorg, Weed-whacker128, Tbhotch, Rarut, DARTH SIDIOUS 2, RjwilmsiBot, Lopifalko, EmausBot, Timtempleton, Nuujinn, Noloader, Super48paul,

Dewritech, Tommy2010, FlippyFlink, Bxj, MajorVariola, Chris conlon, Akebinho, Cf. Hay, Tomato86, Arman Cagle, Yadirh, Sbmeirow, Itahmed, Donner60, SBaker43, Guthrg007, Jjplaya209, Tijfo098, Nill smith, Pyav, Jabbany, Greatwhitesharkbear, 28bot, Sara Wright, Mikhail Ryazanov, ClueBot NG, Gareth Griffith-Jones, Rafigordon, ForwardChange, Raghith, Chester Markel, Dfarrell07, Nahiyan8, Robertssh, Kasirbot, Chriswaigl, Ercrt, Jpinkerton88, Adrianfd, Btrzupek, Helpful Pixie Bot, BBirke, Boomboombi, Usaguruman, Denovoid, BG19bot, Bklisch, MusikAnimal, Badon, Compfreak7, Nabeeh20, Maartenvanrooijen, Yowanvista, SSLcertificatesecurity, Tony Tan, Unixman83, Chmarkine, Mirrakor, RealSebix, Morning Sunshine, Trailspark, Tcc8, Sbose7890, Fylbecatulous, Pizzamancer, Meowimasexycat, BattyBot, Cyberbot II, Makerofthings7, Thulasi.goriparthi, Codename Lisa, SoledadKabocha, Mogism, Cerabot~enwiki, Will Faught, JustAMuggle, Palmbeachguy, Jcsouthworth, Wikivhz, Faizan, SolarStarSpire, Marker Way, Ruby Murray, Codelux, JakeWi, PhantomTech, Praemonitus, Zwodrei, Yellow Lilt, ArmbrustBot, 08af9a09, Comp.arch, Diyoev, Bugorsky, Domswaine, Ginsuloft, Wikisuzan, Someone not using his real name, Meteor sandwich yum, Thompor, Imshingu, Mattghali, Claw of Slime, Monkbot, Chrisdeguara, F3ndot, Jason3221, WikiGopi, Guse1234, Phantom gamer 1993, Infinite0694, Educemail, Brchelmo, Noncombatantorg, Zimzamyap and Anonymous: 879

- **Twinkle (software)** *Source:* https://en.wikipedia.org/wiki/Twinkle_(software)?oldid=680849463 *Contributors:* Lquilter, Greenman, Jeroen, Fo0bar, Stesmo, Dialectric, Rearden9, SmackBot, Frap, Guyjohnston, Thijs!bot, ChrisiPK, Hans Persson, Kgfleischmann, Kbrose, Triwbe, Editore99, ترجمان05, Ndgp, Kl4m-AWB, Leonard^Bloom, XLinkBot, Addbot, Xp54321, Grandscribe, LubosD, Santosga, Luckas-bot, AnomieBOT, Götz, Emil.ivov, LucienBOT, PigFlu Oink, Ondra.pelech, Mekeor, Alisha.4m, Vanished 1850, BattyBot, Pai Walisongo, Kephir, Kulandru mor, Therockam2, Qwertyxp2000 and Anonymous: 14

- **Typhoid adware** *Source:* https://en.wikipedia.org/wiki/Typhoid_adware?oldid=665373227 *Contributors:* Bruce1ee, Eastlaw, Cydebot, Samker, DanielPharos, Callinus, Tinton5, XJDHDR, Buggie111, Bar-abban, Staszek Lem, Pastore Italy, Codename Lisa, Julietdeltalima and Anonymous: 3

- **User Activity Monitoring** *Source:* https://en.wikipedia.org/wiki/User_Activity_Monitoring?oldid=658934864 *Contributors:* Dthomsen8, Meatsgains, OccultZone, Rubbish computer and TimLudy

- **Vanish (computer science)** *Source:* https://en.wikipedia.org/wiki/Vanish_(computer_science)?oldid=618156555 *Contributors:* Gracefool, RussBot, LodeRunner, Tktktk, Falcon8765, DanielPharos, SensuiShinobu1234, Mootros, Haqui11a, H3llBot, Oodri3 and Anonymous: 3

- **Verifiable computing** *Source:* https://en.wikipedia.org/wiki/Verifiable_computing?oldid=620774336 *Contributors:* DavidCary, Rjwilmsi, Bhny, Malcolma, Racklever, Nuesken, R'n'B, SchreiberBike, Dekart, Yobot, WikiDan61, Jonesey95, BG19bot, Aialali, ScienceContentCreator, Wikipatrol94, Solarlifesolarlife and Anonymous: 5

- **Voice over IP Security Alliance** *Source:* https://en.wikipedia.org/wiki/Voice_over_IP_Security_Alliance?oldid=543984155 *Contributors:* Merovingian, MarkSweep, Dyork, Hooperbloob, Jemiller226, SmackBot, TJJFV, Gogo Dodo, Alaibot, Jim.henderson, VolkovBot, Addbot, Lightbot, 123microsoft123 and Anonymous: 2

- **Whisper Systems** *Source:* https://en.wikipedia.org/wiki/Whisper_Systems?oldid=670810154 *Contributors:* Bearcat, Rpyle731, Cydebot, Pro crast in a tor, Gigi head, I JethroBT, Funandtrvl, Electron100, Mortense, PabloCastellano, Jonesey95, RjwilmsiBot, BG19bot, Johnny Squeaky, Robinwest76, BattyBot, BurritoBazooka, Moxiemoxie, Bslackr, Dodi 8238 and Anonymous: 3

- **Wireless intrusion prevention system** *Source:* https://en.wikipedia.org/wiki/Wireless_intrusion_prevention_system?oldid=654444116 *Contributors:* RedWolf, Auric, David Haslam, Brandon, Petri Krohn, SmackBot, Xaosflux, Sepa, Bluebot, TheGerm, Alaibot, Nick Number, Dawnseeker2000, Tqbf, RushJohnson, B. Wolterding, R'n'B, Public Menace, Canaima, Adam.J.W.C., Jojalozzo, Cliffordc5, Akshaymathur, Kbdavis07, Yobot, Backscatter, Xqbot, Vmp2009, Bar-abban, Randyicrew, KLBot2, Lesser Cartographies, Ghcko and Anonymous: 22

- **X.1035** *Source:* https://en.wikipedia.org/wiki/X.1035?oldid=649808951 *Contributors:* Edward, Yobot, Itusg15q4user, ArmbrustBot and Anonymous: 1

- **Zooko's triangle** *Source:* https://en.wikipedia.org/wiki/Zooko'{}s_triangle?oldid=683782563 *Contributors:* Michael Hardy, 6birc, Ciphergoth, Mike Linksvayer, David Latapie, Mattflaschen, Matthäus Wander, TonyW, Domster, DrummondReed, Dionyziz, TheAnarcat, Bayle Shanks, Shepazu, MSJapan, Deku-shrub, SmackBot, McGeddon, 127, Xaosflux, Psiphiorg, JonHarder, Iridescent, Cydebot, Neustradamus, Widefox, Donagle, PGibbons, Zooko, Gwern, Gwen Gale, X-Fi6, Ossguy, Jonathanstray, Niceguyedc, MystBot, Addbot, Yobot, Iangfc, Erik9, Sanpitch, Yutsi, Dewritech, Your Lord and Master, BG19bot, Cincinatis, Cliff12345, Plaintive plaintiff, Browncoyote, Someone not using his real name, MARIODOESBREAKFAST, Spiderjerky, Nemesis0618 and Anonymous: 12

67.5.2 Images

- **File:Ambox_important.svg** *Source:* https://upload.wikimedia.org/wikipedia/commons/b/b4/Ambox_important.svg *License:* Public domain *Contributors:* Own work, based off of Image:Ambox scales.svg *Original artist:* Dsmurat (talk · contribs)

- **File:ArmLogo.png** *Source:* https://upload.wikimedia.org/wikipedia/en/8/82/ArmLogo.png *License:* CC-BY-SA-3.0 *Contributors:* ? *Original artist:* ?

- **File:Arm_partial_screenshot.png** *Source:* https://upload.wikimedia.org/wikipedia/commons/8/86/Arm_partial_screenshot.png *License:* CC BY-SA 3.0 *Contributors:* Transferred from en.wikipedia to Commons by SreeBot. *Original artist:* Atagar at en.wikipedia

- **File:Bus_icon.svg** *Source:* https://upload.wikimedia.org/wikipedia/commons/c/ca/Bus_icon.svg *License:* Public domain *Contributors:* ? *Original artist:* ?

- **File:CA_Browser_Forum_Logo.jpg** *Source:* https://upload.wikimedia.org/wikipedia/en/0/08/CA_Browser_Forum_Logo.jpg *License:* Fair use *Contributors:* https://www.cabforum.org/ *Original artist:* ?

- **File:Commons-logo.svg** *Source:* https://upload.wikimedia.org/wikipedia/en/4/4a/Commons-logo.svg *License:* ? *Contributors:* ? *Original artist:* ?

- **File:Computer-aj_aj_ashton_01.svg** *Source:* https://upload.wikimedia.org/wikipedia/commons/d/d7/Desktop_computer_clipart_-_Yellow_theme.svg *License:* CC0 *Contributors:* https://openclipart.org/detail/105871/computeraj-aj-ashton-01 *Original artist:* AJ from openclipart.org

- **File:Crypto_key.svg** *Source:* https://upload.wikimedia.org/wikipedia/commons/6/65/Crypto_key.svg *License:* CC-BY-SA-3.0 *Contributors:* Own work based on image:Key-crypto-sideways.png by MisterMatt originally from English Wikipedia *Original artist:* MesserWoland

- **File:Crystal_Clear_app_browser.png** *Source:* https://upload.wikimedia.org/wikipedia/commons/f/fe/Crystal_Clear_app_browser.png *License:* LGPL *Contributors:* All Crystal icons were posted by the author as LGPL on kde-look *Original artist:* Everaldo Coelho and YellowIcon

- **File:Crystal_Clear_app_network.png** *Source:* https://upload.wikimedia.org/wikipedia/commons/4/49/Crystal_Clear_app_network.png *License:* LGPL *Contributors:* All Crystal Clear icons were posted by the author as LGPL on kde-look; *Original artist:* Everaldo Coelho and YellowIcon;

- **File:Crystal_Clear_device_cdrom_unmount.png** *Source:* https://upload.wikimedia.org/wikipedia/commons/1/10/Crystal_Clear_device_cdrom_unmount.png *License:* LGPL *Contributors:* All Crystal Clear icons were posted by the author as LGPL on kde-look; *Original artist:* Everaldo Coelho and YellowIcon;

- **File:Data_breach_average_cost_germany.svg** *Source:* https://upload.wikimedia.org/wikipedia/commons/3/3c/Data_breach_average_cost_germany.svg *License:* CC0 *Contributors:* Own work *Original artist:* Bananenfalter

- **File:Edit-clear.svg** *Source:* https://upload.wikimedia.org/wikipedia/en/f/f2/Edit-clear.svg *License:* Public domain *Contributors:* The *Tango! Desktop Project*. *Original artist:*
  The people from the Tango! project. And according to the meta-data in the file, specifically: "Andreas Nilsson, and Jakub Steiner (although minimally)."

- **File:Electrospace-mlp1a.jpg** *Source:* https://upload.wikimedia.org/wikipedia/commons/4/48/Electrospace-mlp1a.jpg *License:* CC BY-SA 3.0 *Contributors:* Own work *Original artist:* Paul2

- **File:Emoji_u1f4f1.svg** *Source:* https://upload.wikimedia.org/wikipedia/commons/1/17/Emoji_u1f4f1.svg *License:* Apache License 2.0 *Contributors:* https://code.google.com/p/noto/ *Original artist:* Google

- **File:Emoji_u1f510.svg** *Source:* https://upload.wikimedia.org/wikipedia/commons/3/35/Emoji_u1f510.svg *License:* Apache License 2.0 *Contributors:* https://code.google.com/p/noto/ *Original artist:* Google

- **File:Encryption_-_decryption.svg** *Source:* https://upload.wikimedia.org/wikipedia/commons/b/bf/Encryption_-_decryption.svg *License:* CC-BY-SA-3.0 *Contributors:* based on png version originally uploaded to the English-language Wikipedia by mike40033, and moved to the Commons by MichaelDiederich. *Original artist:* odder

- **File:EtherApeTorScreenShot.png** *Source:* https://upload.wikimedia.org/wikipedia/commons/7/72/EtherApeTorScreenShot.png *License:* CC BY-SA 4.0 *Contributors:* Own work *Original artist:* Kencf0618

- **File:Firefox_3.0_error_on_svn.boost.org.png** *Source:* https://upload.wikimedia.org/wikipedia/commons/b/be/Firefox_3.0_error_on_svn.boost.org.png *License:* MPL 1.1 *Contributors:* I took this screenshot myself. *Original artist:* Mozilla Foundation

- **File:Flag_of_the_United_States_National_Security_Agency.svg** *Source:* https://upload.wikimedia.org/wikipedia/commons/5/51/Flag_of_the_United_States_National_Security_Agency.svg *License:* Public domain *Contributors:* This vector image includes elements that have been taken or adapted from this: <a href='//commons.wikimedia.org/wiki/File:National_Security_Agency.svg' class='image'><img alt='National Security Agency.svg' src='https://upload.wikimedia.org/wikipedia/commons/thumb/0/04/National_Security_Agency.svg/20px-National_Security_Agency.svg.png' width='20' height='20' srcset='https://upload.wikimedia.org/wikipedia/commons/thumb/0/04/National_Security_Agency.svg/30px-National_Security_Agency.svg.png 1.5x, https://upload.wikimedia.org/wikipedia/commons/thumb/0/04/National_Security_Agency.svg/40px-National_Security_Agency.svg.png 2x' data-file-width='718' data-file-height='718' /></a> National Security Agency.svg. *Original artist:* Fry1989

- **File:Folder_Hexagonal_Icon.svg** *Source:* https://upload.wikimedia.org/wikipedia/en/4/48/Folder_Hexagonal_Icon.svg *License:* Cc-by-sa-3.0 *Contributors:* ? *Original artist:* ?

- **File:Free-speech-flag.svg** *Source:* https://upload.wikimedia.org/wikipedia/commons/f/fd/Sample_09-F9_protest_art%2C_Free_Speech_Flag_by_John_Marcotte.svg *License:* Public domain *Contributors:* Badmouth, (Archived link) *Original artist:* John Marcotte

- **File:Free_Software_Portal_Logo.svg** *Source:* https://upload.wikimedia.org/wikipedia/commons/6/67/Nuvola_apps_emacs_vector.svg *License:* LGPL *Contributors:*

- Nuvola_apps_emacs.png *Original artist:* Nuvola_apps_emacs.png: David Vignoni

- **File:Geographies_of_Tor.png** *Source:* https://upload.wikimedia.org/wikipedia/commons/4/41/Geographies_of_Tor.png *License:* CC BY-SA 4.0 *Contributors:* Own work *Original artist:* Stefano.desabbata

- **File:Gretacoder_210-IMG_0576-white.jpg** *Source:* https://upload.wikimedia.org/wikipedia/commons/5/5a/Gretacoder_210-IMG_0576-white.jpg *License:* CC BY-SA 2.0 fr *Contributors:* Own work *Original artist:* Rama

- **File:HTTPS_Everywhere_logo.png** *Source:* https://upload.wikimedia.org/wikipedia/commons/c/c5/HTTPS_Everywhere_logo.png *License:* CC BY 3.0 *Contributors:* https://www.eff.org/https-everywhere *Original artist:* Electronic Frontier Foundation

- **File:Integrated_circuit_icon.svg** *Source:* https://upload.wikimedia.org/wikipedia/commons/8/8c/Integrated_circuit_icon.svg *License:* LGPL *Contributors:* All Crystal icons were posted by the author as LGPL on kde-look *Original artist:* Everaldo Coelho (YellowIcon);

- **File:Internet2.jpg** *Source:* https://upload.wikimedia.org/wikipedia/commons/d/da/Internet2.jpg *License:* GFDL *Contributors:* Internet1.jpg by Rock1997 modified. *Original artist:* Fabio Lanari

- **File:LampFlowchart.svg** *Source:* https://upload.wikimedia.org/wikipedia/commons/9/91/LampFlowchart.svg *License:* CC-BY-SA-3.0 *Contributors:* vector version of Image:LampFlowchart.png *Original artist:* svg by Booyabazooka

- **File:Letsencrypt_screenshot_2_domain_choice.png** *Source:* https://upload.wikimedia.org/wikipedia/commons/b/bb/Letsencrypt_screenshot_2_domain_choice.png *License:* CC BY 3.0 *Contributors:* https://media.ccc.de/browse/conferences/camp2015/camp2015-6907-let_s_encrypt.html#download, ~31:44 *Original artist:* Peter Eckersley

- **File:Linomalogo.gif** *Source:* https://upload.wikimedia.org/wikipedia/en/d/d1/Linomalogo.gif *License:* Fair use *Contributors:* The logo may be obtained from Linoma Software.
  Original artist: ?

- **File:Logo_of_CSipSimple,_Android_SIP_application_released_under_GPL_license.png** *Source:* https://upload.wikimedia.org/wikipedia/commons/2/26/Logo_of_CSipSimple%2C_Android_SIP_application_released_under_GPL_license.png *License:* CC BY 3.0 *Contributors:* Development of CSipSimple
  Previously published: http://code.google.com/p/csipsimple/ *Original artist:* R3gis3r

- **File:Monitor_padlock.svg** *Source:* https://upload.wikimedia.org/wikipedia/commons/7/73/Monitor_padlock.svg *License:* CC BY-SA 3.0 *Contributors:* Transferred from en.wikipedia; transferred to Commons by User:Logan using CommonsHelper.
  Original artist: Lunarbunny (talk). Original uploader was Lunarbunny at en.wikipedia

- **File:Nathan_Freitas_at_Unlike_Us.jpg** *Source:* https://upload.wikimedia.org/wikipedia/commons/9/9b/Nathan_Freitas_at_Unlike_Us.jpg *License:* CC BY 2.0 *Contributors:* https://secure.flickr.com/photos/networkcultures/8581857091 *Original artist:* Martin Risseeuw

- **File:National_Security_Agency.svg** *Source:* https://upload.wikimedia.org/wikipedia/commons/0/04/National_Security_Agency.svg *License:* Public domain *Contributors:* www.nsa.gov *Original artist:* U.S. Government

- **File:Nuvola_apps_emacs_vector.svg** *Source:* https://upload.wikimedia.org/wikipedia/commons/6/67/Nuvola_apps_emacs_vector.svg *License:* LGPL *Contributors:*

- Nuvola_apps_emacs.png *Original artist:* Nuvola_apps_emacs.png: David Vignoni

- **File:Open_WhisperSystems_logo.png** *Source:* https://upload.wikimedia.org/wikipedia/en/4/4f/Open_WhisperSystems_logo.png *License:* Fair use *Contributors:* https://whispersystems.org/ *Original artist:* ?

- **File:Orbot-logo.svg** *Source:* https://upload.wikimedia.org/wikipedia/commons/8/8b/Orbot-logo.svg *License:* CC BY-SA 4.0 *Contributors:* Own work *Original artist:* André Loconte

- **File:PersonalStorageDevices.agr.jpg** *Source:* https://upload.wikimedia.org/wikipedia/commons/8/87/PersonalStorageDevices.agr.jpg *License:* CC-BY-SA-3.0 *Contributors:* I took this photograph of artifacts in my possession *Original artist:* --agr 15:53, 1 Apr 2005 (UTC)

- **File:Phone_icon_rotated.svg** *Source:* https://upload.wikimedia.org/wikipedia/commons/d/df/Phone_icon_rotated.svg *License:* Public domain *Contributors:* Originally uploaded on en.wikipedia *Original artist:* Originally uploaded by Beao (Transferred by varnent)

- **File:Phoner.png** *Source:* https://upload.wikimedia.org/wikipedia/commons/b/b0/Phoner.png *License:* Copyrighted free use *Contributors:* Self-made screenshot by Patrick87. *Original artist:* Heiko Sommerfeldt (phoner.de)

- **File:PhonerLite.png** *Source:* https://upload.wikimedia.org/wikipedia/commons/f/f9/PhonerLite.png *License:* Copyrighted free use *Contributors:* Self-made screenshot by Patrick87. *Original artist:* Heiko Sommerfeldt (phonerlite.de)

- **File:PhonerLite_Logo.png** *Source:* https://upload.wikimedia.org/wikipedia/commons/1/1e/PhonerLite_Logo.png *License:* Copyrighted free use *Contributors:* phonerlite.de/index de.htm *Original artist:* Heiko Sommerfeldt

- **File:Phoner_Logo.png** *Source:* https://upload.wikimedia.org/wikipedia/commons/9/9f/Phoner_Logo.png *License:* Copyrighted free use *Contributors:* phoner.de/index.htm *Original artist:* Heiko Sommerfeldt

- **File:Portal-puzzle.svg** *Source:* https://upload.wikimedia.org/wikipedia/en/f/fd/Portal-puzzle.svg *License:* Public domain *Contributors:* ? *Original artist:* ?

- **File:Question_book-new.svg** *Source:* https://upload.wikimedia.org/wikipedia/en/9/99/Question_book-new.svg *License:* Cc-by-sa-3.0 *Contributors:*
  Created from scratch in Adobe Illustrator. Based on Image:Question book.png created by User:Equazcion *Original artist:* Tkgd2007

- **File:RedBlack.png** *Source:* https://upload.wikimedia.org/wikipedia/commons/3/3d/RedBlack.png *License:* CC-BY-SA-3.0 *Contributors:* ? *Original artist:* ?

- **File:RedPhone_icon_2014.svg** *Source:* https://upload.wikimedia.org/wikipedia/commons/e/e7/RedPhone_icon_2014.svg *License:* GPLv3 *Contributors:* https://github.com/WhisperSystems/RedPhone/blob/master/artwork/icon.ai *Original artist:* Jake McGinty

- **File:Retroshare-symbol.png** *Source:* https://upload.wikimedia.org/wikipedia/en/5/5b/Retroshare-symbol.png *License:* Fair use *Contributors:* http://retroshare.sourceforge.net/ *Original artist:* ?

- **File:SSL_Certificate_Info_Box_In_Firefox.png** *Source:* https://upload.wikimedia.org/wikipedia/commons/0/00/SSL_Certificate_Info_Box_In_Firefox.png *License:* CC BY-SA 3.0 *Contributors:* Own work *Original artist:* BoobllaAu

- **File:STE_telephone.nsa.jpg** *Source:* https://upload.wikimedia.org/wikipedia/commons/6/61/STE_telephone.nsa.jpg *License:* Public domain *Contributors:* ? *Original artist:* ?

- **File:STU-I.Young.jpg** *Source:* https://upload.wikimedia.org/wikipedia/commons/0/00/STU-I.Young.jpg *License:* CC-BY-SA-3.0 *Contributors:* ? *Original artist:* ?

- **File:STU-I.jpg** *Source:* https://upload.wikimedia.org/wikipedia/commons/b/bd/STU-I.jpg *License:* CC-BY-SA-3.0 *Contributors:* ? *Original artist:* ?

- **File:STU-II.jpg** *Source:* https://upload.wikimedia.org/wikipedia/commons/c/c5/STU-II.jpg *License:* CC-BY-SA-3.0 *Contributors:* ? *Original artist:* ?

- **File:STU-IIcabinet.a.jpg** *Source:* https://upload.wikimedia.org/wikipedia/commons/6/66/STU-IIcabinet.a.jpg *License:* CC-BY-SA-3.0 *Contributors:* ? *Original artist:* ?

67.5.3 Content license